From Millwrights to Shipwrights to the Twenty-first Century

Explorations in a History of Technical Communication in the United States

WRITTEN LANGUAGE
SERIES

Marcia Farr, senior editor

Perspectives on Written Argument
Deborah Berrill (ed.)

From Millwrights to Shipwrights to the Twenty-First Century
R. John Brockmann

Collaboration and Conflict: A Contextual Exploration of Group Writing and Positive Emphasis
Geoffrey Cross

Subject to Change: New Composition Instructors' Theory and Practice
Christine Farris

Literacy: Interdisciplinary Conversations
Deborah Keller-Cohen (ed.)

Literacy Across Communities
Beverly Moss (ed.)

Artwork of the Mind: An Interdisciplinary Description of Insight and the Search for it in Student Writing
Mary Murray

Word Processing and the Non-Native Writer
Martha C. Pennington

Twelve Readers Reading: Responding to College Student Writing
Richard Straub and Ronald Lunsford

Validating Holistic Scoring for Writing Assessment: Theoretical and Empirical Foundations
Michael Williamson and Brian Huot (eds.)

in preparation

Writing on the Plaza: Mediated Literacy Practice Among Scribes and Clients in Mexico City
Judy Kalman

Contexts, Intertexts, and Hypertexts
Scott DeWitt and Kip Strasma (eds.)

Self-Assessment and Development in Writing: A Collaborative Inquiry
Jane Bowman Smith and Kathleen Blake Yancey

From Millwrights to Shipwrights to the Twenty-first Century

Explorations in a History of Technical Communication in the United States

R. John Brockmann
University of Delaware

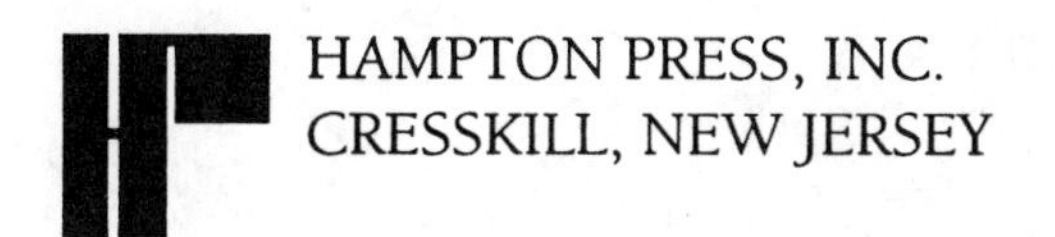

Printed in the United States of America

Library of Congress Cataloging-in-Publication Data

Brockmann, R. John
From millwrights to shipwrights to the twenty-first century : explorations in a history of technical communication in the United States / R. John Brockmann.
p. cm. -- (Written language series)
Includes bibliographical references and index.
ISBN 1-57273-076-5. -- ISBN 1-57273-077-3
1. Communication of technical information--United States--History.
I. Title. II. Series.
T10.5.B75 1998
808'.066--dc21 97-45651
CIP

Hampton Press, Inc.
23 Broadway
Cresskill, NJ 07626

Was du ererbt von deinen Vätern hast
Erwirb es, um es zu besitzen.

You must earn what you inherit from your fathers;
You must make it your own.

Goethe

Contents

LIST OF ILLUSTRATIONS

LIST OF TABLES

Acknowledgements

I would like to gratefully acknowledge:

- Carla Koss for her extensive research and editorial work on this project over many years
- The Honors Department for funding Ms. Koss's original work and research as part of the Undergraduate Research Program
- Professors Robert Day, Deborah Andrews of the University of Delaware English Department, and Barbara Mirel of DePaul University for reviewing this manuscript in an early draft
- Dr. Anne Amsler of ACIT of the University of Delaware and Hans van der Meij of the University of Twente in the Netherlands for reviewing this manuscript in an early draft
- Help from the archivists at the many divisions of the National Museum of American History, Smithsonian Institution; Hagley Museum and Library; Winterthur Library; Historical Society of Chicago; and the University of California at Davis Agricultural Archives and Library
- Mr. William West, Archivist at the NCR Corporation headquarters in Dayton, Ohio, for his aid in obtaining a copy of the 1892 *Sales Primer* and a copy of its later appearance in the first sales manual as the "Demonstration"

- Primary sources materials by and about Oliver Evans have been made available through the efforts of the members of the Oliver Evans Chapter of the Society for Industrial Archeology, especially the late John Kebabian. Information on the SIA can be obtained by writing to 204 West Rose Valley Road, Wallingford, PA 19086-6801 U.S.A.
- Professor Eugene Ferguson and Dr. Robert Multhauf for their helpful suggestions with the Oliver Evans material
- Catherine Kursack who reviewed this manuscript in an early draft and whose helpful suggestions from New Mexico over the Internet always seemed to lead in fruitful directions
- My wife Sarah who put up with my brainstorms from the shower, reviewed this manuscript in an early draft, and was a constant source of encouragement . . . and challenge
- All praise also to God whose spirit of creation, adventure, and insight sustained me on this long, long, process

R. John Brockmann, TSSF

Illustration Acknowledgements

Courtesy of the Hagley Museum and Library, Wilmington, DE

Courtesy of the Massachusetts Historical Society, Boston, MA

Courtesy of the Oliver Evans Press

Courtesy of Carillion Historical Park, Dayton, OH

Courtesy of the Department of Special Collections, University of California Library, Davis, CA

Courtesy of the Lincoln Publishing Co., Lockport, NY

Courtesy of the Charles Babbage Institute, University of Minnesota, Minneapolis..

Courtesy of the National Museum of American History, Smithsonian Institution

Courtesy of The Institute of Electrical and Electronics Engineers, Inc.— Cuthbert C. Hurd. "Sales and Customer Installation." ***Annals of the History of Computing*** 5(2) April 1983.

Part One

Introduction
The Role of Clio, the Muse of History, in Technical Communication*

INTRODUCTION

[T]o know history is go grow up.
—Simon Schama[1]

Technical communicators need to know their profession's forebears. Only by knowing their forebears can technical communicators develop a historical perspective that can reveal, for instance, whether a contemporary standard in technical communication is enduring or transitory. Without a historical perspective, technical communicators have little with which to answer questions about standards, education, or future directions for the professional. In many ways, the profession of technical communication currently is like the spaceship in Einstein's relativity experiment: a solitary, free-floating body knowing neither its speed nor direction because it lacks fixed, external orienting points.

*Portions of this chapter appeared as "Does Clio Have a Place in Technical Writing? Considering Patents in a History of Technical Communication," *Journal of Technical Writing and Communication* 18(4) (1988); 297-304.

At the same time as technical communicators create new "documents" such as online hypertexts for the World Wide Web, they need, more than ever, to appreciate the style and techniques of their forebear's "documents."[2] In the field of biology, Barbara McClintock did not win the Nobel prize solely for finding a breed of corn from the past; she won the award because, in finding that breed, she replenished the contemporary corn gene pool and recovered qualities lost through generations of inbreeding.[3] In a similar way, appreciating older technical communication styles and techniques could revitalize contemporary styles and techniques and recover qualities lost through decades of reusing styles and techniques from only the recent past or proximate environment—"[Variety] is important in a day when more and more technical writing is being turned out according to a formula."[4]

Yet, even if technical communicators do not reuse past styles and techniques, they can perhaps learn from the inadequacies of past styles and techniques. Discarded models and paradigms in science or technology, for example, are always more easily analyzed than currently accepted ones that are too deeply interwoven with the data they organize. So, too, observing the problems of older technical communication styles and techniques may be easier than seeing the problems of present ones. Furthermore, observing the inadequacies of the past may keep technical communicators from being condemned to repeat such problems in the future.

Having a historical perspective in technical communication can help to create a better sense of professional self-identity and tradition—prerequisites, for example, to the establishment of professional ethics which are so "tradition dependent."[5] Having such a perspective can bring the profession up to speed with its kin communication professions—general composition, science writing, and business communication—which are already busily engaged in constructing their own histories.[6] Finally, having a historical perspective in technical communication is even more necessary than in its kin professions because many technical communicators still consider World War II military manual mills to be the birthplace of technical communication.[7]

AN ECLECTIC HISTORY

In creating this book I gathered scattered threads of scholarship from a variety of technological fields and attempted to weave them together. The research methods were eclectic, combining, for example,

- the statistical studies of a large number of anonymously authored manuals with the chronicling of specific individual's writing processes that produced important single books or manuals; and

- the exploits of famous technical communicators with those whose work has been largely forgotten.

I was also catholic in the type of documents examined, which range from patents to cash register-drawer quick reference cards and punched-card calculator instruction manuals.

One of the newest trends in technical communication research is ethnographic research. Such research typically examines technical communication within an organizational context to understand particular communication activities and techniques. For example, one recent ethnographic study of technical communication looked at how the differing communication practices of managers and engineers at Morton Thiokol contributed to the Challenger shuttle disaster.[8] However, one shortcoming of such ethnographic contextualizing is that it has usually ignored the longer history of the organization or technology and their communications practices. By ignoring a longer, historical glimpse of organizations, it's as if most ethnographic studies look at a single frame of film, a short span of time in the life of an organization. In this book, on the other hand, I describe the motion picture frames that led up to that single frame as well as those that followed. For example, I look at the evolution of specific technologies (e.g., antebellum shipbuilding), various corporations (e.g., the Eckert-Mauchly Computer Corporation from 1945 to 1954 and IBM from 1920 to 1990), and individuals (e.g., Oliver Evans's life as inventor and technical gadfly from 1780 to 1819). In turn, these technological, organizational, and individual evolutionary histories give a much deeper and richer context to specific communication designs and techniques.[9]

Research on technical communication from a historical perspective has appeared in a number of articles in technical communication journals and anthologies over the past 30 years.[10] Thus, what I examine in this book is nothing new. However, I do examine three of the larger threads in a tapestry of American technical communication history as well as how these threads can knot to produce contemporary problems in technical communication.[11] I hope this book excites technical communicators to join in the search to find new threads and complete the warp and woof comprising the history of the profession of technical communication. There are many artifacts in many fields still waiting to be examined and studied; many threads to be followed. Clio, the muse of history, is waiting and pointing to an important way for the profession of technical communication to grow up.[12]

THE ORGANIZATION OF THE BOOK

My exploration of the history of American technical communication is divided into three themes and a final demonstration chapter. The time frames of the three themes overlap, producing a multidimensional sense of technical communication in different eras of American history. The three themes are: the important role of visual communication (1791-1887), the power of genera (1791-1980), and the role of technical communicators as innovators within constraints (1948-1954). The final demonstration chapter examines one very specific contemporary dilemma in technical communication and illustrates how a historical perspective can offer important options for a solution.

Theme 1: The Tradition of Visual Communication in 19th-Century American Technical Communication (1791-1887)

In Chapters 1, 2, and 3, I explore the central role that illustrations within texts, three-dimensional models, and dramatized demonstrations had in introducing new technology in the United States prior to 1900. My exploration is based on a close examination of 19th-century texts written or dictated by "mechanicians"—persons "skilled in the operation, repair, and creation of machinery"[13]—and technology salespeople. One of the common events in the careers of each of these 19th-century mechanicians or technology salespeople was that each turned to visual communication only after having their initial attempts at purely verbal or textual communication rebuffed or misunderstood.

Theme 2: The Power of Genre in U.S. Technical Communication (1791-1980)

In Chapters 4, 5, and 6, I explore the role that the constraints of genera and subgenera has had on American technical writers and suggest how these genera and subgenera have evolved. In contrast to the qualitative examination of famous individual's work in Theme 1, in Theme 2 I use quantitative studies of hundreds of patents and dozens of instruction manuals for sewing machines, agricultural implements, and automobiles. The patent authors are known, but the instruction manual authors are more often "anonymous" technical communicators.

Theme 3: The Role of American Technical Communicators—Innovators Within Constraints (1947-1955)

In Chapters 7 and 8, I revisit the observations and analytical techniques of the earlier themes in order to examine closely the development of texts by two technical communicators. Whereas the texts examined in Theme 1 are primarily books written by famous American mechanicians that were reprinted many times, the texts examined here are written by individuals little known outside their own companies, and whose texts were seen only by their companies' customers. Additionally, whereas the texts examined in Theme 2 are large in number and often anonymously authored, the texts examined here are rather few in number. There are only three manuals from two of the first computer companies, and they are clearly identifiable in their authorship (Joseph Chapline and Sidney Lida are the authors). Furthermore, whereas only the final published documents are examined in Themes 1 and 2, in Theme 3 I examine the document development process captured in Chapline's and Lida's early drafts, source material, developmental memos, and oral reminiscences. Thus, in the chapters comprising Theme 3, I observe the communication processes as well as the communication artifacts of the past.

An Application of a Historical Perspective in Technical Communication

My goal in this book is to initiate, not complete, a historical perspective on technical communication in the United States. Thus, I offer the final chapter as a demonstration of how a historical perspective can be usefully focused on a contemporary problem in technical communication in order to offer specific solutions .

ENDNOTES

1. Schama, Simon: "Clio Has A Problem." *The New York Times Magazine*, 9/8/91 :32.
2. Eiler, Mary Ann: "From Screen to Page: A Usability Taxonomy for Displaying Electronic Information in a User Document Manual." In (Brenda Sims, ed.) *Studies in Technical Communication*. Denton, TX: North Texas State University, 1991.: 135—"written documentation is as yet a fairly unmapped genre and may remain so as the industry rushes even at this writing to go 'online'." One of the few occasions when historians of written communication have looked back in order to anticipate possible changes that could occur with online communication was in an article written by JoAnne Yates and Wanda J. Orlinkowski: "Genres of Organizational Communication: A Structurational Approach to Studying Communication and Media." *Academy of Management Review* 17 (2) (1992): 299-326. They describe the development of internal business communication letters into memos and suggest what may occur as the genre moves to electronic mail.
3. Fox, Evelyn K.: *A Feeling for the Organism: The Life and Work of Barbara McClintock*. San Francisco, CA: Freeman, 1983.
4. Miller, Walter James: "What Can the Technical Writer of the Past Teach the Technical Writer of Today?" In (Donald H. Cunningham and

Herman A. Estrin, eds.) *The Teaching of Technical Writing*. Urbana, IL: National Council Teachers of English, 1975: 209.

5. Hauerwas, Stanley, and Willimon, William H.: *Resident Aliens: Life in the Christian Colony*. Nashville, TN: Abingdon Press, 1989: 71-2.
6. In business communication, see Mary T. Carbone: "The History and Development of Business Communication Principles: 1776-1916" *Journal of Business Communication* 31(3) (1994): 173-93; JoAnne Yates: *Control through Communication: The Rise of System in American Management*, Baltimore: Johns Hopkins University Press, 1989; as well as the Association for Business Communication's *Studies in The History of Business Writing* (George H. Douglas and Herbert W Hilderbrandt, eds.). Urbana, IL: American Business Communication, 1985; in science writing see Charles Bazerman: *Shaping Written Knowledge: The Genre and Activity of the Experimental Article in Science*, Madison, WI: University of Wisconsin Press, 1988; and in general composition see *A Short History of Writing Instruction From Ancient Greece to Twentieth-Century America* (James J. Murphy, ed.) Davis, CA: Hermagoras Press, 1990.
7. Gresham, Stephen L.: "Harvesting the Past: The Legacy of Scientific and Technical Writing in America." In *Proceedings of the 24th International Technical Communication Conference, 1977*. Washington, DC: Society for Technical Communication, 1977: 306. Killingsworth and Gilbertson also came close to designating the birth of technical communication as occurring during the 1940s and 1950s: "This crystallization of institutions in the 1940s and 1950s produced the needs that eventually resulted in the formal profession of technical communication." (*Signs, Genres, and Communities in Technical Communication*. Amityville, NY: Baywood Publishing Company, Inc., 1992: 145).
8. Barabas, Christine: *Technical Writing in a Corporate Culture: A Study of the Nature of Information*. Norwood, NJ: Ablex Publishing, 1990. See also Herndl, Carl G., Fennell, Barbara A., and Miller, Carolyn R.: "Understanding Failures in Organizational Discourse: The Accident at Three Mile Island and the Shuttle Challenger Disaster" in (Charles Bazerman and James Paradis, eds.) *Textual Dynamics of the Professions: Historical and Contemporary Studies of Writing in Professional Communities*. Madison, WI: University of Wisconsin Press, 1991; as well as Raven, Mary Elizabeth: "What is Ethnographic Research, and How Can It Help Technical Communicators." In *Proceedings of the 37th International Technical Communication Conference, May 20-23, 1990, Santa Clara California*. Washington DC: Society for Technical Communication, 1990: RT-40-RT-43.
9. Yates and Orlikowski (op. cit., 1992) in their historical approach to business communication genre development also used this deeper and richer contextual investigation methodology

10. See Brockmann, R. John: "Bibliography of Articles on the History of Technical Writing," *Journal of Technical Writing and Communication* 13 (1983): 155-65; Michael G. Moran, "The History of Technical and Scientific Writing," in (Michael G. Moran and Debra Journet, eds.) *Research in Technical Communication: A Bibliographic Sourcebook*, Westport, CT: Greenwood Press, 1985: 25-38; and Rivers, William: "Studies in the History of Business and Technical Writing: A Bibliographical Essay." *Journal of Business and Technical Communication* 8(1) January 1994: 6-57.
11. I have consciously limited myself to exploring only United States technical communication, not because there are uninteresting artifacts in France and England, nor because they are unimportant. Nevertheless, if Rivers's 1994 bibliography of some 200 articles on the history of business and technical communication is accurate, there is a dearth of studies on 19th- and 20th-century American technical communication, and thus the consciously limited focus of this book covers those areas previously left unexplored.
12. If you do not like the idea of exploring the warp and woof of technical communication, consider being a cartographer of technical communication's history: "Instead of an explicator of texts, or historian of literary taste, or literary scientist, or one of the other roles critics sometime play, it may be more revealing for the critic to assume the guise of the mapmaker, a kind of cartographer of literary texts. In this role, the critic would describe the boundaries of a text, in what genre it participates, and how it deforms or conforms to the conventions of that genre. She would describe the contours of meaning within a text and how these contours have changed from time to time. The literary cartographer would give us maps that describe both the structure of a text and a text's evolution. . . . In pursuing these activities, this critic would be engaged in a kind of modeling activity that lays bare both the synchronic and diachronic elements of the literary text." Kent, Thomas: *Interpretation and Genre: The Role of Generic Perception in the Study of Narrative Texts*. Lewisburg, PA: Bucknell University Press, 1986: 27-8.
13. Hawke, David Freeman: *Nuts and Bolts of the Past: A History of American Technology, 1776-1860*. New York: Harper & Row, 1988: 31.

THEME 1

The Tradition of Visual Communication in 19th-Century American Technical Communication

> The dialectic of word and image seems to be a constant in the fabric of signs that a culture weaves around itself. What varies is the precise nature of the weave, the relation of warp and woof.
>
> —Mitchell, *Iconology: Image, Text, Ideology*[1]

In a 1980 article, I lamented the state of graphics instruction in technical communication: "There is little or no attention given to the student's initial steps in developing graphics—the same steps of composition and creation that are now emphasized in the instruction of pre-writing techniques."[2] Yet, if anything, the state of graphics instruction in technical communication deteriorated through the decade of the 1980s. Rubens wrote in 1989: "At present, we seem to take visual literacy as a given despite the fact that our entire education process aims at verbal literacy at the expense of the visual."[3] Moreover, Lee Ellen Brasseur wrote in 1992:[4]

> As teachers of technical writing and communication, we know the importance of designing good graphs, charts and tables; yet we have few good models of the best method for teaching them in our classes.

> Our problems result from the fact that much of the pedagogy which is reflected in the graph and chart chapters of technical writing textbooks has been rhetorically and theoretically deficient, especially when compared to the writing chapters in the same texts.

This ascendancy of verbal over visual literacy in technical communication in the 1980s was, however, only one episode in the longstanding cultural tug-of-war between words and pictures. W. J. T. Mitchell described this tug-of-war in the following way: "The history of culture is in part the protracted struggle for dominance between pictorial and linguistic signs, each claiming for itself certain proprietary rights on a nature to which only it has access."[5]

The ascendancy of verbal over visual literacy in 1980s technical communication was quite similar to the situation in the first decade of this century.[6] A series of papers by English teachers about the relative merits of words and pictures were published between 1908 and 1912 in the *Proceedings of the Annual Meeting of the Society for the Promotion of Engineering Education.*[7] The ascendancy of text over pictures during this time can be observed in statements such as these from the English teachers:

> In the first place, much of the thinking in applied science is done by means of visual images and symbols, and not in words. But thought and language are so nearly inseparable that, at least in the case of the ordinary undergraduate, when the thought has not been carried to actual expression in words there is always ground for questioning the accuracy or the completeness of the thought.[8]

> The immediate lessons, then, are those of accurate and thorough observation, clear perception of purpose, and adaptation to purpose These are, of course, the essential elements of all study; they belong conspicuously to rhetoric because the verbal statement is the direct embodiment of the thought [9]

The message from these teachers at the beginning of 20th century was quite clear: To think clearly and accurately, write; to think inaccurately or incompletely, use pictures.[10]

However, even these English teachers occasionally acknowledged the effectiveness of visual communication:[11]

> For instance, the English instructor, in talking over a character sketch in which the student has given simply the external appearance, is trying to lead the writer to see for himself other possibilities. The student does not catch the idea, when the instructor in drawing, who has overheard a part of the conference, asks, "How about the dotted lines?" and the student sees at once.

What the drawing instructor sought to remind the student by saying, "How about the dotted lines," was that aspects of a drawn object that are hidden when an object is viewed from one perspective could be indicated in perspective drawings by the presence of dotted lines. Thus the drawing instructor hinted in visual terms that the student's character sketch only revealed one aspect or one view—the "external appearance"—of the character; the student needed to suggest or explore the "hidden" aspects of the character. With such a hint, the student was given the opportunity to see and solve the problem "at once." As Roland Barthes noted: "Pictures . . . are more imperative than writing, they impose meaning at one stroke, without analyzing or diluting it.'"[12]

The need to "see at once" was necessary for the engineering student in that early 20th-century scene as well as for the engineers—the "mechanicians"—in the 19th century because mechanicians primarily used visual and tactile communication:[13]

> The kind of thinking involved in designing machine systems was unlike that of linguistic or mathematical thinking in its emphasis on sequence as opposed to classification. To the linguistic and mathematical thinker (for mathematics is merely very formal language), the grammar that embodies the rule of sequence is given; what is crucial is the choice of the correct word or phrase, denoting a class of concepts, and the collection of attributes appropriate to describe the thing in mind. To the mathematical thinker, the grammar of the machine or mechanical system is the successive transformation of power—in quantity, kind, and direction—as it is transmitted from the power source. . . . If one visualizes a piece of machinery, however, and wishes to communicate that vision to others, there is an immediate problem. Speech (and writing) will provide only a garbled and incomplete translation of the visual image. One must make the thing—or a model, or at least the drawing—in order to ensure that one's companion has approximately the same visual experience as oneself.

Visual thinking and communication such as that suggested in the drawing instructor's hint and as used by the mechanicians were a strength of 19th-century American technical communication.[14] For example, in his 1981 book, *Emulation and Invention*,[15] Brooke Hindle carefully plotted the strong relationship between the visual thinking and communication of artists and early American invention. Hindle pointed out that Fulton, for instance, studied painting with Benjamin West, and Samuel Morse, the inventor of the telegraph, exhibited his own paintings at the British Royal Academy of Art. Even the inventor of various types of locks, Linius Yale, began as a portrait and landscape artist[16] Consider, for example, the case of the colt revolver:[17]

> The first truly practical and affordable rapid-fire weapon was the famous Colt revolver. It is said . . . that the idea for it came to Samuel Colt in 1830, when he was a sixteen-year-old seaman on the brig Corvo: by observing the rotation of the ship's wheel (or perhaps the windlass), he conceived of a pistol with a rotating breech. Whatever his source of inspiration, he had carved a rough model before the voyage was over.

Colt conceived of his idea by seeing either the ship's wheel or windlass, and he first communicated it by carving a model. Such American visual thinking and communication was, of course, a continuation of the tradition of visual technological thinking and communication that stretched back to the machine books of Ramelli in 1588 and Agricola in 1556.[18]

Luckily, in the 1990s, with many more powerful tools for creating visual communication, the battle for dominance between words and pictures is different than at the beginning of the century. Not only have computers offered new tools for rendering graphics than were unavailable to those early English teachers in the *Society for the Promotion of Engineering Education*, but the manuals used to explain computer hardware and software, computer user documentation, and the computer controls themselves from windows to icons have given rise to a new appreciation of the crucial role of visual communication.[19] Moreover, technical communicators have eagerly read books focusing on the visual element, such as William Horton's *Illustrating Computer Documentation: The Art of Presenting Information Graphically on Paper and Online* (1991)[20] and his *The Icon Book: Visual Symbols for Computer Systems and Documentation* (1994).[21] Similarly, nonlinear hypertext systems of multimedia information with links and nodes have turned authors at the close of this century to maps rather than lists to indicate organization to audiences.[22] Furthermore, in recent years, gatherings of technical communicators at conventions such as SIGDOC (Association of Computing Machinery, Special Interest Group on Computer Documentation) have devoted large portions of their meetings to discussing and analyzing visual computer interfaces.

In order to put all the new 1990s emphases on visual communication in a larger historical context, I examine in the three chapters within this theme the crucial role of visual communication in 19th century American technical communication. More specifically, I explore three works of mechanician writers and one by an expert technology salesman.

In Chapter 1, I investigate the work of Oliver Evans. Evans was a prolific creator of inventions and filer of patents. He obtained federal patents for improvements in such things as flour-mills run with waterpower (1790) as well as steam engines (1804). Evans was also prolific at suing people to protect his patents. Jousting with such luminaries as Thomas Jefferson in suits and court battles, Evans created an extensive, published record of his creative thinking in a number of pamphlets.[23] In Chapter 1, I focus on his two books, *Young Mill-wright and Millers Guide* (1795) and *The Abortion of the Young Steam Engineer's Guide* (1805), and suggest that his success in communicating to antebellum mechanicians was directly related to the presence of illustrations in his communication.

In Chapter 2, I explore the work of Lauchlan McKay and John W. Griffiths, two fellow, New York, apprentice ship designers and shipwrights, who wrote two of the first American books about ship design: Lauchlan McKay's *The Practical Shipbuilder* (1839) and John W. Griffiths's *Treatise on Marine and Naval Architecture* (1851). What is most interesting about these two books is that, although both of these authors used pictures and three-dimensional models, they did so quite differently despite the similarity of their training from the same master shipwright. McKay used graphics and models to depict reality directly,[24] while Griffiths used graphics and models to depict reality metaphorically.

In the final chapter of this theme, Chapter 3, I examine 19th-century oral technical communication demonstrations that used dramatic audience involvement. This oral technical communication was captured in the pages of NCR *Sales Primer* (1887)[25] a document crucial to NCR's success as a company because it showed how to involve the audience physically in demonstrations of cash registers.

ENDNOTES

1. Mitchell, W.J.T.: *Iconology: Image, Text, Ideology.* Chicago, IL: University of Chicago Press, 1986: 43.
2. "Taking a Second Look at Technical Communication Pedagogy." *Journal of Technical Writing and Communication* 10(4) 1980: 264.
3. Rubens, Phillip. "Online Information, Hypermedia, and the Idea of Literacy." In (Edward Barrett, ed.) *The Society of Text: Hypertext, Hypermedia, and the Social Construction of Information.* Cambridge, MA: MIT Press, 1989: 7.
4. Brasseur, Lee Ellen: "The Visual Composing Process of Graph, Chart, and Table Designers: Results of an Empirical Study." In (Brenda Sims, ed.) *Studies in Technical Communication: Selected Papers from the 1992 CCCC and NCTE Meetings*. Denton, TX: University of North Texas, 1992: 163. See also Katz, Carl: "Desktop Publishing Education and Training: A Critical Challenge." In *1992 Proceedings of the 39th Annual Conference, May 10-13, 1992*. Arlington, VA: Society for Technical Communication, 1992: 351; and Couse, Mary M., Graham, W J. F., and Stokes, Louis W.: "Words into Pictures: Applying Visual Thinking to Online

Documentation". In 1993 *Proceedings of the 40th Annual Conference, June 6-9, 1993*. Arlington, VA: Society for Technical Communication, 1993: 530—"A literature search on the topic 'visual literacy' revealed that it is being discussed in a variety of disciplines: technical writing, journalism, education, psychology, user interface design, and visual arts. Within these disparate fields, a number of major themes emerge, as follows: . . . Current education in visual skills is inadequate. Educators are recognizing the lack of training in visual literacy compared to the effort and emphasis given verbal literacy."

5. Mitchell, W.J.T.: *Iconology: Image, Text, Ideology*. Chicago, IL: University of Chicago Press, 1986: 43. Qtd. in Ben F. and Marthalee S. Barton: "Postmodernism and the Relation of Word and Image in Professional Discourse." *The Technical Writing Teacher* 17(3) Fall 1990: 256-70. Mitchell suggested that the battle between text and graphic reflected "within the realm of representation, signification, and communication, the relations we posit between symbols and the world, signs and their meanings. We imagine the gulf between words and images to be as wide as the one between words and things, between (in the largest sense) culture and nature" (43).
6. Albert Biderman's article in the inaugural issue of *Information Design Journal* had the following to say about the ascendancy of the verbal over visual literacy: "The evolution of graphic methods as an element of the scientific enterprise has been handicapped by their adjunctive, segregated, and marginal position. The exigencies of typography that moved graphics to a segregated position in the printed work have in the past contributed to their intellectual segregation and marginality as well. There was a corresponding organizational segregation, with decisions on graphics often passing out of the hands of the original analyst and communicator into those of graphic specialists—the commercial artists and designers of graphic departments and audio-visual aids shops, for example, whose predilections and skills are usually more those of cosmeticians and merchandisers than of scientific analysts and communicators." "The Graph as a Victim of Adverse Discrimination and Segregation: Comment Occasioned by the First Issue of *Information Design Journal*." *Information Design Journal* 1(1) 1980: 238. See also Olsen, Gary R.: "Eideteker: The Professional Communicator in the New Visual Culture." *IEEE Transactions on Professional Communication* (34)1 March 1991: 13-9.
7. See Tellen, J. Martin: "The Courses in English in Our Technical Schools," in *Proceedings of the 16th Annual Meeting of the Society for the Promotion of Engineering Education, Detroit, MI, June 24-25-26, 1908*, Lancaster, PA: New Era, 1909: 61-73; Kent, William: "Results of an Experiment in Teaching Freshman English," in *Proceedings of the 16th Annual Meeting of the Society for the Promotion of Engineering Education, Detroit, MI, June*

24-2~26, 1908. Lancaster, PA: New Era, 1909: 74-83; "Discussion," in *Proceedings of the 16th Annual Meeting of the Society for the Promotion of Engineering Education, Detroit, MI, June 24-25-26, 1908*, Lancaster, PA: New Era, 1909: 84-97; Clark, John J: "Clearness and Accuracy in Composition," in *Proceedings of the 18th Annual Meeting of the Society for the Promotion of Engineering Education, Madison, WI, June 23-24-25, 1910*. Lancaster, PA: New Era, 1911: 333-42; "Discussion," in *Proceedings of the 18th Annual Meeting of the Society for the Promotion of Engineering Education, Madison, WI, June 23-24-25, 1910*. Lancaster, PA: New Era, 1911: 342-57; Earle, Samuel C.: "English in the Engineering School at Tufts College," in *Proceedings of the 19th Annual Meeting of the Society for the Promotion of Engineering Education, Pittsburgh, PA, June 27-28-29, 1911*. Lancaster, PA: New Era, 1912: 33-47; Raymond, F N.: "The Preparation of Written Papers in Schools of Engineering," in *Proceedings of the 19th Annual Meeting of the Society for the Promotion of Engineering Education, Pittsburgh, PA, June 27-28-29, 1911*. Lancaster, PA: New Era, 1912: 48-54; "Discussion," in *Proceedings of the 19th Annual Meeting of the Society for the Promotion of Engineering Education, Pittsburgh, PA, June 27-28-29, 1911*. Lancaster, PA: New Era, 1912: 54-90. See also Kynell, Teresa: "English as an Engineering Tool: Samuel Chandler Earle and The Tufts Experiment." *Journal of Technical Writing and Communication* 25 (1) 1995: 85-92.

8. Earle, op. cit., 1912: 38.
9. Raymond, op. cit., 1912: 50.
10. See Owens, Larry: "Chapter One: Virgins and Dynamos: The White City" in *Straight-Thinking: Vannevar Bush and the Culture of American Engineering*. Diss. Princeton University, January 1987 (University Microfilms, Ann Arbor Michigan): 2-24.
11. Earle, op. cit., 1912: 44.
12. Barthes, Roland. *Mythologies*. Translated by Annette Lavers. New York: Hill and Wang, 1970: 10. Qtd. in Ben F. and Marthalee S. Barton: "Modes of Power in Technical and Professional Visuals." In Sims, op. cit., 1992: 188.
13. Wallace, Anthony F C.: *Rockdale, The Growth of an American Village in the Early Industrial Revolution*. New York: W. W. Norton, 1980: 238. See also Ferguson, Eugene S.: "The Mind's Eye: Nonverbal Thought in Technology." *Science* 197 (4306) August 26, 1977: 827-36—"If we are to understand the development of Western technology, we must appreciate this important, if unnoticed, mode of thought. It has been nonverbal thinking, by and large, that has fixed the outlines and filled in the details of our material surroundings for, in their innumerable choices and decisions, technologists have determined the kind of world we live in" (827). Also see Ferguson's book-length work on this topic, *Engineering and the Mind's Eye*. Cambridge, MA: MIT Press, 1993.

14. Calhoun, Daniel: *The Intelligence of a People*. Princeton, NJ: Princeton University Press, 1973: 244—"Visual intuition, whether it transcended the verbalism of tradition or bypassed the need for quantitative skill, facilitated the emergence of simple, clean-line styles within a new culture. Even when the elements of design were old, as in shipbuilding, the break in cognitive styles fostered recombination within simplicity."
15. Hindle, Brooke: *Emulation and Invention*. New York: New York University Press, 1981. See also Anthony, Gardener: *An Introduction to the Graphical Language; the Vocabulary, Grammatical Construction, Idiomatic Use, and Historical Development with Special Reference to the Reading of Drawings*, Boston: D. C. Heath & Co., 1922.
16. "New and Useful Improvements: Nineteenth-Century Patent Models." Exhibition at the Hagley Museum and Library, Greenville DE. April-October 1994. See also Root-Bermsteom Robert S.: "Visual Thinking: The Art of Imagining Reality." *Transactions* (American Philosophical Society) 75(part 6) 1985: 50-67.
17. Sweeney, Daniel: "When America was Last in the Arms Race." *American Heritage of Invention & Technology* 10(4) Spring 1995: 23.
18. Ferguson, op. cit., 1977: 828.
19. See for example, Kostelnick, Charles: "Developing a Course in Visual Communication" In *Proceedings of the 37th International Technical Communication Conference, May 20-23, 1990, Santa Clara California*. Washington DC: Society for Technical Communication, 1990: ET-124-ET-127; and Powley, William: "Technical and Scientific Illustrations: From Pen to Computer" In *Proceedings of the 42th International Technical Communication Conference, May 1995*, Washington, DC. Arlington, VA: Society for Technical Communication, 1995: 349-352.
20. New York: John Wiley & Sons, Inc.
21. New York: John Wiley & Sons, Inc. These two books by Horton are not the only fruit of a current resurgence in nonverbal communication in technical communication; see Nancy Allen's review of a number of 1993 and 1994 technical communication textbooks that integrate "visual and verbal tools to enhance communication." *Technical Communication Quarterly* 3(3) Summer 1994: 351-6.
22. Kremer, Rob and Gaines, Brian R.: "Groupware Concept Mapping Techniques." In *Proceedings of SIGDOC '94~ Conference, Banff, Alberta, October 1-2*. New York: Association of Computing Machinery, 1994.
23. *Patent Right Oppression EXPOSED, or Knavery Detected* (1813); *A Trip Made By A Small Man in a Wrestle with a Very Great Man* (1813); *Exposition of Part of the Patent Law* (1816); and *Oliver Evans to His Counsel Who Are Engaged in Defense of his Patent Rights* (1817). This last work functioned much like a memoir. This extensive published record, although unusual for a mechanic ("much writing ill suits a Mechanic"

said the celebrated David [Rittenhouse in 1772 (Rittenhouse to Thomas Barton, February 3, 1772, in Barton, William: *Memoirs of Rittenhouse.* (Philadelphia, 1813): 231] is not the only one. John Fitch, one of Evans's competitors, also left a number of argumentative pamphlets (see Hindle, Brooke: *Emulation and Invention.* New York: New York University Press, 1981: 34, 39.)

24. Chapelle, Howard I.: *The National Watercraft Collection.* Washington, DC: Smithsonian, 1960: 8-12. See also Cutler, Carl C.: *Greyhounds of the Sea*, Annapolis, MD: Naval Institute, 1930: 28-9.
25. For further information see: Johnson, Roy W. and Lynch, Russell W.: "Chapter 29. The Silver Dollar Demonstration" In *The Sales Strategy of John H. Patterson (Sales Leaders Series).* Chicago and New York: The Dartnell Corp., 1932; Bernstein, Mark: "John Patterson Rang Up Success with the Incorruptible Cashier," *Smithsonian* 20(3) June 1989: 150-66; Crandall, Richard L. and Robbins, Sam: *The Incorruptible Cashier, Vol. 1. The Formation of an Industry, 1876-1890.* Vestal, NY: Vestal Press, 1988; and *1884-1922 The Cash Register Era.* Dayton, OH: NCR, 1984 (one of a five part company centennial celebration publication): 7.

Chapter 1

Oliver Evans and the Weave of Text and Graphics for Antebellum Millwrights

Mr. Evans made no pretensions to literature; he considered himself, as he really was, a plain, practical man; and the main object of his writing his work was to introduce his invention to public notice.

—Preface to *The Young Millwright and Miller's Guide* (1826 edition)

At the beginning of his career as an inventor in the early 1780s, Oliver Evans described how the manual labor involved in the process of flour milling[1] had changed very little since Roman times:[2]

THE OLD PROCESS OF MANUFACTURE OF FLOUR

If the grain be brought to the Mill by land carriage, the Miller took it on his back, a sack generally 3 bus., carried it up one story by stair steps, emptied it in a tub holding 4 bus., this tub was hoisted by a jack moved by the power of the Mill which required one man below and another above to attend to it, when up the tub was moved by hand to the granary, and emptied. All this required strong men. From the granary it

> was moved by hand to the hopper of the rolling screen, from the rolling screen by hand to the Millstone hopper, and as ground it fell in a large trough, retaining its moisture, from thence it was with shovels put into the hoist tubs which employed 2 men to attend, one below, the other above, and it was emptied in large heaps on the Meal loft, and spread by shovels, and raked with rakes, to dry and cool it, but this necessary operation could not be done effectually, by all this heavy labour. It was then heaped up over the bolting hopper, which required constant attendance, day and night, and which would be frequently overfed, and cause the flour to pass off with the bran, at other times let run empty, when the specks of fine bran passed through the cloth, which with the great quantity of dirt constantly mixing with the meal from the dirty feet of every one who trampled it, trailing it over the whole Mill and wasting much, caused great part to be condemned, for people did not even then like to eat dirt, if they could see it. After it was bolted it required much labour to mix the richest and poorest parts together, to form the standard quality, this lazy millers would always neglect, and great part would be scrapped or condemned, while others were above the standard.[3]

However, after Evans invented an automated mill to manufacture flour at the end of the 1780s, he could describe the new flour manufacturing process in a much shorter fashion:[4]

> Where the Waggoner standing in his wagon empties the grain from his bags into a spout leading into the machinery moved by the power which drives the Mill, and which carry it through every operation without the least manual labor, until the flour be completely manufactured and packed in the barrel without waste and performing every operation to perfection with mathematical certainty finishing it as it goes. . . . The Miller frequently shuts his door, blows out his lights and goes to his bed to rest while every operation is performed most faithfully, by servants who never sleep, lie, or shirketh from their duty, nor do they ever become weary and require to be relieved, but perform his task faithfully and certainly.

All of the mill work could be done automatically without any human labor—"by the power which drives the Mill, and which carry it through every operation without the least manual labor." Additionally, Evans claimed that his invention would save costs to such a degree that, as he later noted in a letter to Congress:[5]

> If they [his inventions] had been promptly and extensively put into operation, and the savings or gains by the use of them collected into the

> public treasury, it would have been sufficient to have discharged the public debt, defrayed the expense of government, and freed the people of the United States from taxes.

However, the significance of Evans's automated flour mill went far beyond saving labor and money in the late 18th century. John Kasson wrote of Evans:[6]

> Well over a century before Henry Ford and his associates applied the continuous automatic conveyor to automobile production in 1914, Evans stood as the period's most dramatic example of the way in which new technological developments promised to transform American manufactures while avoiding the creation of a degraded proletariat.

Moreover, Siegfried Giedion claimed that Evans's automated flour-milling process "opened up a whole new chapter of mankind":[7]

Figure 1.1. Portrait of Oliver Evans

> For Oliver Evans, hoisting and transportation have another meaning. They are but links within the continuous production process: from raw material to finished goods, the human hand shall be replaced by the machine. At a stroke, and without forerunner in this field, Oliver Evans achieved what was to become the pivot of later mechanization.

Yet, it was not easy for Evans to convince his fellow millwrights and millers to use his pivotal breakthrough, or, more importantly, to pay him a license fee for its use. In fact, they resisted his message at every turn. According to Evans, "The human mind seems incapable of believing anything that it cannot conceive and understand. . . . I speak from experience . . . "[8] The millwrights' and millers' resistance to Evans's message was so great that their first recorded reaction was: "It will not do! it cannot do!! it is impossible that it should do!!!"[9]

Nevertheless, Evans persisted in his attempts at persuasion and eventually found a way to help the millwrights and millers understand his invention. Thus, Matthew Carey, editor of the *American Museum or Universal Magazine*, sang of Evans's eventual success in persuading the millwrights and millers when he wrote in 1792: "We are happy to learn that mr. Evans is now likely to reap the fruits of his labour; upwards of 100 mills have adopted his improvements, and they are daily increasing in public estimation."[10]

Much of Evans's success in persuading the millwrights and millers can be attributed to the unique qualities of his technical communication style captured in his first book, *The Young Mill-wright and Millers' Guide* (1795), and its 15 successive editions, which collectively made *The Guide* the most reprinted book about technology written before the Civil War.[11] What made the *The Guide* so successful in antebellum America was Evans's dense weave of text and graphics.[12]

THE "VISIONARY PROJECTOR"

However, before Evans could write *The Guide*, he had to invent the milling machines. Evans's 1817 "memoirs" describe how he invented the flour-manufacturing machine in the early 1780s just after the completion of the War of Independence.[13] Evans recalled: "This arrangement [of milling machines] I so far completed before I began the mill that I have in my bed viewed the whole operation with much mental anxiety."[14] A later biographical sketch of Evans's early life also observed that Evans "would lie for hours, cogitating upon his plans and improvements."[15] Furthermore, five years before his death in March 1814, Evans wrote that he was again lying in bed—this time overcome by exhaustion—when he began to think about his country being devastated by raiding British naval vessels during the War of 1812. Evans wrote a letter to

President Madison in which he noted that "his mind runs constantly on devising the means of destroying enemies ships in our waters." For all these dreams and plans, Evans remembered in his 1817 memoir that he was called by his fellows a "visionary projector."[16]

In the late 18th century, a "visionary projector" was one who formed a fantastic or speculative project based on knowledge obtained from other than ordinary sight.[17] By emphasizing the noun, "projector," such a title could have been meant in derision. For example, "projector" could have meant "worthlessness" such as when Evans's contemporary, Hezekiah Niles recalled that in their boyhood days: "adults had agreed that Oliver Evans 'would never be worth any thing, because he was always spending time on some contrivance or another.'"[18] Indeed, Evans himself may have felt it was meant derisively because when he wrote a letter in 1794 requesting a loan from a sponsor, he cautioned his sponsor: "I will observe to you that I having the name of a Projector will probably find difficulty in obtaining money from most men."[19]

However, the word "visionary projector" also had an adjective, and by emphasizing the adjective "visionary," Evans joins the ranks of those whom contemporary historians of technology define as having used their "mind's eye":[20]

> Many features and qualities of the objects that a technologist thinks about cannot be reduced to unambiguous verbal descriptions; they are dealt with in his mind by a visual, non-verbal process. His mind's eye is a well-developed organ that not only reviews the contents of his visual memory, but also forms such new or modified images as his thoughts require.

Although it is impossible now to peer into Evans's "mind's eye" as he lay for hours inventing his machines, one can catch a glimpse of this in the creative thinking process described in his memoir (1817):

> Discovery of the Elevator
>
> My next step was to ascertain if any machinery was in use for other purposes which might be applied to this, but could not find any until some time in the spring or summer of 1783, when I recollected to have seen a drawing of a chain pump for raising water from the holds of ships, the principles of which I hope to be able to apply so as to raise the grain and meal.[21]
>
> Discovery of the Hopper-Boy
>
> . . . until the idea struck my mind of laying the meal by a machine, first in a large circular ring around the bolting hopper, and then gathering it from the ring to the center. I perceived with infinite pleasure, I had discovered the principles . . .[22]

> Discovery of the Conveyor
>
> My first conception in the invention of this machine was the application of the endless screw to revolve in a trough . . . but changed the principle of the machine to that of many ploughs, by flights in the shaft, set in a spiral line, to follow each other.[23]

It was a "drawing" that Evans "recollected to have seen" when he discovered the principle of the elevator. In like fashion, it was a "large circular ring" that gave him "infinite pleasure" in discovering the principles of the hopper-boy. Finally, he conceived of the conveyor in visual terms, such as "an endless screw in a trough" and "many ploughs, by flights in the shaft, set in a spiral line."

The visual quality of Evans's "mind's eye" is even more obvious in Figures 1.2, 1.3, and 1.4 drawn from his portfolio of design drawings and specifications.[24] This portfolio contains the only record of his rough drafts; he burned all his other design drawings and sketches in a fit of pique in 1809.[25] Although these particular portfolio excerpts reveal Evans's thoughts about a steam engine rather than a flour mill, they are relevant because both inventions were progressing simultaneously in Evans's considerations in the 1780s.[26]

Figure 1.2 from the portfolio is a textual presentation of a steam engine much like Evans may have submitted with his patent applications. In the accompanying figure, Evans carefully labeled and sequentially explained the progression of action from the source of power A ("The furnace containing the fire") to F ("A little forcing pump"), which was turned by the power of the furnace. The picture controlled the organization of the words by moving the reader from the center of the picture (A) to the outside of the picture (F).

In Figure 1.3 from the portfolio, the role of text and pictures is not as linear or disciplined as in Figure 1.2. Instead, Figure 1.3 suggests the precedence of pictures over text and mathematics in Evans's mind's eye—the page seems to indicate that Evans first drew the boiler and connecting pipes and then proceeded to write the text and mathematics over them.

Figure 1.4 from the portfolio resembles Leonardo da Vinci's notebooks in the interpenetration of pictures and text. For example, it seems that once again Evans first sketched the boiler and then wrote the text. Sometimes Evans wrote text right over the picture, and, at other times, the pictures compartmentalized the text into sections and forced Evans—on the extreme right for example—to write vertically. Thus, the pictures appear to give rise to Evans's words as well as later to control them and organize them.

As effective as this weave of text and graphics was for helping Evans create his inventions in the early and mid-1780s, they only remained within Evans's mind's eye as a "visionary projector." Evans did not initially use the same weave of text and graphics when he needed to communicate his automated flour-mill to others.

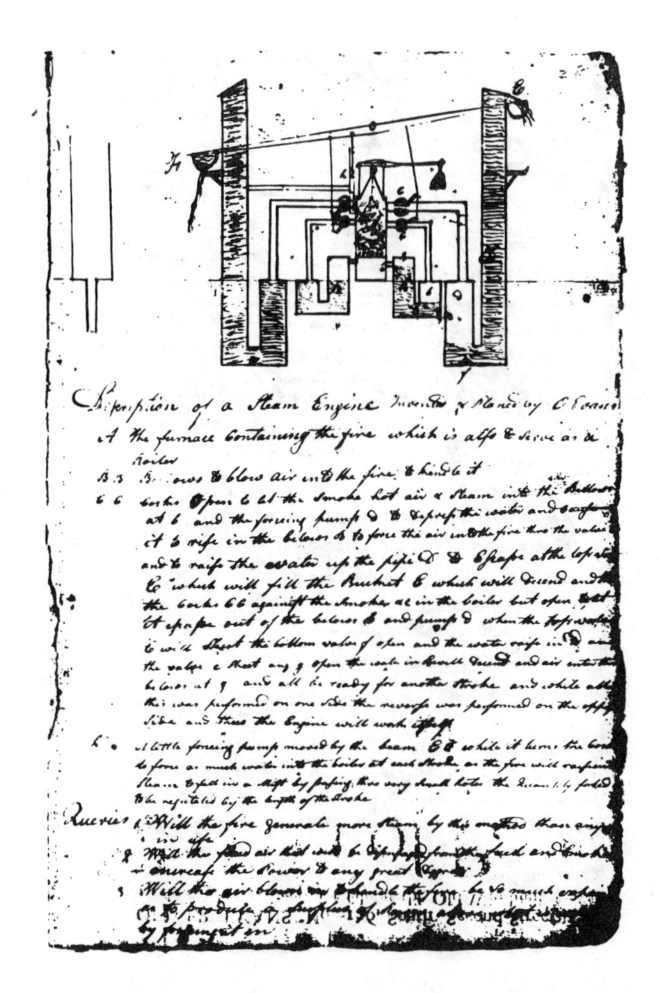

Figure 1.2. A text and graphic combination, page 33 from "The PORTFOLIO" (Hagley Museum and Archives, Wilmington, DE, Accession 1538)

FROM THE "MIND'S EYE" TO COMMUNICATION WITH OTHERS: EVANS'S INITIAL COMMUNICATION ATTEMPTS

In 1785, Evans approached Thomas Lea, a miller on the Brandywine River in Delaware, with an oral request for financial help in building his automated mill:[27]

> the descender, transferring down an incline; the elevator for raising vertically; the conveyor and the drill for moving horizontally, and the hopper-boy [hopper], whose function was to spread and cool the meal and feed it regularly into the bolting hopper.

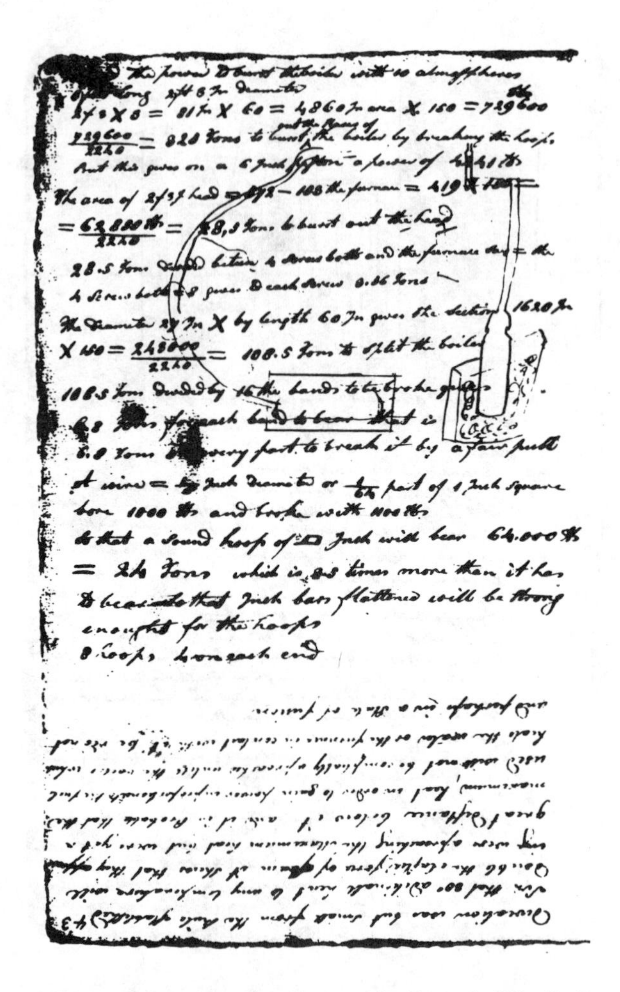

Figure 1.3. The graphic precedes the text in Evans's Mind's Eye, page 50 from "The PORTFOLIO" (Hagley Museum and Archives, Wilmington, DE, Accession 1538)

Lea's response was "No."[28]

Next, Evans approached James Latimer, a flour merchant and justice of the New Castle (Delaware) County Court of Common Pleas, with a similar oral request for help. However, Latimer's response was even less encouraging: "Ah! Oliver, you cannot make water run up hill, neither can you make boys without women."[29]

Perhaps because Evans's description of his mill was so at odds with the preconceived notions of milling in the 1780s, his spoken words alone failed to persuade either Lea or Latimer to loan him funds. Nonetheless, by

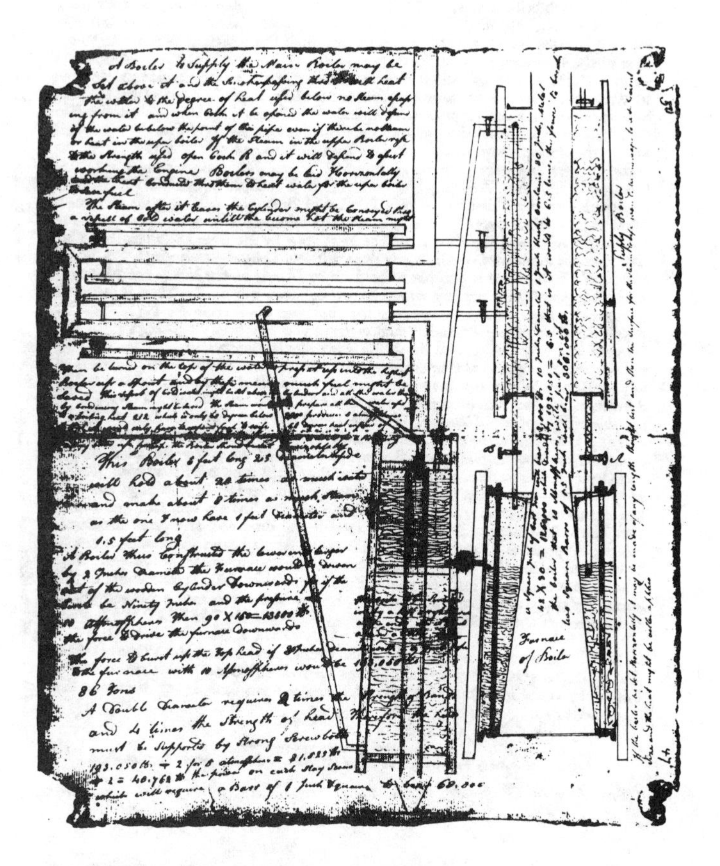

Figure 1.4. Textual-graphical interpenetration in Evans's Mind's Eye, pages 54-7 from "The PORTFOLIO" (Hagley Museum and Archives, Wilmington, DE, Accession 1538)

1787, the legislatures of Delaware, Pennsylvania, Maryland, and New Hampshire assigned Evans patent privileges for his milling machinery.[30] Armed with this new legal foundation, Evans created his first communication describing his inventions in a poster—a broadside (Figure 1.5)—that was printed on December 19, 1787. However, in Evans's lengthy description of his inventions, he did not use any of the visual techniques he had already used in his own internal process of invention. There were no pictures in the broadside

Evans described his inventions in the broadside and then attempted to persuade the millers to purchase his "directions" and "rights" by pointing

To the Millers.

THE Subſcribers have a Merchant-Mill on Redclay Creek, 3 Miles above Newport, Newcaſtle County, Delaware, with Evans's new-invented Elevators and Hopperboys erected in her, which does the principal Part of the Work. One of the Elevators receives the Wheat at the Tail of the Waggon, and carries it up into Garners, out of which it runs through Spouts into the Screen and Fan, through which it may be turned as often as neceſſary, till ſufficiently cleaned; [illegible] into a Garner over the Hopper which feeds the Stones regularly.—Another Elevator receives the Meal when ground and carries it up, and it falls on the Meal-loft, where the Hopperboy receives it and ſpreads it abroad thin over the Floor, and turns it over and over perhaps an hundred Times and cools it completely, then conveys it into the Boulting-Hopper, which it attends regularly. [illegible] Elevator alſo carries up the Tail Flour with a Portion of Bran, and mixes it with the ground Meal to be boulted over, by which means the Boulting is done to the greateſt Perfection poſſible, and the Cloths will be kept open by the Bran in the hotteſt Weather without Knockers.—All this is done without Labour, with much leſs Waſte, and much better than is poſſible to be done by Hand, as the Miller has no need to trample in the Meal, nor any way to handle or move it from the Time it leaves the Waggoner's Bag, until it comes into the ſuperfine Cheſt ready for Packing.—The [illegible] Expence of the Materials and erecting ſaid Machinery will not exceed from Twenty to Forty Dollars, as the Mills may differ in Conſtruction.—[illegible] now do the Work that uſed to employ two or three, two Hands are [illegible] Mill with two Waterwheels and two Pair of Stones ſteady running, [illegible] Aſſiſtance, if the Machinery be well applied—They are ſimple and [illegible] not ſubject to get out of Repair. If Millers will think on this when they are [illegible] carrying heavy Bags, or with hoiſting their Wheat or Meal, ſpreading [illegible] attending the Boulting-Hopper, Screen and Fan, and when they ſee the [illegible] tered over the Stairs, &c. waſting, or when they hoiſt their tail Flour with [illegible] to boult over—and when their Flour is ſcraped for neglect in Boulting, and when the ſuperfine is let run into the Middlings by overfeeding, &c, &c. and conſider that theſe Machines will effectually remedy all this, and ſave great Expence in Wages, Proviſions, Bruſhes and Candles—[illegible] he may conclude that it is [illegible] to continue in the old Way, while ſuch excellent Improvements are extant. Those who chooſe to adopt them, may have Permiſſion, with full Directions for [illegible] them, by applying to OLIVER EVANS, the Inventor, who has an excluſive [illegible] or to either of the Subſcribers.

JOHN THEOPHIL[illegible]
OLIVER EVANS

N. B. Farmers and others may have Wheat ground during the Winter Seaſon at ſaid Mill (on good Burrs and all Things in the beſt Order) with great Care and Diſpatch, at the low Rate of Thirty Shillings per 100 Buſhels, or Eighteen Shillings per Load.

Redclay Creek, Dec. 19, 1787.

Lancaſter: Printed by STEINER, ALBRIGHT & LAHN, *a few doors ſouth of the Court-Houſe.*

Said Elevators will hoist Water to any desired height for the purpose of Watering Meadow at a very small expense Oliver Evans

Figure 1.5. First public textual description of Evans's inventions, 1787 (Massachusetts Historical Society)

out that such machinery would require less work, produce less waste, and be inexpensive while being durable. Evans even offered his invention, license- and cost-free, to the first miller in each district who would use his machinery. Yet, like Lea and Latimer before, there were no takers. Evans wrote in his journal that this lack of an affirmative response to his 1787 broadside was "truly discouraging."[31]

Perhaps one reason for Evans's inability to convince millers and millwrights to consider his invention and pay him his fee was that distant millers who did not personally know Evans could not believe that his concepts were either real or practical. Their suspicion only grew when they discovered that Evans's own neighboring millers on the Brandywine River were not using the machines:[32]

> This opposition cost Mr. Evans thousands of dollars, in the travelling of many miles, in fruitless attempts to establish his invention; for, wherever his agents went, the enquiry was "have the Brandywine millers adopted it?" the answer was "no!" which was generally followed by this pertinent reply: "if those who are so much more extensively engaged in the manufacture of flour, do not think it worth their attention, it certainly cannot be worth ours."

Additionally, there were also economic reasons that may have caused millers and millwrights not to adopt Evans's invention in these early days of the republic. For example, to acquire the necessary equipment for Evans's automated mill, a miller would have to make a large capital investment just at a time of economic disarray and depression following the end of the War for Independence. Such an investment would have been problematic, however, at any time both to the small mill owners without the capital as well as to the large mill owners who were already fully invested in the established milling technology.[33]

However, surely another reason for such discouraging results was Evans's failure to communicate effectively. In his 1787 broadside, Evans bluntly presented himself as a "projector" who was selling plans for profit: "Those who choose to adopt them, may have Permission, with full Directions for erecting them, by applying to Oliver Evans, the Inventor, who has an exclusive Right." Surely, such a presentation of his persona in the broadside, as someone owning exclusive rights and giving others permission, did not help to convince millers who had just finished fighting a war for independence and equality. Nor, did Evans's purely textual descriptions reach the mind's eye of the millers and millwrights.

Even when Evans eventually had something for millwrights and millers to see—working equipment installed by late 1790 in his own mill on Red Clay Creek in Delaware—Evans still "experienced great difficulty in teaching millers to understand the utility of his improvement and the millwrights how to construct and install them properly."[34] One day, for example, when he noticed a group of millers inspecting his automatic mill while he was out in the fields, he stayed away from the mill long enough to give the millers time to see for themselves that the machinery required no human intervention. Nevertheless, the millers were not convinced:[35]

> But to his astonishment, he soon heard that on their return they had reported to their neighboring millers that the whole contrivance was a set of rattle-traps,[36] not worthy the attention of men of common sense.

A MIX OF TEXT AND PICTURES—THE UNIVERSAL ASYLUM AND COLUMBIAN MAGAZINE AND A NEW BROADSIDE

In January 1791, Evans decided to try a different communication approach from the one he had used in the broadside of 1787. To describe this new approach, Evans wrote a letter to William Young,[37] who was publishing a journal in nearby Philadelphia called the *Universal Asylum and Columbian Magazine*. In this new approach, Evans made pictures the key to his communication:*

> Wilmington Jany 1791
>
> Sir
>
> I will thank you very much if you will get introduced onto my Draught a Figure of a Miller standing at the Spout A. viewing the wheat as it runs from the Waggoners Bag—stooping a little resting himself with his left hand on the side of the Spout, with his right hand full of wheat looking attentively at it this is the Millers Business in real Practice and in my opinion will greatly adorn the Draught and convey a strikin idea—I wish also to have introduced after the explanation of the Process the following words or to the same purport—Viz Thus from the time the wheat is emitted from the Waggon until it is compleatly manufactured into superfine flour the whole is done by machinery without any part thereof being moved by manual labour—This will fourcibly convey the idea which many readers cannot take up unless thus expressed. I think I shall want at least 10 of this number to send to Different Places—I have in contemplation to furnish you with a drawing of a packing machine for packing the flour into barrels next month—and ditto of the manner of laying out Millstones with observations and reasons thereon for the month following Providing you approve and will let me have the plates afterwards—
>
> Several drawings of different kind of water wheels with explanations of the true principles of each shewing the advantages and disadvantages of Double and Single geered Mills on different heads the whole combined might form a pamphlet that might convey much useful knowledge and greatly tend to the improvement of the important Branch of our Manufacture of flour.
>
> I am Sir with much Esteem
> Your Humble servt
> Oliver Evans
>
> To William Young
> Bookseller at the Corner
> of Chestnut and Second Street
> Philadelphia

---

*The errant spelling and punctuation in this letter, such as they are, are all Evans's.

Years after failing to convince Thomas Lea with his purely oral presentation, and one year after failing to convince any miller with his purely textual broadside,[38] Evans finally concluded that the only way to "fourcibly convey" his invention was through a number of "draughts." Figures 1.6 to 1.10 show the result of this letter—a second broadside circulated all along the East Coast that reprinted his article from the January 1791 issue of the *Universal Asylum and Columbian Magazine*.[39]

The most striking element of this second broadside was the preeminence of the "plate"—a word Evans used the describe the new broadside's five figures because they were all etched on one copper*plate*. Evans made the plate preeminent by

- positioning it on the top left of the broadside—the reader's entry point to the broadside,[40]
- making 42% of all the paragraphs refer to the plate, and
- organizing the text in the broadside in a graphical manner moving from left to right in Figure I and in a zigzag fashion from top to bottom. Once Evans completed his textual description of the central portion of the plate, Figure I, he gave a more detailed textual description of each component (the elevator, the conveyer, and the hopper box) using Figures II, III, IV, and V above and below Figure I. He covered the components second because Figures II, III, IV, and V were visually placed on the borders of the plate.

In switching from using only words to a mix of words and pictures, Evans acquired in this second broadside a number of rhetorical advantages. First, instead of bluntly projecting an image, a persona, as a "projector" owning exclusive rights and giving others permission, Evans moved his persona into the background and let his pictures convey much of his message. Thus the broadside appears more objective. Such an appearance of objectivity allayed distrust and was an especially important advantage to Evans, who felt that "having the name of a Projector will probably find difficulty in obtaining money from most men."[41]

A similar rhetorical ploy of using pictures to bolster believability and remove the focus on a foreground persona had been used a century earlier by Boyle in his records of the initial experiments of the Royal Society. The pictures accompanying Boyle's reports "served to announce that 'this was really done' and that it was done in the way stipulated; they [the pictures] allayed distrust and facilitated virtual witnessing."[42]

A second rhetorical advantage of the newly added pictures in this second broadside was the replacement of Evans blunt appeal to pathos—the appeal to a reader's emotions—found in the first broadside ("If Millers will think on this when they are fatigued carrying heavy Bags..."). This blunt

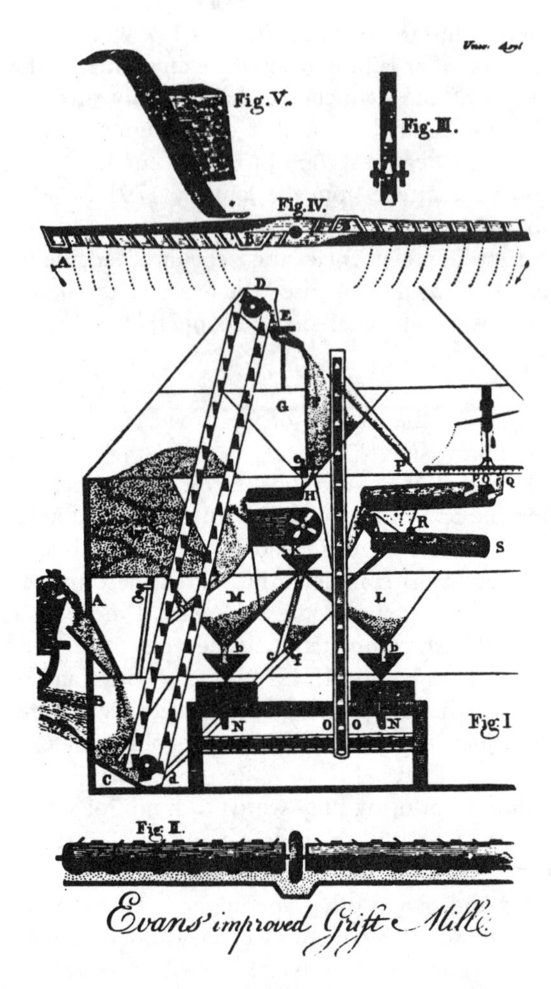

Figure 1.6. The opening graphic from *The Universal Asylum and Columbian Magazine* broadside, January 1791. (Germantown Historical Society & Oliver Evans Press, Wallingford, PA, 1990)

appeal was replaced with a more subtle pictorial appeal to pathos as Evans envisioned in his description of the drawing sent to the publisher Young:[43]

> if you will get introduced onto my Draught a Figure of a Miller standing at the Spout A. viewing the wheat as it runs from the Waggoners Bag—stooping a little resting himself with his left hand on the side of the Spout, with his right hand full of wheat looking attentively at it this is the Millers Business in real Practice and in my opinion will greatly adorn the Draught and convey a strikin idea . . .

By O L I V E R E V

BY theſe improvements, for which he has lately obtained a Patent, the grain and meal are carried from one ſtory to another, or from one part of the ſame ſtory to another; the meal is cooled; and the boulting-hoppers are attended by machinery, which is moved entirely by the power of the mill, and leſſens the expenſe of attendance at leaſt one half.

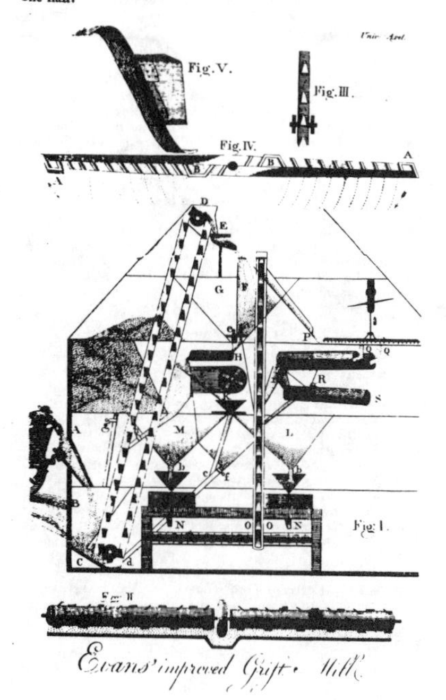

at the upper pully. See C D. in Fig. 1ſt. Fig. 3 exhibits a view of the pully of the meal elevator, as it is ſupported on each ſide, with the ſtrap and buckets deſcending to be filled.—Fig. 5, is a perſpective view of the bucket of the wheat-elevator; and ſhows the manner in which it is faſtened, by a broad piece of leather which paſſes under the elevator-ſtrap, and is nailed to the ſides with little tacks. It is alſo tacked to the under ſide of the ſtrap to keep it in its place. Buckets made of ſheet iron rivetted to the ſtrap ſuit beſt for wheat.—Beſides the uſes of the Elevator, that have already been mentioned; it may be employed to unlade ſhips, thus—Annex a conveyer to the axis of the lower pully; let it paſs through the wall of the mill, extend to the ſide of the ſhip, and convey the wheat into the elevator. When large quantities of wheat are to be received, turn the crane ſpout over the great Wheat Granary W. out of which it may be taken, as wanted, by drawing the gate g, which lets it fall into the Garner B, or into the Elevator at any other convenient place.

The CONVEYER is an eight-ſided ſhaft, ſet on all ſides with pins, in an oblique manner. It is put in motion in a trough; and, performing its office on the principle of a continued ſcrew, conveys the grain, or meal, from one end of the trough to the other, whether its direction be aſcending, deſcending or horizontal. See N N, Fig. 1ſt.—Fig. 2 is a perſpective view of the Conveyer, as it lies in its trough, at work; ~~and ſhows the manner in which it is joined to the pullies, at each ſide of~~ the elevator.

The HOPPER-BOY is an upright ſhaft, ſet in motion and carrying round with it an horizontal piece called the arms, ſet with flights, in the under ſide, in an oblique direction; ſo conſtructed as to ſpread the meal thinly over the floor, to cool; and, at the ſame time, to gather it into the boulting-hopper. Theſe arms are balanced by a weight, hung to a rope paſſing over a pully, at the head of the upright ſhaft; ſo that it plays lightly on the meal, and will riſe or fall as the heap increaſes or decreaſes. See it in fig. 1 over Q Q. Fig. 4 is a perſpective view, of the under ſide of the arms of the *hopper-boy*, with flights complete. The dotted lines ſhow the track of the flights of one arm; thoſe of the other following, and tracking between them. A A are the ſweepers. Theſe carry the meal round in a ring, trailing it regularly all the way, the flights drawing it to the centre, as already mentioned. B B are the ſweepers that drive it into the boulting hoppers.

Brief DIRECTIONS *for conſtructing the ſeveral* MACHINES.
Of the WHEAT-ELEVATOR.

Let the *pullies* be 17 inches in diameter, 4 inches in thickneſs, and half an inch higher in the middle than in any other part. The motion about 47 revolutions in a minute, which will move the ſtrap 200 feet, in the ſame time. The *ſtrap* is to be made of good harneſs, 4 inches wide; ~~and to be furniſhed with a~~ *bucket*, of the ~~ſame width, 5~~ inches in length, for every foot the ſtrap is in length. This *elevator* will hoiſt 3375 buſhels in 12 hours.

To prevent the machine from being over loaded, the wheat muſt be let in at the bottom, to meet the buckets. The Caſes ought to be a little crooked, to ſuit the motion of the ſtrap, that the buckets may not touch them in deſcending.

Figure 1.7. From *The Universal Asylum and Columbian Magazine* broadside, January 1791 (Germantown Historical Society and Oliver Evans Press, Wallingford, PA, 1990)

Readers tend to feel manipulated if authors use pathos in too blunt a fashion, and perhaps this was how the millers felt when Evans presented his direction—"If Millers will think on this when they are fatigued"—in his first broadside. However, Evans gained new persuasive subtlety by letting his readers come to this same interpretation in the second broadside and article with the "stooping" figure and by letting them recall their own fatigue involved with flour manufacturing.

DESCRIPTION *of the* PLATE.

THE grain is emptied into the ſpout A, by which it deſcends into the garner B ; whence, by drawing the gate at C, it paſſes into the elevator C D, which raiſes it to D, and empties it into the Crane-Spout E, which is ſo fixed on gudgeons that it may be turned to any ſurrounding granaries, into the Screen-Hopper F, for inſtance (which has two parts, F and G) out of which it is let into the Rolling Screen, at H, by drawing the ſmall gate a. It paſſes through the Fan I, and falls into the little Sliding-Hopper K, which may be moved, ſo as to guide it into either of the Hanging-Garners, over the ſtones L or M ; and it is let into the Stone-Hoppers by the little bags bb as faſt as it can be ground. When ground it falls into the Conveyer N N, which carries it into the Elevator O O, this raiſes and empties it into the Hopper-Boy at P, which is ſo conſtructed as to carry it round in a ring, gathering it gradually towards the center, till it ſweeps it into the Bouting-Hoppers Q Q.

The tail flour, as it falls, is guided into the Elevator, to aſcend with the meal , and, that a proper quantity may [illegible] elevated, there is a regulating board R, ſet under the ſuperfine cloths, on a joint x, ſo that it will turn towards the head or tail of the Reel, and ſend more or leſs into the Elevator, as may be required.

There is a piece of coarſe cloth or wire put on the tails of the ſuperfine reels, that will let all paſs through except the bran, which falls out at the tail, and a part of which is guided into the elevator with the tail flour, to aſſiſt the boulting in warm weather the quantity is regulated by a ſmall board r, ſet on a joint under the ends of the reels. Beans may be uſed to keep the cloths open, and ſtill be returned into the elevator to aſcend again. What paſſes through the coarſe cloth, or wire, is guided into the cloth S, to be boulted.

To clean Wheat, ſeveral times.

Suppoſe the grain to be in the ſcreen-hopper F. Draw the gate a ; ſhut the gate e ; move the ſliding hopper K over the ſpout K c d ; and let it run into the elevator to be raiſed agan. Turn the crane ſpout over the empty hopper G, and the wheat will be all depoſited there nearly as ſoon as it is out of the hopper F. Then draw the gate e, ſhut the gate a, and turn the crane ſpout over F; and ſo on alternately, as often as neceſſary. When the grain is ſuficiently cleaned, ſlide the hopper K over the holes that lead into the ſtones.

The ſcreenings fall into a garner, hoppervide. To clean them draw the gate f, and let them run into the elevator, to be elevated into the ſcreen hopper F. Then proceed with them as with the wheat, till ſufficiently clean.—To cle[illegible] the fannings, dra[illegible] the little gate h, and let them into the elevator, [illegible] as before.

Deſcription of the ELEVATOR, CONVEYER, *and* HOPPER-BOY.

The ELEVAT[illegible]thern ſtrap, revolving round two pullies, with buckets fa[illegible], which are filled at the lower and emptied

Of the MEAL-ELEVATOR.

Pullies 16 inches in diameter and 3 1-2 thick, higheſt in the middle. Buckets the width of the ſtrap ; their number in proportion to the work they have to do ; if to hoiſt meal, tail-flour and bran, for two pairs of ſtones, put one for every foot. The motion 24 revolutions of the pully in a minute, which will move the ſtrap about 100 feet. It will diſcharge the better if made to lean a little.

Of the CONVEYER.

After the *deſcription* that has already been given of the Conveyer, we need make but few remarks on the manner of conſtructing it.---The eight-ſided ſhaft ought to be 5 inches in diameter : the flights 3 inches long, and 2 1-2 wide. Set the firſt flight in an oblique manner, on one ſide of the ſhaft, at the end next the elevator ; turn the ſhaft the way in which it is to go when at work, and at the diſtance of 1 1-2 inch, ſet a flight on the next ſide, in the ſame oblique manner as the other ; and thus go on to turn the ſhaft, and to inſert a flight at every 1 1-2 inch diſtance, till the ſpiral line be completed to the other end of the ſhaft. Then the ſpire of the ſcrew will appear, at firſt view, to be turned the wrong way, the flights ſtanding acroſs the ſpiral line.

Beſides the *conveying flights*, above-mentioned, half their number of other flights are to be ſet with their broad-ſide foremoſt. Theſe are called *lifting-flights*. Their uſe is, to lift the meal from one ſide of the ſhaft, and let it fall through the air to the other ſide. This is of great uſe in cooling the meal.

Of the DESCENDER.

When there is not ſufficient deſcent from the elevator to the hopper-boy for the meal to run, the Deſcender becomes neceſſary. This is a broad ſtrap of thin canvaſs or linen, ſet to revolve round two pullies, hung on nice pivots, in a trough or caſe, one end of which is lower than the other ; ſo that the weight of the meal let fall from the elevator on the upper ſide of the ſtrap, near the higher pully, ſets the ſtrap in motion, by which the meal is diſcharged at the lower pully into the hopper-boy. Here a ſmall bucket, extending acroſs the ſtrap, is uſed, for the purpoſe of bringing up the waſte meal out of the caſe.

The *pullies* ought to be one foot in length ; 9 inches diameter at the ends, and 10 in the middle : the *ſtrap* 12 inches wide ; and the trough 14 inches wide, and 15 deep. As this machine will ſeldom be neceſſary, no repreſentation of it is given in the Plate.

Of the HOPPER-BOY

Length of the arms, 14 feet. Their width to be 7 inches at the centre, and 5 at the end[illegible] their thickneſs 2 inches and a half, the fore ſide ſloping upwards[illegible] Flights are to be ſet four inches aſunder ; their inclination to be a[illegible]t 2 1-4 inches for five in length near the cen-

Figure 1.8. From *The Universal Asylum and Columbian Magazine* broadside, January 1791. (Germantown Historical Society & Oliver Evans Press, Wallingford, PA, 1990)

FROM ARTICLE AND BROADSIDE TO BOOK

In 1791, after the article and broadside were published, Evans evidently felt that his broadside could use some further visual aid. He therefore accompanied the broadside in Wilmington and Philadelphia with a three-dimensional model of his mill. This idea of using a model as a way to introduce new technology was quite popular in America throughout the 19th century. For exam-

A N S.

ter; 1 inch in 5 at the ends, and 2 inches in depth, that in moving round, they may collect the meal towards the center. One Sweeper muſt be placed at each extemity of the arm, and one over the hopper. The upright ſhaft to be four feet and a half, made round, that the arm may have liberty to riſe about three feet. Let the pully for the balance be 8 inches in diameter.

Several Uſes to which the ELEVATOR *may be applied, beſides thoſe repreſented in the Plate, viz.*

To elevate Wheat, or any other grain, from ſhallops into mills, or ſtore-houſes; either by having it firſt brought, by a Conveyer, into the Elevator, or by a ſhort Elevator that may bring it in through a door or window, and diſcharge it into the main Elevator. This ſhort Elevator may be put up, and taken down, as required.

If two pair of Stones be turned by one water-wheel, at the ſame time, one to chop and the other to grind down, as it paſſes through one pair it may be elevated into the Hopper of the other: by this method, much more may be done by one wheel in the ſame time.

If the Screen and Fan be on the ſecond floor, ſo that a large quantity of Wheat cannot be cleaned, and laid over the Stones, at a time, to ſupply them regularly, it may be elevated from the tail of the Fan to a hanging Garner or any floor over the Stones, from whence it can be let in by a Spout.

If there be no convenience to clean Wheat over the Stones, then a Garner may be made hopperwiſe under the Fan, and a ſmall Elevator ſet to run conſtantly, to elevate the Wheat from ſaid Garner into a Spout, that will ſupply two or three pair of Stones regularly. This Elevator muſt be ſo conſtructed as to hoiſt a little faſter than to ſupply the Stones, always keeping the Spout full, the ſurplus will fall back.

If the clean Wheat ſuits beſt to be at a diſtance from the Stones, and near on a level with the Hopper, the Elevator may be ſet to rake it into the Hopper. If it brings too much it will carry it back again.

If the Wheat cannot be laid over the Rolling Screen, but on the ſame floor, or one below it, either under it, or at a diſtance, the Elevator may be ſet to bring it in; the ſurplus it will carry back.

If the Wheat paſſes through a Rubbing or Shelling Mill, it may be elevated to the cleaning machines.

If the Wheat be damp, lying in a Garner, it may be ſpouted into the Elevator, and elevated into another Garner, this will dry it much, in a dry day.

A Kiln may be contrived to dry Corn with, a Hopperboy to ſtir it, the grain let in at the center and ſpread ſo ſlowly that it will be ſufficiently dried before it comes to the extremes, which are to deliver it into the Elevator, to be elevated into the mill and depoſited over the Stones.

If applied to elevate grain, or the like, from ſhips to ſtorehouſes, and wrought by a man who applies his weight as the power, he will be able to elevate as much as two men can carry; but it may be moved by a horſe, ſteam, wind, or tide.

If the tail flour or bran will not run into the Elevator, it may be ſet to rake it in, and to elevate the bran and ſhorts to another ſtory, that it may be ſpouted into a ſhallop, or to take it in any direction to a more ſuitable place.

Figure 1.9. From *The Universal Asylum and Columbian Magazine* broadside, January 1791. (Germantown Historical Society & Oliver Evans Press, Wallingford, PA, 1990)

ple, in the 1776 petition to the Pennsylvania Assembly by the United Company of Philadelphia for Promoting American Manufactures, the petitioners asked for £7 for each model of the spinning jenny, to hasten its use and "so contribute to the Establishment of American Manufactures."[44] The U.S. government required working scale-models as part of all patent applications until very late in the 19th century. [45] Further, evidence of the importance of three-dimensional models to such diverse groups as New York shipwrights and cash register salesmen will be discussed in the next two chapters of this book.

If Wheat or Meal will not run from the Elevator as diſcharged, to the place deſired, it may be applied to rake it to the place.

If there is no room for a Hopper-Boy the meal may be cooled by Conveyers, and let fall over the Boulting-Hopper.

The Conveyer will anſwer many of the above purpoſes, as well as the Elevator.---The Hopper-Boy may be varied many ways, to ſuit different ſituations, and to attend one, two, or three, Boulting-Hoppers at once. So that it is evident, that the improvements will apply to all flour mills, however conſtructed.

When applied to mills that grind for toll only---the meal is elevated as it is ground, and in order that it may cool it is let run down a long broad Spout into the Boulting-Hopper. The Shoe is ſmall that it may clean out quicker. The Cheſt made ſteep and to come to a point like a funnel, under which are hung two Bags, one for meal, the other for the bran; it is ſo contrived as to keep the fine and coarſe meal ſeparate, if deſired. When the miller ſees the firſt griſt all into the Elevator, he ſtops the other from running in, until all the firſt is in the cloth, he then lets the ſecond run in, and ſtops the knocking of the Shoe by pulling a cord that is faſtened to it, and paſſing over a Pully, hangs near the meal Spout, until all the firſt griſt is in the Bags, and others put in their place, he then lets the Shoe knock again.

The following are two of many certificates of the utility of theſe improvements, which have been obtained from thoſe who have applied them in practice.

THESE are to certify that we have erected *Oliver Evans's* new invented mode of elevating, conveying and cooling Meal, &c. As far as we have experienced, we have found them to anſwer a valuable purpoſe, well worthy the attention of any perſon concerned in Merchant or even extenſive country Mills, who wiſheth to leſſen the labour and expenſe of manufacturing wheat into flour.

Ellicotts'-Mills. Baltimore County—State of Maryland. Auguſt 4th, 1790.

JOHN ELLICOTT,
JONATHAN ELLICOTT,
GEORGE ELLICOTT,
NATHANIEL ELLICOTT.

WE the ſubſcribers do hereby certify, that we have introduced *Oliver Evans's* improvements into our Mills at Brandywine; and have found them to anſwer as repreſented by the above plate and deſcription; alſo to be a great ſaving of waſte, labour, and expenſe and not ſubject to get out of order.—We therefore recommend them, as well worthy the attention of thoſe concerned in manufacturing grain into flour.

Brandywine Mills 3d Month 28th, 1791.

JOSEPH TATNALL,
THOMAS LEA,
SAMUEL HOLLINGSWORTH,
THOMAS SHALLCROSS,
CYRUS NEWLIN.

☞ Perſons deſirous of obtaining the privilege of uſing theſe improvements, may apply to Oliver Evans, the proprietor, at *Wilmington (Delaware* State) to Robert Leſlie in *Philadelphia*, to Jonathan Ellicott, at *Ellicotts'-Mills*, near *Baltimore*, or to Elias Ellicott, in *Baltimore*, to Nathaniel Ellicott, *Peterſburg, Virginia;* Samuel Reynolds, *Albany;* William Byrnes, *New-Windſor*, on *Hudſon River*, State of *New-York*.

*** The following low price is eſtabliſhed, and will not be raiſed before the firſt day of April 1792, viz. For the privilege of uſing ſaid improvements in merchant mills, the ſum of 33 dollars for each water wheel, to which the ſame is applied—and for mills that grind for toll only, the ſum of 16 dollars.

Figure 1.10. From *The Universal Asylum and Columbian Magazine* broadside, January 1791. (Germantown Historical Society & Oliver Evans Press, Wallingford, PA, 1990)

Because his communication pieces—the article, broadside, and model—finally spoke the language of the millers, Evans overcame their suspicions.[46] A favorable response to the article, broadside, and model was quick. On March 28, 1791, the millers on the Brandywine River—Evans's home territory—signed a testimonial recommending Evans's machines as "well worth the attention of those concerned in manufacturing grain into flour."[47] No longer were Evans's machines "rattle-traps," as the millers had earlier claimed. With such a signed testimonial, Evans also began to overcome the suspicions of distant millers who needed proof that Evans's own neighbors supported him. As a result of Evans's new more visual approaches to communicating his ideas, Matthew Carey, the editor of *American Museum*, wrote the following in May 1792:[48]

> Such a mill, ten years ago, would have been ranked with the perpetual motion, or philosopher's stone, as the dreams of a visionary projector; but we have lived to see it constructed; and are at a loss which most to admire; the extent of the improvement, or the simplicity, and if I may so express myself, the harmony with which it is applied. . . . We are happy to learn that mr. Evans is now likely to reap the fruits of his labour; upwards of 100 mills have adopted his improvements, and they are daily increasing in public estimation.

No longer were the dreams and plans of Evans's mind's eye dismissed as those of a "visionary projector." Now millers and millwrights understood and began to pay him his license fees. Yet, Evans's campaign on behalf of his invention was not complete. In the final sentence of his 1791 letter to Young, Evans had made a proposition for creating a "pamphlet" that never came to fruition through Young's The *Universal Asylum and Columbian Magazine*:[49]

> Several drawings . . . with explanations of the true principles of each...the whole combined might form a pamphlet that might convey much useful knowledge and greatly tend to the improvement of the important Branch of our Manufacture of flour.

Nevertheless, this pamphlet proposition did plant the seed for the creation of the book-length description of Evans's machines in *The Young Mill-wright and Miller's Guide*.

In advertising *The Guide*, Evans originally wrote that the book would be modest in size (350 pages and 20 plates) and price ($1.50).[50] However, as Evans warmed to his subject, the book grew to over 500 pages and 26 plates, and, consequently, the price doubled. This unexpected increase in size and price eventually even took a physical toll on Evans, for he later wrote in 1813 that: "this cost him nearly three years of the most intense study and application which had seriously impaired his constitution and reduced him to a state of real indigence."[51] Moreover, in another 1817 tract, Evans described this "indigence" in vivid details:[52]

> I was reduced to such abject poverty that my wife sold the tow clothe which she had spun with her own hands for clothing for her children, to get bread for them, my head was covered with many gray hairs and I required spectacles.

The reasons why Evans risked such financial and physical ruin in creating *The Guide* were both altruistic as well as mercenary.

Risking Financial And Physical Ruin?

Among the more altruistic explanations for why Evans faced personal disaster in expanding his five-page article to a book of 500 pages was one offered by Ferguson in his biography of Evans. Ferguson wrote that Evans had a "zeal to publish technical information for the guidance of young men."[53] He may be correct because his explanation echoes a conversation Evans was reported to have had a year before his death:[54]

> Mr. Evans had much to say on the difficulties inventive mechanics labored under for want of published records of what had preceded them, and for works of reference to help the beginner. In speaking of his own experience, he said that everything he had undertaken he had been obliged to start at the very foundation; often going over ground that others had exhausted and abandoned; leaving no record.

Another altruistic explanation for Evans's risk may have been the patriotism he shared with many others in those early years after the War for Independence. Recall that five years before his death in 1814, Evans had written to President Madison to say that when he had begun to think about the assault on his country by British naval vessels during the War of 1812 "his mind runs constantly on devising the means of destroying enemies ships in our waters." Evans struck a similar patriotic chord in the conclusion of the original edition of *The Guide* when he wrote:[55]

> Sensible of the expense, time, labor, and thought that this (though small) work has cost me, and hoping that it may be well received by, and prove serviceable to my country—I wait to see its fate; and feel joy at being ready to say FINIS.

However, Evans's most powerful motive for taking such risks was probably mercenary. He wanted to leave no stone unturned in his efforts to sell and popularize his designs. Thomas Jones,[56] who edited *The Guide* after Evans's death, added a preface in 1834 in which he noted that "the main object of Evans's writing this work was to introduce his invention to public notice . . . " That certainly echoed what Evans himself wrote in his 1817 memoir:[57]

> I commenced the work, intending only to explain my own improvements, but being advised to add something respecting building mills, as

> well as on the manufacture of flour, to induce millwrights and millers to purchase and read it, I was led into the investigation of the principles of water acting on wheels to drive mills, and of the construction of mills.

Moreover, this mercenary explanation was further illustrated by the fact that all the licenses (Figure 1.11) that were later issued by Evans at $32 each contained Plate VIII (Figure 1.17) from the book.[58]

Whether the motive was altruistic, mercenary, or a mixture of the two, Evans wrote *The Guide* for very practical considerations because, as he noted, "I am not writing so particularly to men of science as to practical mechanics."[59] In a similar fashion, the first published review of *The Guide* in the *American Monthly Review* (September 1795) noted: "Part II. and III. entitled, The Young Mill-Wright's Guide, seem particularly intended for the instruction of practical mill-wrights."[60] Hence, Evans's original creative strategies, book, pictures, and financial success of his licenses were all inextricably intertwined. Evans's ideas began in his "mind's eye," which eventually led him to communication with both text and pictures, and the process concluded with visual designs purchased with license fees by mechanicians. However, between is mind's eye and the mechanicians lay *The Guide*; a very curiously organized book.

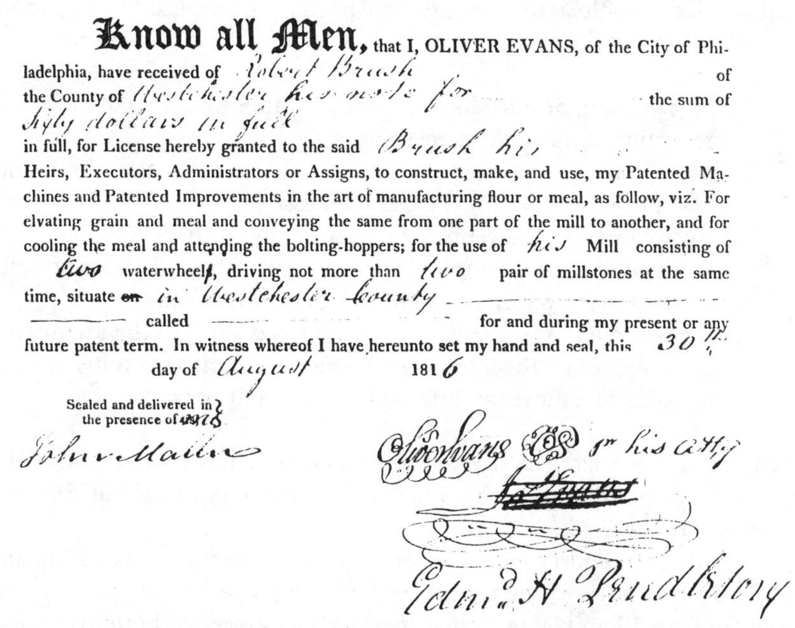

LICENSE.

Know all Men, that I, OLIVER EVANS, of the City of Philadelphia, have received of Robert Brush of the County of Westchester his note for the sum of Fifty dollars in full in full, for License hereby granted to the said Brush his Heirs, Executors, Administrators or Assigns, to construct, make, and use, my Patented Machines and Patented Improvements in the art of manufacturing flour or meal, as follow, viz. For elvating grain and meal and conveying the same from one part of the mill to another, and for cooling the meal and attending the bolting-hoppers; for the use of his Mill consisting of two waterwheels, driving not more than two pair of millstones at the same time, situate in Westchester County called for and during my present or any future patent term. In witness whereof I have hereunto set my hand and seal, this 30th day of August 1816

Sealed and delivered in the presence of us

Figure 1.11. Front of blank license for Evans's design, page 161 from "the PORTFOLIO" (Hagley Museum and Archives, Wilmington, DE, Accession 1538.

THE YOUNG MILL-WRIGHT AND MILLER'S GUIDE, 1795: VISUAL COMMUNICATION AND VISUAL COMPREHENSION

On first glance, *The Guide* appears very disjointed. There are six parts with various lengths and weaves of text and graphics. Some parts were written wholly by others; for example, Part V by Thomas Ellicott. Some parts were taken verbatim from other published books; for example, much of Part 1 and the Appendix were either from John Smeaton's *Experimental Enquiry Concerning the Natural Powers of Wind and Water to turn Mills and other machines*[61] or William Waring's "Observations on the Theory of Water Mills".[62] Some parts reprinted Evans's own *Universal Asylum and Columbian Magazine* article and second broadside. Some parts seemed extraneous; for example, flour-manufacturing and milling were interrupted by discussions of mechanics, hydraulics, and a consideration of the invention process itself ("The True Paths to Inventions"). Furthermore, some parts repeat information presented in other parts; the material on the overshot waterwheel appeared three times, in Part I, II, and V.[63]

However, this disjointedness disappears when Evans's intentions underlying the various parts and their placement in *The Guide* are clarified. In fact, the sense of disjointedness disappears completely by observing how Evans carefully organized his book by the method of rhetorical organization that was fashionable at the time, the "dispositio" of classical Greek and Roman rhetoric. Classical dispositio organization includes five divisions:[64]

1. an *exordium*, or introduction, whose purpose is to make the readers receptive to the author's message.
2. a *narratio*, the statement of background facts that lays the foundation for understanding the central message.
3. the *confirmatio*, the central statement of the author's message.
4. the *refutatio*, the section in which the author would seek to discredit any opposing views in a kind of preemptive strike.
5. the *peroratio*, or conclusion, which would sum up the argument as a whole, move the audience to action, and often amplify the conclusions to a more important plane of significance.

Table 1.1 summarizes the parts of *The Guide*, how Evans's parts fit in classical rhetorical dispositio, and what their positioning suggests about their various purposes in *The Guide*.

Parts III and IV were the core of Evans's message, his confirmatio, and where Evans described the inventions he hoped to sell to the public. The centrality of Parts III and IV to Evans's message was suggested both by the fact that Part III almost completely repeated his already successful 1791 *Universal Asylum and Columbian Magazine* article and second broadside,[65] as well as by

Table 1.1. Seven Parts of *The Guide*.

Title	Preface	Part I Mechanics and Hydraulics	Part II Young Millwright's Guide	Part III Containing Evans's Patented Improvements	Part IV Young Miller's Guide	Part V The Practical Millwright	Appendix Rules for Discovering New Improve-ments
Number of pages	2	154	68	51	38	90	10
% of total pages in the book	1	37	17	12	9	22	2
% of pages taken verbatim from other authors	0	27	0	0	0	100	33
% of pages with references to graphics	0	14	36	80	16	50	50
Dispositio Functions	Exordium	Narratio	Narratio	Confirmatio	Confirmatio	Refutatio	Peroratio

the fact that the licenses later issued by Evans reprinted a graphic from Part III (Plate VIII). It was also in Part III and IV that Evans took the least amount of material from other authors. Consequently, it is almost as if the work from other authors in Parts I, V, and the Appendix were used by Evans as a persuasive setting for his own jewels in Parts III and IV. (Plate VIII and the 10 pages of explanation in Part III, Articles 89 and 90, are shown in Figures 1.12–1.16.) Not only were Parts III and IV at the core of Evans's message, but the manner in which Evans composed them, their weave of text and graphics, best exemplified the work of Evans's mind's eye. Nearly 80% of the pages in Part III have references to graphics. Such a dense weave of text and graphics was so successful in conveying information to 18th-century mechanicians that it even occasionally worked to Evans's disadvantage. For example, during one of Evans's many patent lawsuits,[66] the defendant claimed that the $40 fee for a license to use Evans's hopper-boy, referenced in Article 89, was sufficient; Evans's set price for a license was, however, $100. The defendant claimed that the $40 fee was sufficient because the defendant had already gotten enough information to make his own hopper-boy from "a plate in the Millwright's Guide."[67]

78 DESCRIPTION OF MACHINES. *Chap. I.*

Art. 88. a confiderable diftance, with a fmall defcent. Where a motion is eafily obtained from the water, it is to be preferred to that of working itfelf, it being eafily ftopped, is apt to be troublefome.

The Crain Spout is hung on a fhaft to turn on pivots or a pin, fo that it may turn every way, like a crane; into this fpout the grain falls from the elevator, and, by turning, it can be directed into any garner. The fpout is made to fit clofe, and play under a broad board, and the grain is let into it through the middle of this board, near the pin, fo that it will always enter the fpout. See it under B, fig. 1. L is a view of the under fide of it, and M is a top view of it. The pin or fhaft may reach down fo low, that a man may ftand on the floor and turn it by the handle x.

CHAPTER II.

Art. 89.

APPLICATION OF THE MACHINES, IN THE PROCESS OF MANUFACTURING WHEAT INTO SUPERFINE FLOUR.

PLATE VIII, is not meant to fhew the plan of a mill; but merely the Application and Ufe of the patented Machines.

Chap. II. APPLICATION OF THE MACHINES. 79

Art. 89. Of receiving the wheat.

The grain is emptied from the waggon into the fpout 1, which is fet in the wall, and conveys it into the fcale 2, that is made to hold 10, 20, 30, or 60 bufhels, at pleafure.

There fhould, for convenience of counting, be weights of 60lbs. each; divided into 30, 15 and 7 1-2lbs. then each weight would fhew a bufhel of wheat, and the fmaller ones halves, pecks, &c. which any one could count with eafe.

When the wheat is weighed, draw the gate at the bottom of the fcale, and let it run into the garner 3; at the bottom of which there is a gate to let it into the elevator 4—5, which raifes it to 5, and the crane fpout being turned over the great ftore garner 6, which communicates from floor to floor, to garner 7, over the ftones 8, which fuppofe to be for fhelling or rubbing the wheat, before it is ground, to take off all duft that fticks to the grain, to break fmut or fly-eaten grain, lumps of duft, &c. As it is rubbed it runs, by the dotted lines, into 3 again; in its paffage it goes through a current of wind blowing into the tight room 9, having only the fpout a, through the lower floor, for the wind to efcape; all the chaff will fettle in the room, but moft of the duft paffes out with the wind at a. The wheat again runs into the elevator at 4, and the crane fpout, at 5, is turned over the fcreen hoppers 10 or 11, and the grain lodged there, out of which it runs into the rolling fcreen 12, and defcends through the current of wind made by the fan 13, the clean heavy grain defcends, by 14, into the conveyer 15—16;

Figure 1.12. Article 89 from Part III, *The Guide*

In contrast to Evans's dense weave of text and graphics in Parts III and IV, which echoed the visual quality of his own mind's eye, Part I, (the narratio, see Table 1.1), revealed a quite different weave of text and graphics. Only 14% of the total pages in this part had any references to graphics (see Figure 1.17). Such a contrast in the weaves of text and graphics suggested that Part I and Part III in *The Guide* were, in fact, based on different ways of thinking. Part I was based on the scientific-mathematical thought, whereas Part III was based on the visual thought of Evans's mind's eye.

Part 1, written in such a scientific-mathematical manner, functioned in *The Guide* as Evans's "first principles." Initiating a book with "first principles" was a widely used practice in the 18th century: For example, when Thomas Price wrote his history of America, he initiated it with the story of Adam and Eve. Evans's "Adam and Eve" was the exposition on mechanics and hydraulics derived from Smeaton, Waring, Martin, and Ferguson.

80 APPLICATION OF THE MACHINES. *Chap. II.*

Art. 89.

which conveys it into all the garners over the ſtones 7—17—18, and theſe regularly ſupply the ſtones 8—19—20, keeping always an equal quantity in the hoppers, which will cauſe them to feed regularly; as it is ground the meal falls to the conveyer 21—22, which collects it to the meal elevator, at 23, and it is raiſed to 24, whence it gently runs down the ſpout to the hopper-boy at 25, which ſpreads and cools it ſufficiently, and gathers it into the bolting hoppers, both of which it attends regularly; as it paſſes through the ſuperfine cloths 26, the ſuperfine flour falls into the packing cheſt 28, which is on the ſecond floor: If the flour is to be loaded on waggons, it ſhould be packed on this floor, that it may conveniently be rolled into them; but if the flour is to be put on board a veſſeſ, it will be more convenient to pack on the lower floor, out of cheſt 29, and roll it into the veſſel at 30. The ſhorts and bran ſhould be kept on the ſecond floor, that they may be conveyed by ſpouts into the veſſel's hold, to ſave labour.

The rublings which fall from the tail of the 1ſt reel 26, are guided into the head of the 2d reel 27; which is in the ſame cheſt, near the floor, to ſave both room and machinery. On the head of this reel is 6 or 7 feet of fine cloth, for tail flour, and next to it the middling ſtuff, &c.

The tail flour which falls from the tail of the 1ſt reel 26, and head of the 2d reel 27, and requires to be bolted over again, is guided by a ſpout, as ſhewn by dotted lines 31—22, into the conveyer 22—23, to be hoiſted again

Chap. II. APPLICATION OF MACHINES. 81

with the ground meal; a little bran may be let in with it, to keep the cloth open in warm weather—But if there be not a fall ſufficient for the tail flour to run into the lower conveyer, there may be one ſet to convey it into the elevator, as 31—32. There is a little regulating board, turning on the joint x under the tail of the firſt reels, to guide more or leſs with the tail flour.

Art. 89. Tail flour & middlings hoiſted and bolted over.

The middlings, as they fall, are conveyed into the eye of either pair of millſtones by the conveyer 31—32, and ground over with the wheat; which is the beſt way of grinding them, becauſe the grain keeps them from being killed, and there is no time loſt in doing it, and they are regularly mixed with the flour. There is a ſlanting ſliding board, to guide the middlings over the conveyer, that the miller may take only ſuch part, for grinding over, as he ſhall judge fit; and a little regulating board between the tail flour and middlings, to guide more or leſs into the ſtones or elevator.

Middlings ground over with the wheat.

Which ſaves labour and time.

The light grains of wheat, ſcreenings, &c. after being blown by the fan 13, fall into the ſcreenings garner 32; the chaff is driven further on, and ſettles in the chaff-room 33; the greater part of the duſt will be carried out with the wind through the wall. For the theory of fanning wheat, ſee art. 83.

To clean the Screenings.

Draw the little gate 34, and let them into the elevator at 4, and be elevated into garner

Screenings cleaned.

M

Figure 1.13. Article 89 from Part III, *The Guide*

Evans not only drew his requisite "first principles" from Smeaton, Waring, Martin, and Ferguson, but, by including them in Part 1 in his text, Evans also bolstered his own unknown inventions by using their established names and widely respected books.[68] Evans probably expected to reprint the material (some 27% of Part I appears verbatim from these established authors) because reprinting was entirely permissible at that time:[69]

> The customary approach to compiling a handy reference work was to copy, with or without acknowledgment, sections of earlier books on mechanics and hydraulics. Such plagiarism, as we think it, was neither forbidden by law nor frowned upon by custom.

However, Evans's reprinting was inconsistent: Sometimes he reprinted his source material verbatim, and sometimes he rewrote the material in his own words. Most often, the material Evans reprinted verbatim was originally written by authors in a mathematical and verbal style (Smeaton and Waring), whereas the material Evans paraphrased was written by authors who used a

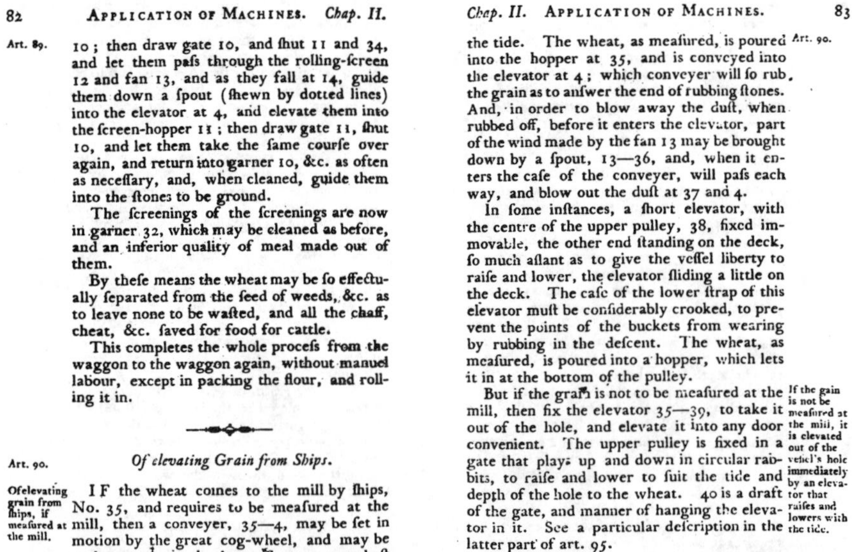

82 APPLICATION OF MACHINES. *Chap. II.*

Art. 89.

10; then draw gate 10, and ſhut 11 and 34, and let them paſs through the rolling-ſcreen 12 and fan 13, and as they fall at 14, guide them down a ſpout (ſhewn by dotted lines) into the elevator at 4, and elevate them into the ſcreen-hopper 11; then draw gate 11, ſhut 10, and let them take the ſame courſe over again, and return into garner 10, &c. as often as neceſſary, and, when cleaned, guide them into the ſtones to be ground.

The ſcreenings of the ſcreenings are now in garner 32, which may be cleaned as before, and an inferior quality of meal made out of them.

By theſe means the wheat may be ſo effectually ſeparated from the ſeed of weeds, &c. as to leave none to be waſted, and all the chaff, cheat, &c. ſaved for food for cattle.

This completes the whole proceſs from the waggon to the waggon again, without manuel labour, except in packing the flour, and rolling it in.

Art. 90.

Of elevating Grain from Ships.

Of elevating grain from ſhips, if meaſured at the mill.

IF the wheat comes to the mill by ſhips, No. 35, and requires to be meaſured at the mill, then a conveyer, 35—4, may be ſet in motion by the great cog-wheel, and may be under or above the lower floor, as may beſt ſuit the height of the floor above high water. This conveyer muſt have a joint, as 36, in the middle, to give the end that lays on the ſide of the ſhip, liberty to raiſe and lower with

Chap. II. APPLICATION OF MACHINES. 83

Art. 90.

the tide. The wheat, as meaſured, is poured into the hopper at 35, and is conveyed into the elevator at 4; which conveyer will ſo rub the grain as to anſwer the end of rubbing ſtones. And, in order to blow away the duſt, when rubbed off, before it enters the elevator, part of the wind made by the fan 13 may be brought down by a ſpout, 13—36, and, when it enters the caſe of the conveyer, will paſs each way, and blow out the duſt at 37 and 4.

In ſome inſtances, a ſhort elevator, with the centre of the upper pulley, 38, fixed immovable, the other end ſtanding on the deck, ſo much aſlant as to give the veſſel liberty to raiſe and lower, the elevator ſliding a little on the deck. The caſe of the lower ſtrap of this elevator muſt be conſiderably crooked, to prevent the points of the buckets from wearing by rubbing in the deſcent. The wheat, as meaſured, is poured into a hopper, which lets it in at the bottom of the pulley.

If the gain is not be meaſured at the mill, it is elevated out of the veſſel's hole immediately by an elevator that raiſes and lowers with the tide.

But if the grain is not to be meaſured at the mill, then fix the elevator 35—39, to take it out of the hole, and elevate it into any door convenient. The upper pulley is fixed in a gate that plays up and down in circular rabbits, to raiſe and lower to ſuit the tide and depth of the hole to the wheat. 40 is a draft of the gate, and manner of hanging the elevator in it. See a particular deſcription in the latter part of art. 95.

By the ſtrength of one man.

This gate is hung by a ſtrong rope paſſing over a ſtrong pulley or roller 41, and thence round the axis of the wheel 42; round the rim of which wheel there is a rope, which paſſes

Figure 1.14. Article 89 and 90 from Part III, *The Guide*

much more visual style of presentation (Ferguson and Martin). This variation in the handling of his source materials can actually help to describe the role of Evans's mind's eye in his own comprehension of new technology.

The Mathematical and Verbal Styles of Smeaton and Waring

In 1795, reviewers noted that Part I of *The Guide* was "science," and it is clear that the science came primarily from the work of John Smeaton.[70] On May 3, 10, and 24, 1759, Smeaton addressed the Royal Society concerning the scientific understanding of water and wind mills. He later published in their *Philosophical Transactions* a lengthy exposition entitled *Experimental Enquiry Concerning the Natural Powers of Wind and Water to turn Mills and other machines depending on a circular motion and an experimental examination of the quantity and proportion of mechanic power necessary to be employed in giving different degrees of velocity to heavy bodies from a state of rest also new fundamental experiments upon the collision of bodies*. For his address and subsequent article, Smeaton received the Royal Society's Gold Medal in 1759.[71]

84 APPLICATION OF MACHINES. *Chap. II.*

Art. 90.

round the axis of wheel 43, round the rim of which wheel is a ſmall rope, leading down over the pulley P, to the deck, and faſtened to the cleet q; a man by pulling this rope can hoiſt the whole elevator; becauſe if the diameter of the axis be 1 foot and the wheels 4 feet, the power is increaſed 16 fold, by art. 20. The elevator is hoiſted up, and reſted againſt the wall, until the ſhip comes too, and is faſtened ſteady in the right place, then it is ſet in the hole on the top of the wheat, and the bottom being open, the buckets fill as they paſs under the pulley; a man holds by the cord, and lets the elevator ſettle as the wheat ſinks in the hole,, until the lower part of the caſe reſts on the bottom of the hole, it being ſo long as to keep the buckets from touching the veſſel; by this time it will have hoiſted 1, 2 or 300 buſhels, according to the ſize of the ſhip and depth of the hole, at the rate of 300 buſhels per hour. When the grain ceaſes running in of itſelf, the man may ſhovel it up, till the load is diſcharged.

Lifting the wheat up itſelf.

200 buſhels per hour.

The elevator diſcharges the wheat into the conveyer at 44, which conveys it into the ſcreen-hoppers 10—11, or into any other garner, from which it may deſcend into the elevator 4—5, or into the rubbing-ſtones 8.

Into a conveyer that conveys it into any garner, and at the ſame time rubs off the duſt.

This conveyer may ſerve inſtead of rubbing-ſtones, and the duſt rubbed off thereby may be, by a wind-ſpout from the fan 13, into the conveyer at 45, blown out through the wall at p. The holes at 44 and 10—11 are to be ſmall, to let but little wind eſcape any where but out through the wall, where it will carry the duſt.

Chap. II. APPLIÇATION OF MACHINES. 85

A ſmall quantity of wind might be let into the conveyer 15—16, to blow away the duſt rubbed off by it. Art. 90.

The fan muſt be made to blow very ſtrong, to be ſufficient for all theſe purpoſes, and the ſtrength of the blaſt regulated as directed by art. 83.

A Mill for grinding Parcels. Art. 91.

HERE each perſon's parcel is to be ſtored in a ſeparate garner, and kept ſeparate thro' the whole proceſs of manufacture, which occaſions much labour; almoſt all of which is performed by the machines. See plate VI. fig. 1; which is a view of one ſide of a mill containing a number of garners holding parcels, and a ſide view of the wheat elevator.

Application to a mill for grinding merchant work in parcels; the grain elevated from the waggon into any garner, and from them again, into the rolling-ſcreen, &c. by drawing gates; and kept ſeparate.

The grain is emptied into the garner g, from the waggon, as ſhewn in plate VIII; and, by drawing the gate A, it is let into the elevator AB, and elevated into the crane-ſpout B, which being turned into the mouth of the garner-ſpout BC, which leads over the top of a number of garners, and has, in its bottom, a little gate over each garner; which gates and garners are all numbered with the ſame numbers reſpectively.

Suppoſe we wiſh to depoſit the grain in the garner No. 2, draw the gate 2 out of the bottom, and ſhut it in the ſpout, to ſtop the wheat from paſſing along the ſpout paſt the hole, ſo that it muſt all fall into the garner; and thus for the other garners 3-4-5-6-&c. Theſe

Figure 1.15. Article 90 from Part III, *The Guide*

Smeaton's objective in his address and article was the scientific investigation of the principles of mechanics for fellow scientists:[72]

> The principles of mechanics were never so clearly exhibited as in his writings, more especially with respect to resistance, gravity, and the power of water and wind to turn mills.[73]

Thus, Smeaton's focus on the principles and audience of "natural philosophers" led him to use pictures differently than Evans, who had more practical motives and an audience of "practical mechanicians." In Smeaton's work, there were only five plates; in Evans's work, there were five times as many.[74] Furthermore, in most of Evans's book, the text was dependent on the pictures for its message and organization, proceeding from one call-out to the next, spatially left to right through the graphic.

In contrast, Smeaton displayed his plates in the front of his book and then later described them in much the same way as he described his experi-

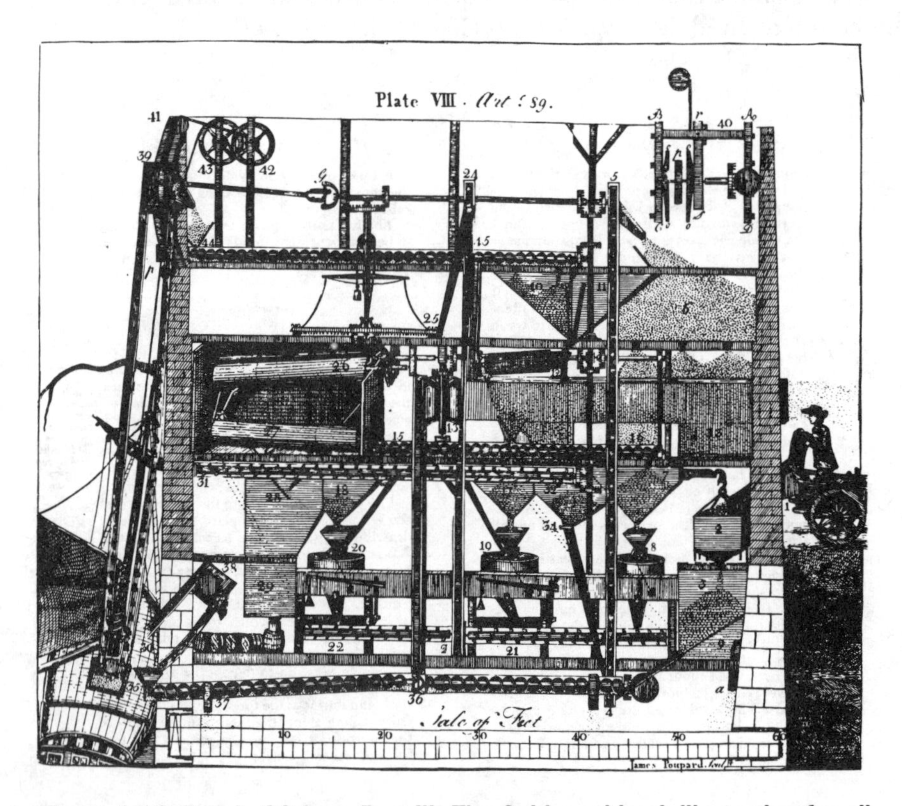

Figure 1.16. Article 90 from Part III, *The Guide* and back illustration for all of Evans's mill design licenses

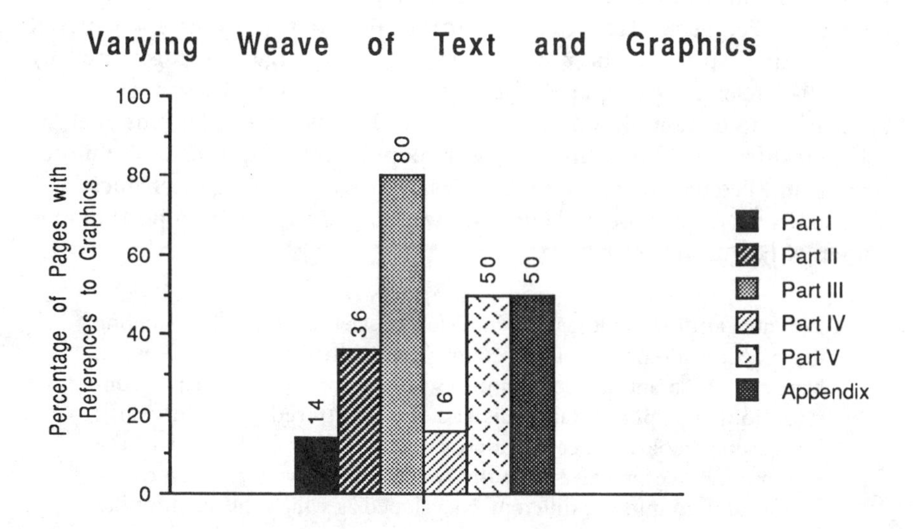

Figure 1.17. Varying weave of text and graphics

mental apparatus, leaving the graphics out entirely and representing them entirely in words and mathematics. For Smeaton, the English scientist, the picture was only a minor means to an end; for Evans, the American mechanician and seller of designs to other American mechanicians,[75] the visual was the alpha and the omega of his entire project. The difference in the use of pictures by Smeaton and Evans was summed up by George Basalla:[76]

> The process of nonverbal thinking is central to the work of engineers and technicians but much less so to scientists, who are more likely to manipulate concepts, mathematical expressions, or hypothetical entities.

Furthermore, although Evans reprinted the largest amount of "science" from Smeaton, he also took five pages of "science" directly from Waring. These five pages also had an uncharacteristic mathematical and verbal style and for many of the same reasons: [77] Waring's 1793 address to the "natural philosophers" of the American Philosophical Society in Philadelphia was written in the same verbal and mathematical style as Smeaton's address before the British Royal Society.

The Visual Styles of Ferguson and Martin

When Evans developed his "first principles" in Part I, he often referred to Ferguson and Martin; however, when he did, he did not excerpt material from them verbatim. Perhaps Evans was able to rephrase and revise their material because in Ferguson's *Lectures* and Martin's encyclopedia Evans encountered more kindred spirits in the use of visuals: 93% of Ferguson's pages used by Evans had references to graphics as did 52% of Martin's. These more visual approaches to communication made sense in Ferguson's and Martin's rhetorical context because, like Evans, they were popularizers of the "natural philosophers," and Ferguson's and Martin's audiences were practicing mechanicians.[78] This similarity of purpose and audience was suggested in a 1806 posthumous edition of Ferguson's Lectures:[79]

> We must attribute that general diffusion of scientific knowledge among the practical mechanics of this country, which has, in a great measure, banished those antiquated prejudices, and erroneous maxims of construction, that perpetually mislead the unlettered artist [artisan]. [Ferguson's book has been] widely circulated among all ranks of the community. We perceive it in the workshop of every mechanic. We find it diffused into the different Encyclopedias which this country has produced; and we may trace it in those popular systems of philosophy which have lately appeared.

Thus, with the more "scientific" mathematical and verbal presentations of material by Smeaton and Waring, Evans repeated the material verbatim; whereas with the more visual approach of the science popularizers like Martin and Ferguson, Evans was able to rephrase the material in his own words.

Perhaps the variations in the way Evans used his source material in Part 1 revealed the similarity of Evans's comprehension and communication style. Evans digested and reformulated the work of Ferguson and Martin because these authors communicated visually—they spoke to Evans's mind's eye, but Evans simply reprinted the work of Smeaton and Waring because he had not successfully digested and reformulated their work—they had not spoken to his mind's eye. Thus, it may be that Evans's choice of integrating text and graphics in his own communications to 18th-century mechanicians was heavily guided by the fact that such a style was the type of communication that Evans himself best understood. It may be that the concluding metaphor in an 1821 biographical sketch of Evans was more real than symbolic:[80]

> Oliver Evans was a born mechanic; He wrote for the workman who had need of an aid to his intelligence and he thus knew how to put himself in such a one's place.

The role of Evans's "mind's eye" not only to invent and communicate but also to comprehend could be glimpsed in his letter of June 16, 1817, addressed to his children:[81]

> George [his son] . . . walks smartly about the works which so far excel ours that we had better travel a whole year to see such before we begin to erect Machinery to facilitate our business in executing or constructing Steam engines I therefore request Mr. Cadwalader [another son] to come immediately up and stay two months and make accurate drawings of every thing he may see from actual measurements and get the whole fixed in his mind...let him bring several sheets of drawing paper.

Evans clearly wanted his son Cadwalader to "get the whole fixed in his mind" and thus he urged him to make "accurate drawings" as a method of comprehending the technological information, Evans echoed Benjamin Franklin's "Proposals relating to the Education of Youth" in which Franklin wrote:[82]

> . . . this was 'the Time to show them Prints of Antient [sic] and modern Machines to explain them, to lete [sic] them be copied' in order to 'fix the Attention on the several Parts, their Proportions, Reasons, Effects, &c . . . With 'Skill of this kind, the Workman may perfect his own Idea of the Thing to be done, before he begins to work.'

Visual invention, visual communication, and visual comprehension, all based on the mind's eye, were the strengths of Evans's work as a mechanician and as an author.

THE LEGACY OF *THE GUIDE*

In his own life time, Evans revised and printed two more editions, in 1807 and in 1818. In 1818,[83] the firm of Matthew Carey and Son of Philadelphia—the same Matthew Carey, editor of the *American Museum*, who had paid Evans laudatory comments 16 years earlier in 1792—purchased the copyright to *The Guide* for $125.[84] and reprinted it 13 times. The last edition appeared in 1860.

Evans wrote two appreciations of *The Guide*. The first was in his 1817 memoir:[85]

> the *Millwrights and Miller's Guide*, containing some eternal truths, true theories, which will, like Euclid's Elements, stand the test of time, and lead no practitioner into an expensive error.

The second was in a letter he penned from Raleigh, NC a year later:[86]

> *The Millwright's Guide* is held in great estimation as true a guide as truth all follow it and it never leads them [w]rong. This you must think gratifying to its author to think that his name will live after those of all his enemies perish.

Hunter in his recent massive book on waterpower in the United States offers this assessment of legacy of *The Guide*:[87]

> Although the *Miller's Guide* had its chief usefulness in the hands of millwrights, it doubtless had its place on the shelves of the rapidly growing professions of engineers, civil and mechanical, on whom the advance of hydraulic engineering in the service of industry chiefly depended. In certain respects the Miller's Guide must have hastened the decline of the traditional technology it reflected. It did this in the first instance simply by committing to the printed page what had hitherto been communicated only by example and oral tradition, undergoing in the process comparison, criticism, and selection. . . . In their [Evans and Ellicott] preoccupation with medium to large-scale flour milling and their emphasis upon continuity of operations and economy in water use, they may be said to had given the first strong impulse in this country to the shift from water milling to waterpower.

It would be a tidy story, indeed, to leave Evans in 1795 with *The Guide* successfully published and the reviews and positive appreciations pouring in. However, at the same time Evans was working on *The Guide*, he was also investigating steam engines and steam-driven machinery. These investigations culminated in 1805 in a book that Evans called a "scientific work"[88] in which he sought to explain the newly discovered principles of high-pressure steam engines. However, this "scientific work" evolved in something "more difficult and abstruse."[89]

THE ABORTION OF THE YOUNG STEAM ENGINEER'S GUIDE, 1805: THE LIMITS OF VISUAL THINKING

In his 1817 memoir, Evans wrote of the difficulty he had in writing this second book, using many of the same words he had earlier used to describe his toil with *The Guide*:[90]

> It brought on grey hairs again, and the use of spectacles, and greatly injured my constitution and health a second time, but which I soon regained upon quitting intense study, and resuming active bodily exercise but the grey hairs and the use of spectacles I could never get rid of.

Furthermore, in a "Postscript" that he included in the original edition of his second book, Evans stated his own disappointment with the book:

> Having read my work in sheets, as it came from the press, I am highly sensible of its deficiencies. It is truly an "Abortion of the Young Steam Engineer's Guide," a work for which I had issued Proposals.

In the short span of 10 years, Evans went from writing a book that in his own estimation was as enduring as Euclid's *Elements*, to a book that, again in his own estimation, was an "abortion."

One reason for such a change might have been the bitter disappointment he encountered in 1804 while in the midst of drafting the *Abortion*. When Congress did not extend the patent rights for his milling machinery (a subject Evans wrote of in the Preface to the *Abortion*), Evans felt psychologically hobbled and unable to expend the time or energy that would have been necessary to do justice to his new subject:[91]

> The time which should have been occupied in examining the different authors who treat of the principles could not be spared. To compare their experiments, results, and observations would have swelled the volume to a large size.

Moreover, Evans was in a precarious financial situation while writing *The Abortion* much like the one he encountered during the drafting of *The Guide*. The engravings alone would have cost $3000.[92] Yet, this time no one rescued Evans:[93]

> I was left in poverty at the age of 50, with a large family of children and an amiable wife to support, for I had expended my last dollar in putting my 'Columbian' steam engine in operation, and in publishing the 'Steam Engineer's Guide'.

With his time, money, and energy depleted, Evans's ability to communicate suffered, and it suffered just at the very time he needed it most because, in *The Abortion*, he was endeavoring to create the very kind of book that he had been unable to achieve with *The Guide*—a work of "science" similar to those by Smeaton and Waring, written in a mathematical-verbal style, which previously gave Evans's mind eye such problems that he reprinted them in verbatim. By choosing to write "science," Evans chose an area in which his assets as a thinker and writer, the visual quality of his "mind's eye," placed him at a disadvantage.

The resulting book was a quarter the size of *The Guide*. Whereas his earlier book had 26 plates with nearly 100 figures, *The Abortion* had just five plates and five figures. Additionally, *The Guide's* subtle persuasion was now replaced by a style resembling a bludgeon.

For example, in the "Preface" of *The Guide*, Evans included the following statement, which he must have hoped would project a humble persona and thus make his audience receptive:

> Wherefore it is not safe to conclude that this work is without error—but that it contains many, both theoretical, practical and grammatical; is the most natural, safe, and rational supposition.

However, Evans's persona in *The Abortion's* preface became accusatory, complaining, and arrogant. Speaking of himself in the third person, Evans wrote:

> His plans have thus proved abortive, all his fair prospects are blasted, and he must suppress a strong propensity for making new and useful inventions and improvements; although, as he believes, they might soon have been worth the labour of one hundred thousand men.

It may be gratifying to have others believe one's work "worth the labour of one hundred thousand men," but it was not very humble or modest for Evans to claim it about himself—even stated in the third person.

Later on in *The Abortion* when Evans needed to refute a critic, he relied heavily on a sharpness of criticism that was completely absent from *The Guide*:[94]

> Have I been half so dexterous as yourself, who sent Dr. Coxe to view my principles, then in operation and use one year, (publicly exhibited and explained to every one who inquired after the principles) and to put a number of questions to me, which drew in answer, a full explanation of the construction and principles of my invention, and which, when you were in possession of, the improvement became obvious to you, and you went and attempted to take out a patent for, and assumed it to yourself; but herein you have failed for want of a competent knowledge; besides you are not the original inventor. . . . Are you sure you are competent to assert, that I have assumed very erroneous principles while you show you do not understand them yourself? . . . Your ignorance of the principles of my invention has caused you to commit and set yourself in the way as an obstacle to the introduction of the most useful improvements ever made on steam engines . . .

Finally, Evans's patriotic chord, which was evident in the conclusion of the original edition of *The Guide*, was replaced in *The Abortion* with indignation and an almost a whining tone: "I now conclude, and renounce all further pursuit of inventions and discoveries, at least until it shall appear clearly to be my interest."

Table 1.2. Five Parts of *The Abortion* With Their Fit in Classical Rhetoric's Dispositio.

		Articles I–XV	Article XVI–XVIII*	Appendix	Conclusion
Title	Preface	Of Steam . . . to Of vibrating motions of machinery	Description of a steam engine on the new principle & Explanation of the screw mill invented by the author Useful inventions by different persons*	Stevens/Evans/ Mitchell debating letters	
Number of Pages	7	64	16	55	1
% of total number of pages in book	5	46	9	39	1
% of pages taken verbatim from other authors	0	0	0	72	0
% of pages referencing graphics	0	0	84	0	0
Dispositio Functions	Exordium	Narratio	Confirmatio	Refutatio	Peroratio

*Article XVIII is deleted from the second American edition along with the three plates contained in it.

A lack of time, money, and energy—along with an uncomfortable style—all undercut the strength of Evans's visual thinking and thus the source of his creative powers. At times, he was able to recover his characteristic dense weave of text and graphics, as can be seen in Article XVI, "Description of a Steam Engine on the New Principle" (see Figures 1.18 to 1.21).

Yet, Evans was able to sustain this style for only 16 pages. Most of The Abortion was written in Smeaton's and Waring's mathematical-verbal style (see Figures 1.22 and 1.23).

The Abortion was rarely noted in Evans's own tracts and pamphlets, but he spoke affectionately and at length about *The Guide*. In *Patent Right Oppression EXPOSED, or Knavery Detected* (1813), Evans described the creation of *The Guide* in six paragraphs but wrote only one about *The Abortion*. Furthermore, in *Oliver Evans to His Counsel Who Are Engaged in Defense of his Patent Rights* (1817), *The Abortion* received only a brief notice.[95] Not only did Evans like *The Guide* better than *The Abortion*, but so did his readers. *The Guide* went through 15 editions; *The Abortion* only three.[96]

Of course, in addition to his awkward style and lack of rhetorical subtlety, there were other problems for the readers of his second book. For example, there was at the time great resistance to the high-pressure steam engine that Evans championed—he advocated in *The Abortion* moving from the customary of 2 to 4 pounds per square inch (psi) of pressure in the established Watt engines of the era to 100, 125, and higher psi.[97] Doolittle spoke of

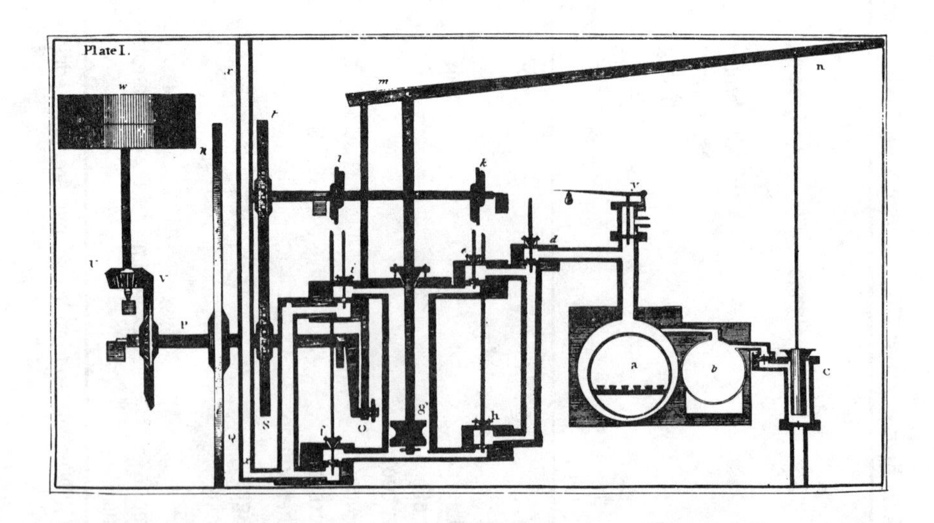

Figure 1.18. Page from Article XVI of *The Abortion of a Young Steam Engineer's Guide* demonstrating Evans's typical tight weave of text and graphics

this resistance in his 1821 French translation of *The Abortion* when he noted that: "For a long time people were doubtful as to the value of high-pressure steam engines because they were believed to be dangerous."[98] A later biographer of Evans also spoke of this resistance:[99]

> In the Emporium of the Arts and Sciences, Vol. 2, published in Carlisle, Pa., 1812, we find quite an extended account of the state of the steam engine at that period, and the feeling against the use of high-pressure steam is well illustrated by an account of the explosion of one of Trevethick's boilers with fatal effect. This fear of the power of high-pressure steam dated from the time of Watt, who thought Richard Trevethick ought to have been hanged for using it, and was a potent factor in the opposition which Evans encountered in his efforts to introduce his engine.

Despite Evans's personal low estimation and the contemporary resistance to the type of steam engine championed in his book, later evaluations of *The Abortion* have been more positive. The June 29, 1866 edition of the British journal *Engineering* noted that *The Abortion* contained:[100]

67

ARTICLE XVI.

DESCRIPTION OF A STEAM ENGINE ON THE NEW PRINCIPLE.

EXPLANATION OF PLATE I.

PLATE I. represents a perpendicular section of the different parts of a steam engine on the new principle explained in this work, but they are differently arranged in the construction.

a The end view of the boiler, consisting of two cylindrical tubes, the best form for holding a great power, the lesser inside of the greater. The fire is kindled in the inner one, which serves as a furnace, the water being between them. The smoke passing to the other end, is turned under the supply boiler, *b*, to heat the water for supplying the waste occasioned by working; *c* the supply pump, which brings water up, and forces it into the supply boiler, at every stroke of the engine.

The steam ascends the pipe, and if the throttle valve *d* be lifted to let the steam into the engine, and valves *e* and *f* be opened, the steam drives the piston *g* to the lower end of the cylinder, as it appears in the plate, the steam escaping before the piston through the valve *f*. As soon as the piston is down the valves *e f* shut and *h i* open, the steam enters at *h* to drive the piston up again, and escapes before the piston through the valve *i*. These 4 valves are wrought by 2 wheels, *k l*, with cams on their sides, which strike against 4 levers, not shown

Figure 1.19. Page from Article XVI of *The Abortion of a Young Steam Engineer's Guide* demonstrating Evans's typical tight weave of text and graphics

68

in the plate, to which the stems of the valves are attached, and which open and shut them at the proper time. The motion of the piston *g* gives motion to the lever *m n*; and the rod *m o*, connected to the crank, puts it in motion, and the fly wheel *q r* keeps its motion regular; the spur wheels *s t*, of equal size, move the valve wheels *l k*; the lever *m n* works the supply pump *c*. Thus the motion is continued, and the cog wheel *v* of 66 cogs going into the tunnel *u* of 23 cogs, gives the stone *w* 100 revolutions per minute, when the piston strikes 35 strokes. This cog wheel may move any other work, or instead thereof a crank may move a pump or saw, as this engine may be made to strike from 10 to 100 strokes per minute, as the case may require; and if the working cylinder be 8 inches diameter, it will drive a pair of 5 feet millstones, or other work requiring an equal power.

The steam, after it leaves the engine, escapes up the pipe *x x*, through the roof of the house, or into a condenser, if one be used, or through the supply boiler to heat the water.

y A safety valve, kept down by a lever graduated like a steelyard, to weigh the power of the steam; this valve will lift and let the steam escape, when its power is too great.

If the pipe of the safety valve be turned into the flue of the furnace, then, by lifting the valve, the ashes may be blown out of the flue.

This engine is of a simple construction, easily executed by ordinary mechanics: the valve seats are formed by simple plates, with holes in them and are easily cast.

Figure 1.20. Page from Article XVI of *The Abortion of a Young Steam Engineer's Guide* demonstrating Evans's typical tight weave of text and graphics

69

In working this engine to drive ten saws, we find that if we put her in motion as soon as she has power to drive one saw, and suffer her to move briskly, she carries off the heat from the boiler nearly as fast as it is generated, and fuel may be consumed and time spent to little purpose; but if we confine and retain the steam in the boiler, until it lifts the safety valve with a power sufficient to drive ten saws, she will start with that load, and carry it all day, and consume but little more fuel.

It takes up but little room in the building. The draught is drawn from half an inch to a foot, except the millstones, and two wheels that move them; they are a quarter of an inch to a foot.

Figure 1.21. Page from Article XVI of *The Abortion of a Young Steam Engineer's Guide* demonstrating Evans's typical tight weave of text and graphics

58

minute, the velocity of the piston. Suppose we take 36 for the number of strokes the engine is to strike per minute; then 220 feet divided by 36 quotes 6.1 feet for the length of the double stroke, say 6 feet; that is 3 feet length of stroke, 36 down and 36 up strokes per minute to make the piston pass about 2½ miles per hour.

Suppose again the piston to carry an average load equal to 50 pounds to the inch instead of 80 pounds, to make allowances, (see article 6), then every 3 inches area of the piston is equal to a horse's power.

The side of a square being 1, the diameter of a circle of equal area is $1\frac{128}{1000}$: therefore to find the diameter of the cylinder for any number of horse's power take the following

RULE.

Multiply the number of horses by 3, extract the square root and multiply by $1\frac{128}{1000}$: the product will be the diameter of the cylinder, 3 feet length of stroke; 36 strokes per minute.

The diameter of a circle being 1, the side of a square of equal area will be $\frac{7854}{10000}$: therefore to find the area of any cylinder, multiply the square of the diameter by .7854 of a decimal and the product is the area.

To produce the power of a horse, a piston 3 inches area must move 220 feet, or 2640 inches per minute, with a load of 150 pounds; and 2640 multiplied by 3 is equal to 7920 cubic inches of space the piston forms into a perfect vacuum behind it per minute to be filled by the heat, (see article 12.) Therefore to find the power of any engine, multiply the area of the piston by the length of stroke in inches and by the number of

Figure 1.22. Page from Article XIII of *The Abortion of a Young Steam Engineer's Guide* demonstrating Evans's mathematical-verbal scientific style

> a very complete and generally correct amount of the properties of steam and a clear and practical statement of the advantages of the high presure engine . . . making allowances for the state of knowledge sixty years ago, it was a book every way worthy of a careful and practical engineer.

One modern historian called *The Abortion* "one of the most durable technical handbooks of the century,"[101] whereas another claimed that it "no doubt found use in many engine shops."[102]

There was some evidence that Evans continued to revise *The Abortion* from the time of its first publication in 1805 until his death in 1819. At his death, Evans bequeathed to his son Cadwalader a copy of *The Abortion* into which Evans had interleaved 214 pages: "I also give to my son Cadwalader The Abortion, bound with blank leaves, in which I have written for his instruction."[103]

In this specially printed book, Evans discussed various engineering subjects, amended and corrected some of the book's tables, and included two more fully developed plates and a number of rough sketches. Although many

59

strokes per minute, and the product is the space it passes through, or the vacuum it forms behind it; which divided by 7920 quotes the number of horses' power, when the piston carries an average load of 50 pounds to the inch area.

To find the number of strokes an engine must strike per minute to produce any number of horses' power, multiply the number of horses' power by 7920, and divide by the cubic inches of space the piston passes through at one stroke, the quotient is the number of strokes the engine must strike per minute, carrying 50 pounds to the inch area.

A horse can work only 8 hours steadily in 24, therefore 3 relays are necessary, and an engine of 10 horses' power will do the work of 3 times 10, equal to 30 horses.

Bolton and Watt's best steam engines, on what I call the old principle, require 1 bushel of the best New-Castle coals, from Walker's pits, (England) to do the work of a horse per day. It has been shown (article 6) that my new principle will produce at least 3 times the effect from equal fuel; one bushel of coals to do the work of 3 horses; and can be built at half the price.

Figure 1.23. Page from Article XIII of *The Abortion of a Young Steam Engineer's Guide* demonstrating Evans's mathematical-verbal scientific style

have accepted this interleaved version of *The Abortion* as just a commonplace book for Evans's notes, it may also be true that to the day he died, Evans continued to use his mind's eye to revise *The Abortion* in an effort to attain a dense weave of text and graphics in his work of "science."

ENDNOTES

1. The total (gross) value of flour milling ("flour and meal") not only led all American manufacturing but was greater than the combined values of the cotton, woolen, and basic iron industries. (U.S. Census Office: *Eighth Census, Manufacturers of the United States in 1860 Washington, DC., 1865*: 733; cited in Hunter, Louis C.: *A History of Industrial Power in the United States, 1780–1930. Volume One: Waterpower in the Century of the Steam Engine* Charlottesville, VA: University Press of Virginia Published for the Eleutherian Mills–Hagley Foundation, 1979: 102.)
2. Stork, John and Teague, Walter D.: *Flour for Man's Bread: A History of Milling*. Minneapolis, MN: University of Minnesota Press, 1952: 94.

3. Oliver Evans letter, James L. Woods's collection of Evans's papers in the Franklin Institute Library.
4. Oliver Evans letter, James L. Woods's collection of Evans's papers in the Franklin Institute Library.
5. Evans, Oliver: "To the honorable Senators" (short pamphlet, self-published) circa 1814, from James L. Woods's collection of Evans's papers in the Franklin Institute Library. Also in Bathe, Greville and Dorothy: *Oliver Evans: A Chronicle of Early American Engineering*. New York: Arno Press, 1972: 214–6. (Reprint of 1935 publication by Historical Society of Pennsylvania.)
6. Kasson, John: *Civilizing the Machine*. New York: Penguin Books, 1976: 28.
7. Giedion, Siegfried: *Mechanization Takes Command*. New York: W. W. Norton & Company, 1948: 85, 86.
8. Evans, Oliver: *The Abortion of a Young Steam Engineer's Guide*. Philadelphia, PA: Fry & Krammerer, 1805. (A facsimile edition is available from the Oliver Evans Press, 204 West Rose Valley Road, Wallingford, PA 19086.)
9. Elisha, Patrick N. I. (pseudonym for Oliver Evans): *Patent Right Oppression EXPOSED, or Knavery Detected*. Philadelphia: R. Folwell, 1813 (cited in Bathe and Bathe, op. cit., 1972: 30).
10. Carey, Matthew: *American Museum or Universal Magazine*, May 1792: 225–6. (Also cited in Ferguson, Eugene S.: Oliver Evans, Inventive Genius of the American Industrial Revolution. Greenville, DE: Hagley Museum, 1820: 27.) Matthew Carey was a key member of a movement around Philadelphia in the 1780s and 1790s that led to the creation of both the Pennsylvania Society for the Encouragement of Manufactures and the Useful Arts as well as to the "creation and dissemination of ideas" in the *American Museum*" (Jeremy, David J.: "British Textile Technology Transmission to the United States: The Philadelphia Region Experience, 1770-1820." *Business History Review* 47(1) (Spring 1973): 29-32; see also Rezneck, Samuel: "The Rise and Early Development of Industrial Consciousness in the United States, 1760-1830," *Journal of Economic and Business History* 4 (1932): 784-811). The point here is that Evans's rhetorical success in conveying his ideas and getting them accepted did not happen in a vacuum but within the carefully cultivated industrial consciousness that he had created with others around Philadelphia in the 1790s. Later, during the 1840s, a similarly inventive mechanic consciousness evolved for the shipwright industry that was centered at New York City's American Institute (see Chapter 2 about John W. Griffiths's shipwright career). Both of these inventive mechanic climates can be equated with the climate for computer inventiveness in Silicon Valley.

11. The 15th edition of *The Guide* appeared in 1860. Portions of *The Guide* appeared in an English journal in 1796 and in a French journal in 1802. See Ferguson, op. cit., 1980: 14.
12. We know, by Evans's remark to a young George Escol Sellers concerning the importance of a school to teach mechanical drawing, that Evans thought graphics and visual thinking were important. See Ferguson, Eugene S. (ed.): "The Philadelphia of Oliver Evans," in *Early Engineering Reminiscences (1815-40) of George Escol Sellers*, Washington, DC: Smithsonian Institution, 1965: 38. Facsimiles of the 1975 original edition of the *Young Mill-wright and Miller's Guide*, the 1792 broadside of the mill developed from the *Universal Asylum* article, and *The Abortion of a Young Steam Engineer's Manual* (1805) are available from the Oliver Evans Press, 204 W. Rose Valley Road, Wallingford, PA 19086.
13. In the early 1780s, Evans completed his apprenticeship and joined his brothers who were operating a small flour mill.
14. Evans, op. cit., 1817: 7-8.
15. Latimer, George A.: "A Sketch of the Life of Oliver Evans, A Remarkable Mechanic and Inventor." Wilmington, DE: Harkness, 1872: 7.
16. Evans, Oliver: *Oliver Evans to His Counsel Who Are Engaged in Defense of his Patent Rights*. Philadelphia, 1817: 8 (this work functioned very much like a memoir for Evans).
17. *Oxford English Dictionary*. Oxford, England: Oxford University Press, 1971: 3642. The term "projector" as a derisive term echoes the dozen or so times the word "project" or "projector" appeared in the fourth, fifth, and sixth chapters of Jonathan Swift's *Gulliver's Travels*, written earlier in the 18th century. In particular, the Academy of Lagado is described as an "Academy of Projectors." See Gies, Joseph: "The Genius of Oliver Evans." *American Heritage of Invention and Technology* 6(2) Fall 1990: 52; Rogers, Pat: "Gulliver and the Engineers." *Modern Language Review* 70 (1975): 260–70; and Keller, Alex: "The Age of the Projectors." *History Today*, July 1966: 467–74. Keller also discussed the use of the word "projector" in Ben Jonson's *The Devil is an Ass* (1631). It is also interesting to note the title of a 1788 pamphlet attacking the steam boat design of Evans's competitor, James Rumsey. This 1788 pamphlet was entitled "The Projector Detected or, Some Strictures on the Plan of Mr. James Rumsey's Steam Boat" and written by Englehart Cruse (Baltimore, MD: John Hayes, 1788). In the pamphlet Cruse suggests the colonial import of the word "projector" in the following: "If you were lost in your profound researches, it is no more than what has happened to other great men—give you time, and time it seems you must have, and your ingenuity will be displayed with fire and smoke" [7] or, "Legislatures of any of the States can be induced thereby to patronize or throw away money on such vapoury exploits, they will justly merit the censure of their constitutents, and

ever man of common sense" [12]. Finally, "the world will consider you as a metaphysician, a builder of castles in the air!" [13].

18. Niles' *Weekly Register* 3 (1812–3), *Addenda*, p. 1—qtd in Ferguson, op. cit., 1980: 13.
19. Letter to John Nicholson, September 12, 1794. Copy of letter in Grenville Bathe papers in the National Museum of American History, Smithsonian Institution; original letter was owned by Simon Gratz.
20. Ferguson, op. cit., 1977: 827.
21. Evans, op. cit., 1817: 5–6.
22. Ibid.: 7.
23. Ibid.: 11–2.
24. "The PORTFOLIO" contains drawings and specifications that were originally purchased from Christopher Evans, son of Oliver Evans, and now housed in the Transportation Library in the University of Michigan's Undergraduate Library. Photostats are available at the Hagley Museum and Archives, Wilmington, DE, Accession 1538. See also "Appendix T" in Bathe & Bathe, 1972. Bathe & Bathe speculate that these undated drawings were developed between 1786 and 1808.
25. Ibid.: 157
26. Evans's 1786–87 patent petitions to Delaware, Maryland, New Hampshire, Pennsylvania, and Virginia mentioned the flour mill as well as his high-pressure steam engine. See Bathe & Bathe, op. cit., 1972: 14–5. See also *The Guide*, Part II: 34–5. In 1786, Evans described both inventions to the legislatures.
27. Sellers, Coleman: "Oliver Evans and His Inventions." *Journal of the Franklin Institute* 122(1) July 1886: 6. See also Hawke, David Freeman: *Nuts and Bolts of the Past: A History of American Technology, 1776—1860.* New York: Harper & Row, 1988: 78—Evans "first incorporated the three basic types of conveyor, as still used today, into a continuous production line…The 'endless belt' (belt elevator), the 'endless screw' (screw conveyor), and the 'chain of buckets' (bucket conveyor), which he used from the very start, constitute to the present day the three types of conveyor systems. Later these three elements became exhaustively technified in their details, but in the method itself there was nothing to change."
28. Evans, op. cit., 1817: 10. Interestingly, however, this same Thomas Lea of Brandywine mills was one of those listed as subscribers who later underwrote the publication of Evans's *The Guide* in 1795 (see 1795 edition, list of subscribers in endpages).
29. Evans, op. cit., 1813: 186.
30. Evans had sent a model and a drawing to the Maryland legislature to obtain his patent (Evans, op. cit., 1795, Part II: 34).
31. Evans, op. cit., 1813 (cited in Bathe and Bathe, op. cit., 1972: 21).
32. Bathe and Bathe, op. cit., 1972: 30.

33. Hunter suggests that if one reads between the lines of Evans's 1795 book—where detailed directions are given for the construction of an impressive four-story masonry building—one can surmise that Evans was primarily addressing large merchant mill owners. Large continuously-processing mills only comprised at most a third of the water mills present in the U.S. at the time; most were small, seasonally-run mills producing for the local populace. See Hunter, op. cit., 1979: 100–1.
34. Evans, op. cit., 1817: 15.
35. Evans, Oliver: *A Trip Made By a Small Man in a Wrestle with a Very Great Man*. Philadelphia, 1813: 159–60.
36. According to the *Oxford English Dictionary*, in 1766 "rattle-traps" meant something like "nick-nacks, trifles, odds and ends, curiosities, small or worthless articles."
37. Bathe & Bathe, op. cit., 1972: 24.
38. Evans may not have used visuals in his first broadside because of the cost of copperplates, which as noted later in his book, led him into deep debt. Notice in the letter to Young how Evans wanted to trade his articles for copperplates. Unfortunately, the *Asylum* survived for only one year, so the extended trade never worked out. Instead, as we see in *The Guide*, Evans did receive the copperplate for this one article.
39. Ibid.: xiv. When the article was made into a one-sheet, 17-inch-by-21-inch broadside, Evans added uses for the elevator, a second testimonial from a Brandywine miller—which was important because Evans showed that his neighbors were using his designs—and a small ad explaining how to purchase his plans at $32 each. The plate and the entire textual description were also repeated verbatim in The Guide, Part III: pp. 78–87. The name "Univ. Asylum" and the Plate number are either scratched out or replaced in *The Guide*. So evidently Evans got the plate for this one article.
40. The theory of the "Gutenberg Diagram" of reader eye movement over a page postulates that a reader's eyes begin reading at a position on a page at the top left (point of arrival) and move in a zigzag fashion to the bottom right (point of departure). See Arnold, Edmund C.: *Producing Flyers, Folders, and Brochures*. Chicago, IL: Lawrence Ragan Communications, 1984: 48–9.
41. Letter to John Nicholson, September 12, 1794. Copy of letter in Grenville Bathe papers in the National Museum of American History, Smithsonian Institution; original letter was owned by Simon Gratz.
42. Sharpin, Steven: "Pump and Circumstance: Robert Boyle's Literary Technology." *Social Studies of Science* 14 (1984): 492. See also Bazerman, Charles: *Shaping Written Knowledge: The Genre and Activity of the Experimental Article in Science*. Madison, WI: University of Wisconsin Press, 1988: 71—"The changes in illustrations through the period also express the growing importance of methods. The early issues of the jour-

nal [*Philosophical Transactions of the Royal Society*] frequently illustrate the phenomena being reported on or new technological marvels, but rarely is the apparatus used for an experiment considered worth a picture. However, as experiments become more ingenious, elaborate, or just careful, illustrations follow. . . . The realism of illustration becomes particularly important as the account or story of the experiment becomes the reader's vicarious surrogate for the actual experiment."

43. Bathe & Bathe, op. cit., 1972: 24.
44. Jeremy, op. cit., 1973: 28.
45. Ferguson, Eugene S., and Baer, Christopher: *Little Machines: Patent Models in the Nineteenth Century*. Greenville, DE: Hagley Museum, 1979.
46. Also, the very specific terms in the article and second broadside (e.g., "the pulleys be 17 inches in diameter, 4 inches in thickness, and half an inch higher in the middle") may be why they succeeded in conveying the idea that these machines were real and not imaginary.
47. Bathe & Bathe, op. cit., 1972: 29.
48. *American Museum or Universal Magazine*, op cit.: 225–6.
49. Bathe & Bathe, op. cit., 1972: 24.
50. Bathe & Bathe, op. cit., 1972: 45–7.
51. Evans, op. cit., 1813: 176.
52. Evans, op. cit., 1817: 16.
53. Ferguson, op. cit., 1980: 30.
54. Ferguson, op. cit., 1965: 38.
55. Evans, op. cit., 1795, "Appendix": 8.
56. Jones was a member of the American Philosophical Society, a Professor of Mechanics, the Recording Secretary at the Franklin Institute, and the Superintendent of the Patent office in 1828.
57. Evans, op. cit., 1817: 16–7.—Could he have anticipated the use of the book as evidence in later lawsuits? He instructs his lawyer to show *The Guide* "to impress the Court & Jury, with adequate conceptions, of the great exertions I have made, and the expense I have born to disseminate my improvements." ("Newspapers & Advertisements Certificate Etc." from James L. Woods's collection of Evans's papers in the Franklin Institute Library.)
58. In his screw mill licenses Evans also inserted a plate, the screw mill from *The Abortion of a Young Steam Engineer's Guide*. All subsequent licenses had the same plate on the back side.
59. Ferguson, op. cit., 1980: 31.
60. Bathe & Bathe, op. cit., 1972: 293.
61. Articles 67, 68, and 69 in *The Guide* are from Parts I, II, and III of Smeaton's *Experimental Enquiry Concerning the Natural Powers of Wind and Water to turn Mills and other machines* for a total of 32 pages—the most pages taken from any one author.

62. *Transactions of the American Philosophical Society*. Philadelphia, PA: Aitken & Son, 1793: 144–149. All five pages of Article 38 in *The Guide* were taken verbatim from Waring. Unlike Martin and Ferguson, only two of Waring's seven pages (or 33%) contained references to graphics.
63. Hunter concurs when he noted: "There is overlapping and duplication in the discussion and tables of the collaborators [Evans and Ellicott], which at times may have been a source of confusion to the reader." Hunter, op. cit., 1979: 94–5.
64. Corbett, Edward P. J.: *Classical Rhetoric for the Modern Student* (3rd ed.). New York: Oxford University Press, 1990: Part III, "Arrangement of Material;" see also Brockmann, R. John: "Oliver Evans and His Antebellum Wrestling with Rhetorical Arrangement" in (Kynell, T. and Moran, W., eds.) *History of Technical Communication*. Norwood, NJ: Ablex Publishing, forthcoming.
65. In comparison to Evans's 1791 article and broadside, Part III differed because it more completely described the production cycle by including the input and output of the milling process and by being more specific in some details. Evans went beyond the bolting cloth, where he left off in his 1791 article, and showed the finished flour being loaded onto ships for transport (Figure 1.17). By 1795, he was also much more specific in the terms of his explanation. For example, in his 1791 article and broadside, Evans wrote (Figure 7): "The grain is emptied into the spout A, by which it descends into the garner B..." But in 1795, Evans rewrote this in the following manner (Figure 13): "The grain is emptied from the waggon into the spout I, which is set in the wall, and conveys it into the scale 2, that is made to hold 10, 20, 30, or 60 bushels, at pleasure."
66. Fessenden, Thomas Green: *An Essay on the Law of Patents for New Inventions*. Boston, MA: Charles Ewer, 1822.
67. Ibid.: 248.
68. Day, Robert A.: *How to Write and Publish a Scientific Paper*. Phoenix, AZ: Oryx Press, 1988: 33. Allan G. Gross in *The Rhetoric of Science* (Cambridge, MA: Harvard University Press, 1990: 12) explained this naming dropping and reference to widely-respected books as the use of the rhetorical method of ethos, "the persuasive effect of authority." Gross continued: "Innovation is the *raison d'etre* of the scientific paper; yet in no other place is the structure of scientific authority more clearly revealed. By invoking the authority of past results, the initial sections of scientific papers argue for the importance and relevance of the current investigation; by citing the authority of past procedure, these sections establish the scientist's credibility as an investigator" (p. 13). Moreover, Evans used "name dropping" to bolster his own claims elsewhere in *The Guide*. For example, he included a list of subscribers who had underwritten *The Guide* in the 1795 edition. The list included the names of

such prominent Americans as Washington, Jefferson, Randolph, Morris, and Burr.

69. Ferguson, op. cit., 1980: 31.
70. In a 1951 essay on John Smeaton, Penrose wrote that he "came to be recognized as England's foremost authority on and builder of harbors, bridges, canals, public waterworks, drainage projects involving great land areas, as well as mills and machinery. Smeaton was the first self-proclaimed Civil Engineer, and before the English established their Society of Civil Engineers, the Smeatonian Society for prominent Civil Engineers was established. (See Penrose, Charles: "Two Men—and Their Contributions in Two Countries!" *Newcomen Society Address*, New York: Newcomen Society in North America,. 1951: 11. See also Wilson, P. N.: "The Waterwheels of John Smeaton," *Newcomen Society Transactions 30* (1960); Reynolds, Terry S.: "Scientific Influences on Technology: The Case of the Overshot Wheel, 1752–54" *Technology and Culture* 20 (2) April 1979: 270–95; and Reynolds, Terry S.: *Stronger Than A Hundred Men: The History of Vertical Water Wheels* Baltimore: John Hopkins University Press, 1983.)
71. Smiles, Samuel: *Lives of the Engineers: Smeaton and Rennie.* London: John Murray, 1891: 94. After Smeaton's death, this article was collected and republished in London by I & J Taylor in 1794. Thus it was quite easy for either the early *Transactions* version or the reprint in book form to have found its way into Evans's hands as he wrote his own book in 1793 and 1794.
72. Reynolds claimed Smeaton was not a scientist because his experiments did not emerge from a system of conceptual or theoretical knowledge, but arose from a "technological milieu and were conducted with traditional technological tools." However, this modern reassessment of Smeaton ignored the "natural philosophy," scientific audience, persona, and context of the Royal Society in which Smeaton presented and published his experiments: "The Society of Arts was established in 1754 and it was recognized that this society was concerned with manufactures in a way which the Royal Society was not. In 1771, for example, Sir John Pringle, President of the Society, replying to a communication from David Macbride of Dublin regarding an improved process of tanning, referred this to the Dublin Society and also 'to the Societies which are established in London and Edinburgh, for the purpose of encouraging trade and manufactures; as judging it will be more in their way than in the Royal Society's to extend the utility of this invention'" (Musson and Robinson, op. cit., 1969: 31). In fact, the way and place in which Smeaton chose to present his experiments strongly suggested that he thought he was communicating science to fellow scientists in 1759.
73. Smiles, op cit., 1891: 177.

74. The word "plate" can be misleading. For example, if you look at Figure 5 in *The Guide*, Evans would have called it one plate, despite the fact that it had five separate figures on it. And in some of the plates in *The Guide*, there are as many as 10 figures on each plate. The cost of the copper-plate engravings probably forced Evans to use every available space on the plate.
75. In translating *The Abortion of the Young Steam Engineer's Guide* into French, Mr. I. Doolittle wrote: "Oliver Evans was a born mechanic; He wrote for the workman who had need of an aid to his intelligence and he thus knew how to put himself in such a one's place." Bathe & Bathe, op. cit., 1972: 280.
76. Basalla, George: *The Evolution of Technology* Cambridge: Cambridge University Press, 1988: 97.
77. Waring introduces graphics only on the fifth of his seven pages, and then, almost immediately, he reiterates them in mathematical-verbal language.
78. Interestingly, Evans's competitor in the design of steamboats, John Fitch (see the discussion in the *Abortion of a Young Steam Engineer's Guide*) also made extensive use of Ferguson and Martin. See Hindle, Brooke: *Emulation and Invention*. New York: New York University Press, 1981: 29–30.
79. Musson and Robinson, op. cit., 1969: 103
80. Bathe & Bathe, op. cit., 1972: 280.
81. Oliver Evans letter, James L. Woods's collection of Evans's papers in the Franklin Institute Library. Also in Bathe & Bathe, op cit., 1972: 248.
82. Franklin, Benjamin: "Proposals" (cited in Brooke Hindle: *Emulation and Invention*. New York: New York University Press, 1981, p. 13–4). See also Brockmann, R. John: "Using Emulation to Teach Nonverbal Technical Communication in the 21st Century" In *The Proceedings of the 43rd Technical Communication Conference*. Arlington, VA, 1996: 8-13.
83. Much of the archaic spelling is corrected in the third edition.
84. Ferguson, op. cit., 1980: 59.
85. Evans, op. cit., 1817: 16–7.
86. Bathe & Bathe, op. cit., 1972: 258.
87. Hunter, op. cit., 1979: 104.
88. The words Evans used to advertise *The Abortion* in *The Aurora* newspaper were "scientific work". See Bathe & Bathe, op cit., 1972: 107.
89. Evans, op. cit., 1817.
90. Ibid., 1817: 18.
91. Evans, op. cit., 1805: "Postscript."
92. Evans, op. cit., 1805: "Postscript". It is difficult to describe exactly how much Evans's $3000 would be worth in today's dollars. For example, *The Value of a Dollar, 1860–1989* (Scott Derks, ed.). Detroit, MI: Gale

Research Inc., 1994) does not have data going back far enough. However, prices and salaries of the time can give some sense of the value of $3000 in 1805. For example, when George Washington wrote to Evans in 1798 requesting help in hiring a miller for the Evans's mill he had constructed at Mount Vernon, he wrote that he intended to pay $166 2/3's dollars annually in addition to providing the miller with food, firewood, a milking cow, and a cottage (Bathe & Bathe, op cit., 1972: 62.) Without these, the annual salary for a millwright in 1800, according to Carroll D. Wright (*Industrial Evolution of the United States*, 1895), was approximately $260.00 for six days work a week, 10 hours a day. (By 1860, the yearly salary for a carpenter in Connecticut had risen to $475.20 [Derks, ibid., 1994].) In comparing the cost of objects, recall that the price of *The Guide* with 300–400 pages, hardbound binding with lettering on the spine, was $2.00; one of Evans's letters in 1804 mentions that a single bag of feed cost $1.00 (Bathe & Bathe, op cit., 1972: 98), and, in 1806, when Evans purchased a nearly quarter acre lot inside the city limits of Philadelphia, bounded on four sides by streets, he paid $2000 (Bathe & Bathe, op cit., 1972: 121). (In 1863, however, a 2 1/2 story frame house on a comparable lot near Fifth Avenue in New York in the fashionable Harlem section with gas and water connections cost $7,800 [Derks, ibid., 1994].)

93. Evans, op. cit., 1817: 18.
94. Bathe and Bathe, op. cit., 1972: 105–6.
95. Perhaps Evans's evaluation of *The Abortion* cannot be entirely trusted because, as Hezekiah Niles wrote in the July 5, 1834, edition of the *Niles Register*: "Had Mr. Evans not been rendered almost misanthropic by what he, (as we thought, erroneously), believed was the injustice and ingratitude of the public, we are of opinion that a discovery made by him, as to the application of steam power, would have been proclaimed, which, even at this day, would be regarded wonderful."
96. The first edition was published by the author through Fry & Krammerer of Philadelphia; the second was an abridged U.S. edition published in 1826 by H. C. Carey & I. Lea, and the third was a translation by I. Doolittle into French in 1830. This French edition contains no publication dates and has been variously dated by bibliographic sources. See Bathe & Bathe, op cit., 1972: 345 versus Library of Congress dating of the edition.
97. Hunter, Louis C.: *A History of Industrial Power in the United States, 1780–1930. Volume Two: Steam Power* Charlottesville, VA: University Press of Virginia Published for the Eleutherian Mills–Hagley Foundation, 1985: 47.
98. Bathe & Bathe, op. cit., 1972: 281.
99. Sellers, op. cit., 1886: 13.

100. Qtd in Hunter, op. cit., 1985: 39
101. Ferguson, op cit., 1980: 9
102. Pursell, Carroll: *Early Stationary Steam Engines in America.* Washington, DC: Smithsonian Press, 1969: 47.
103. Bathe & Bathe, op cit., 1972: 340.

Chapter 2

The Changing Weave of Text and Graphics in Antebellum Shipwright Manuals

> Having arranged our investigations to our mind, we shall discover the corresponding model looming up to our inner vision, and appearing in all its proper proportions and true shape. It is with our mental eye we should first view our incipient architecture afar off, and examine it well before we conclude to tow it into port. After we have submitted the distant ideal vessel to the ordeal of nautical criticism, as an intruding stranger advancing to receive our sympathy and paternal regard, we are prepared to adjust its form to the measure of capacity required. The work of "making the model" is now fairly under way, and we should grasp its configuration fast before our mind's eye, until the labors of the hand have produced it tangibly before us. Our task is not merely to fashion a block of wood into the likeness of something that floats the commercial element, but to produce the exact counterpart of the living shapes which genius has configured in the "lofts" of the mind.
>
> —John W. Griffiths, 1854[1]

Oliver Evans's mind's eye initiated his process of invention as well as enabled him to apprehend the technical information of 18th-century British authors.

Evans's mind's eye then became externalized in pictures and models as a method of communicating the machinery of flour mills and steam engines to American millwrights and millers. In a similar way, the mind's eye of a pair of shipwrights authors, Lauchlan McKay[2] and John W. Griffiths,[3] became externalized in pictures as a method of communicating to their audience of American shipwrights. However, whereas Evans had used a three-dimensional model only once in 1791 for introducing his automated flour mill, McKay and Griffiths spent much time and effort in the 1830s–1850s introducing and standardizing a three-dimensional model as a tool for shipwrights. Griffiths also used graphics in a way that had never previously been used by Evans or McKay; he used them as visual metaphors in order to introduce new design techniques to American shipwrights. The three-dimensional models that McKay and Griffiths advocated called half-hull, lift-models (see Figure 2.1) were a very particular type of model for use only in shipyards, and these models were the subject of John W. Griffiths's ode that opened this chapter.

Figure 2.1. A picture of a shipwright and his half-hull lift-model from *Harper's Weekly* 5 May 1877 21(1062): 358.

THE PURPOSE OF A HALF-HULL LIFT-MODEL

A merchant desiring a vessel usually approached a shipwright and gave him the length and gross tonnage of a desired vessel.[4] From these two characteristics, the shipwright would immediately know whether the merchant wanted, in ascending order of size and complexity, a sloop, schooner, brig, brigantine, bark, or ship, and so on.[5] The trick for the shipwright, however, was in the next step—communicating the exact dimensions and design of the vessel to the largely illiterate shipyard workers.

Early designers, who were also the builders, kept the plans in their heads, and they executed the designs in wood and canvas through their mind's eye. As soon as other people became involved in the building of a vessel, designers needed a way to communicate their plans. Probably one of the first attempts to solve this communication problem involved the use of draughts or drawings. For example, a manuscript entitled "Fragments of Ancient English Shipwrightry" allegedly contained the earliest known vessel's "draught"—a 1586 plan by one of Queen Elizabeth's master shipwrights.[6] Such design drawings usually included:[7]

> an outline of the stem, keel, sternpost of the vessel and several transverse "sections" including the transom. The meat of the design lay in the four longitudinal lines called the "narrowings" and "rising" lines. These determined the points of maximum breadth and the contour of the bottom, and thus whether the boat would be wide or narrow, fast or slow, heavy or light.

These "draughts" were kept somewhat simple for the workmen by having nearly all lines in them drawn using portions of circles, arcs, or sections of arcs. By using only arcs or sections of arcs shipwrights maintained regular ratios between the vessel's length, depth, and breadth.[8] So regular were these ratios that British and American tonnage taxes prior to the American Civil War were simply reckoned by putting the length and breadth of a vessel in a formula—the depth was never actually needed to arrive at the vessel's total tonnage.[9]

These draughts, although kept somewhat simple, did require some sophisticated skills from the draftsmen and workers who would translate the initial designs into full-scale chalk outlines for cutting the timber. This "laying off" resembled dressmaker's outlines for a pattern.[10] Not only did the "laying off" method require some sophisticated skills, but the building blocks of the draughts, the very regularity of the arcs themselves, caused problems. What if the arcs and their relationships were not so regular; what if the relationships between the arcs and the dimensions of the arcs progressively changed in order to develop a "streamlined" contour to the vessel's hull? The need to change the regularity of the arcs and their relationships happened for exactly this reason around 1800 with the development of vessels with streamline dimensions or "fast" lines.

With "fast" lines, a vessel cut through the sea like a knife rather than fought it with a blunt bow and stern like a hammer. Americans wanted ships with "fast" lines to elude British naval patrols during the War of 1812, as well as later in peacetime to smuggle, to get the freshest tea from China, and to arrive quickly at the California and Australian gold fields. The Baltimore clipper, developed around 1800, was a ship with "fast" lines;[11] its designs revealed the importance of gradually tapering long lines, a sharp rise of the sides from the keel rather than a symmetric rounded arc rise, and a sharply angled bow and stern. However, the "fast" lines that the Americans wanted required progressively changing dimensions and thus a knowledge of logarithms to find, describe, and interpret the ship's curves in the "laying off" process. Such calculations were difficult to develop, interpret,[12] and represent in the draughts on paper:[13]

> It will be at once apparent to the thinking-man that it is impossible to represent two curves in a single line, or to delineate the shape of a line in two ways without making two lines; the most profound mathematician will admit this, and still further, it is difficult to retain two shapes in the eye at the same time, in all their relative proportions. While the present practice is adhered to, of determining the shape we want by the eye, we can scarcely suit ourselves on a plane, or if we do on the draught we are not suited in the vessel, because she is not exactly what we expected. This discrepancy in the mode can only be remedied by drawing a perspective plan, which must of necessity form a second drawing, the principal objection to which exists in the fact, that it is quite an extensive operation, and unless the work is performed in strict accordance with the laws of perspective, which pertain to geometrical science, a correct idea cannot be given; thus it will be perceived that the draught alone does not furnish an index to rotundity in ships . . .

To be able to represent the "fast" lines as well as to communicate these dimensions without using logarithms, shipwrights turned to the use of a three-dimensional model, the wooden half-hull lift-model. The half-hull lift-model was a three-dimensional embodiment in wood of the designers vision. Essentially, the half-hull lift-model transformed ship design from a mathematical or drawing problem into a woodworking problem.[14] Furthermore, the wood carved and whittled into a three-dimensional half-hull lift-model could "render an infinite variety of logarithmic curves in the bow and stern."[15] Such a model could reveal the entire vessel with its changing dimensions and varying relationships of length, depth, and breadth to the designer in a single glance. Hence the wooden half-hull lift-model was perfect for correcting design problems, and this advantage was pointed out as early as 1831 in the *Encyclopedia Americana*:[16]

> Our American builders have a different mode, very easy and satisfactory. They begin by making a wooden model of the proposed construction, the thing itself in miniature. Here the length, breadth, bulk, all the dimensions, and most minute inflections of the whole, are seen at a single glance; the eye of the architect considers and reconsiders the adaptation of his model to the proposed object, dwells minutely on every part, and is thus able to correct the faults of his future ship, at the mere expense of a few chips, and while yet in embryo.

Moreover, the half-hull lift-model saved designers money:[17]

> A shipbuilder-designer had little time to thoroughly study hull form by means of line drawings alone. The supervision of the yard took up most of his time and energy, and he had personal responsibility for getting materials, payments on completed work, new business, and for hiring and firing. Hence, any shortcut in design was attractive, and the model was a superb tool for preliminary hull design.

So much depended on the half-hull lift-model that in 1855, master shipwright John W. Griffiths wrote in the *Monthly Nautical Magazine and Quarterly Review*:[18]

> It is, doubtless, a conceded fact among commercial men qualified to direct the progressive movements of ship-building at the present day, that the superiority of shipping depends upon the model. The supremacy of the United States, as a commercial nation, rests upon the shape of her ships; and the enterprise of her mechanics must ever keep the lead in perfecting the composition of the superior qualities of her vessels, if they would maintain their present proud position before the world.

THE DEVELOPMENT OF HALF-HULL LIFT-MODELS

A "block" variation of the half-hull lift-model, composed of a solid block of wood, was known and used in England prior to 1700, and a framed variation of the half-hull lift-model—a side-view of a vessel's horizontal ribs and longitudinal lines called the "hawk-nest" or "crow's nest" model—appeared late in the 18th century[19] but fell out of favor with shipwrights by the 1830s.[20] Such archaic types of three-dimensional models had, as Griffiths noted in 1851:[21]

> been made in Europe as early as the middle of the last century; but they were what would be recognized as the skeleton model, made of pieces representing the half-frames, and are neither adapted to the purposes of building, or of exhibiting the lines of flotation.

However, the type of model that had the greatest impact on American ship building was not the skeleton type English model but rather the lift-model. The lift-model was actually a stack of 3/8– to 1/2–inch thick alternating planks of light pine and dark cedar.[22] The planks, or "lifts," were temporarily fastened together by dowels or long triangular wedges and fashioned into the desired form of the hull by mallet and gouge. (The scale of the lift-model was a half-inch to the foot in the 1840s and early 1850s.[23]) When needed, the lifts could be disassembled—as a kind of three-dimensional exploded diagram—for easy enlargement, or "laying off," in the mould loft. "No plans were required and none were drawn."[24] The process of creating and then using the half-hull lift-model was pragmatic rather than scientific. Yet, despite all the obvious benefits of the half-hull lift-model, Murphy claimed, in *American Ships and Ship-building* (1860) that: "Up to 1816 a wooden model was unknown in this city [New York City], every vessel being built from drawings or designs on paper . . . "[25]

Perhaps the fact that the half-hull lift-model was not in the traditional British method could explain why it was unknown in 1816. Half-hull lift-models were not part of the traditional shipwright process as outlined in many of the British shipwright books of the late 18th and early 19th centuries. For example, there was no mention of half-hull lift-models in:

- Stalkaart's *Naval Architecture or Rudiments and Rules of Shipbuilding* (1781),
- Hutchinson's *A Treatise on Naval Architecture* (1791),[26]
- Chapman's *A Treatise on Ship-building with Explanations and Demonstrations* (1820), or
- Steel's *The Elements and Practice of Naval Architecture* (1796, 1805, 1812, 1822, 1848).[27]

Moreover, even if these British books had included material on the development of half-hull lift-models, they were not widely available in America.[28]

However, two American towns, Salem and Newburyport, claimed to have had shipwrights who initiated the use of the American half-hull lift-model in the 1790s.[29] Yet, it really was not until the half-hull lift-model was used in a major shipyard and port like New York by Isaac Webb, "the Father of Shipbuilders," that it became acceptable to shipwrights.[30]

In the late 1810s and 1820s, Isaac Webb had many gifted apprentices working for him including his son, William Webb, the brothers Donald and Lauchlan McKay,[31] and John W. Griffiths.[32] Using half-hull lift-models,

William Webb, Donald McKay, and John Griffiths produced ships with extremely "fast" lines and christened them with names such as *Rainbow* (1844), *Sea Witch* (1846), and *Staghound* (1850). These ships routinely broke speed records when commanded by captains such as Donald and Lauchlan McKay, and the ships entered into the history books under the unforgettable name of "clipper ships."

Vessel design continued to evolve in the 1850s with increases in the dimensions of ships and the use of new materials such as wrought iron. These longer dimensions and new materials required a design precision impossible with wood carving.[33] Furthermore, with the shift from sail power to steam power, the internal stresses of the vessel required as much design work as the outside of the ship's hull; however, the half-hull lift-model was effective only in working out the vessel's exterior hull problems. Consequently, Chapelle claimed in *The Search for Speed Under Sail: 1700–1855* that:[34]

> The builder's model reached its height in popularity about 1851 and then began a slow decline until its use was finally relegated to the more primitive "country yards" and boat shops.

When John W. Griffiths's *Monthly Nautical Magazine and Quarterly Commercial Review* first appeared in 1854, the use of the half-hull lift-model as a shipwright's sole design tool was over. Griffiths himself declared the half-hull lift-model inadequate when he wrote that:[35]

> The modeling of vessels, unaided by the science of numbers, could never raise the art of construction above the standard of doubtful utility. Without calculations, the model is a mere block of wood, the draft a monotonous routine of lines without an expositor, from which no tangible evidence is furnished. The calculations unfold at once the marvels which lie hid, like marble in the quarry.

However, Griffiths made a conscious effort to unfold the marvels of ship design hidden within the half-hull lift-model by publishing calculations of individual ships (displacement, centers of effort, buoyancy, gravity, lateral plane) in his later books and journals.[36] These calculations both encouraged and reflected the growing mathematical and geometric sophistication of the shipwrights and appeased their demands for an exactness that could not be captured solely by the eye:[37]

> [w]ithout the builder's CALCULATIONS, nothing is known of the model beyond the intuitive fancy of the spectator, aided by his unfailing eye, and some friend's opinion. Therefore, without the unerring

deductions of figures, we cannot hope to compare model with model when the fancy has become attached to favorite dimensions, or the eye wedded to threadbare forms, and refuses, perhaps, to acknowledge correct principles if clothed in unfamiliar shapes.

Yet Griffiths did not discard the half-hull lift-model introduced by his master Webb; instead he added a mathematical exactness. Griffiths demonstrated how the half-hull lift-models could be used to investigate such sophisticated design aspects as water resistance and hydrostatic balance. Hence, in adding these more sophisticated dimensions, Griffiths essentially reinvented the meaning and use of the half-hull lift-model with the result that the models acquired an entirely new rhetorical complexion. Nevertheless, Griffiths did not write the first popular American shipwright book using half-hull, lift models; that honor went to his fellow apprentice, Lauchlan McKay.

THE PRACTICAL SHIPBUILDER (1839) AND *A TREATISE ON MARINE AND NAVAL ARCHITECTURE* (1851)

Following his teenage apprenticeship with "the Father of Ship-builders" and fellow apprentice John W. Griffiths, Lauchlan McKay worked at three other New York shipyards[38] until he became a master shipwright. It was while he was working as a master shipwright for the Navy in the late 1830s that he asked for a leave of absence, went to his older brother Donald's house in Newburyport, and wrote his book, *The Practical Shipbuilder*.[39] Lauchlan was 28 when he wrote *The Practical Shipbuilder*, thus not quite 10 years separated him from his apprenticeship days with Webb and Griffiths.

Although there is no specific evidence of collaboration when Lauchlan went to write his book at Donald's house in Newburyport, it is highly probable that such collaboration took place.[40] Not only were both brothers former apprentices with the same master, Isaac Webb, but Donald later wrote a book prospectus that was very similar in design to Lauchlan's *The Practical Shipbuilder*.[41] Moreover, Donald's first wife, Albenia Boole McKay may have collaborated with the brothers on the text with information derived from her own family's work in New York shipbuilding. Furthermore, she may have helped Lauchlan with the book's illustrations because there is recent evidence that she later offered such aid to Donald.[42]

Lauchlan McKay wrote only one book, and *The Practical Shipbuilder* was published in only a single edition. Even though Lauchlan's grandson, Richard McKay, claimed that Lauchlan's book became "a text-book for every shipyard in the United States and was used in the drafting lofts for hundreds of ships,"[43] Lauchlan has largely been recorded in history books as a legendary captain on heroic clipper ship voyages rather than as a widely read author. John W. Griffiths's career, however, was quite different.

Unlike Lauchlan McKay, who almost initiated his professional shipwright career with a single published work, John W. Griffiths wrote his *Treatise on Marine and Naval Architecture* some 20 years after his apprenticeship, after a decade of experience in designing, lecturing, teaching, and writing on various aspects of naval architecture. From such a wealth of experience at 42, Griffiths produced a book that had seven editions in five years and was published in both New York and London. Moreover, this was not his only book. In 1844, Griffiths had published a short work (with a long title), *Marine and Naval Architecture; or, the Science of Ship Building, Condensed into a Single Lecture, and Delivered Before the Shipwrights of the City of New York, in the Rooms of the American Institute, September 20th, 1844.*[44] He published his third book, *The Shipbuilder's Manual and Nautical Referee* in 1853, and, two decades later in 1875, he published his fourth and final book, *The Progressive Ship Builder.*[45] Griffiths did more than write books: From October 1854 to March 1858, he edited a nautical journal, initially titled *U.S. Nautical Magazine and Naval Journal*, which later became *Monthly Nautical Magazine* and *Quarterly Commercial Review.*[46] Lauchlan McKay may have been the heroic clipper ship captain, but Griffiths certainly had the more extensive publishing career.[47]

Griffiths also drew on more than his own personal experience when he wrote his 1851 *Treatise*. Certainly he had gained a wealth of personal experience as a shipwright working at Virginia's Gosport Navy Yard as a draughtsman, as well as with a New York design firm in the 1830s.[48] Yet, it was a largely ignored British book on marine architecture research that spurred Griffiths on to write his own book; the British book was the posthumously published report of water tank studies of hulls done in the 1790s by Mark Beaufoy.[49, 50]

Beaufoy was a British dilettante scientist who had accepted a challenge from the London Society for the Improvement of Naval Architecture. The society offered £100 or a gold medal to anyone "ascertaining the laws of resistance of water".[51] Like fellow Royal Society member John Smeaton in 1759, in the 1790s Beaufoy thought of himself as a scientist writing to other scientists in the Society for the Improvement of Naval Architecture. Consequently, according to the scientific practice of the day, "at no time does Beaufoy make use of graphical methods to illustrate or present data, always preferring the far less suggestive method of numerical tables."[52] Only one plate was included by Beaufoy in his reports, and it was a display of his scientific apparatus (see Figure 2.2).

When Griffiths discussed Beaufoy's studies in *The Treatise*, he wrote that "a more uninteresting scientific work of near 700 pages, was perhaps never published in the English language."[53] (A recent assessment of Beaufoy's work seems to concur, describing Beaufoy's book as a "massive tome totally lacking in any clear train of thought or theoretical structure."[54]) What Griffiths

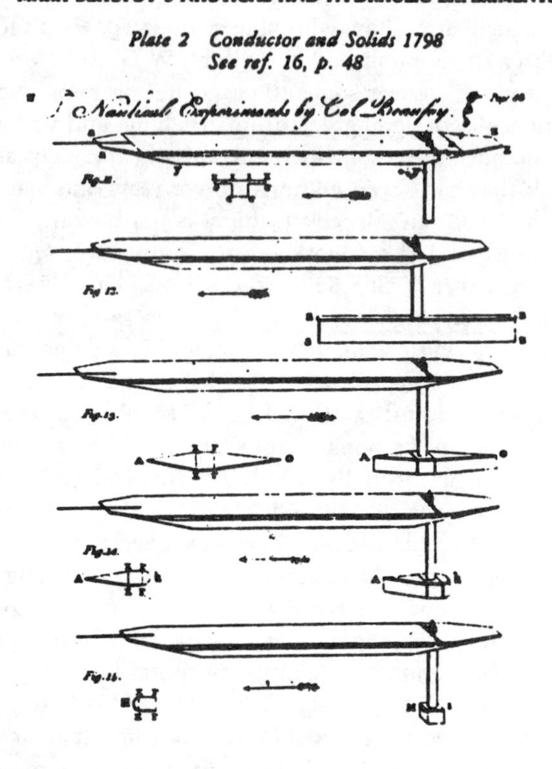

Figure 2.2. Plate from Beaufoy's *Nautical and Hydraulic Experiments* that bears a distinct resemblance to the split apart half-hull lift-model

most objected to was that Beaufoy's scientific experiments were not directly connected to practical shipwright concerns: Beaufoy's scientific apparatuses "have scarcely the least analogy to the form of vessels,"[55] and Beaufoy tested his hulls in a submerged state—hardly the best way to test a sailing vessel's design.

Yet Griffiths probably also noticed something revolutionary about Beaufoy's studies:[56]

> Beaufoy's attention appears to have been directed away from the bow. He seems to have developed an awareness that the after part of the body could have an influence on its resistance characteristics, which represents a distinct departure from the misunderstood Newtonian emphasis on the forepart of the vessel.

What was revolutionary was that Beaufoy was able to improve the movement of a hull through water by changing his design focus to the stern and hull aft of the midpoint from the typical design focus at that time on the bow and hull forward of the midpoint. Griffiths probably was uniquely placed to notice this revolutionary aspect of Beaufoy's research because the only graphic in Beaufoy's entire tome looked surprisingly like a disassembled half-hull lift-model—the same disassembled half-hull lift-model introduced to Griffiths 20 years earlier by his master Isaac Webb.[57] Perhaps in one stroke, Griffiths visually discovered that the half-hull lift-model could not only "produce the exact counterpart of the living shapes which genius has configured in the 'lofts' of the mind,"[58] but could also function, as Beaufoy had demonstrated, as a apparatus for scientific experiments. Griffiths announced this new dual nature of lift-models in his 1851 *Treatise*:[59]

> It is to be regretted that so many mechanics regard the model of a ship as a mere block of wood, like the casual observer who looks upon the marble in the quarry, without being able to discover the statue of the philosopher or the statesman. With such a glance mechanics will never be able to rend the veil that seems to hide nature's laws from their careless vision. But the man in whose mind's eye the surplus of the material itself melts away before his eager gaze, and leaves the ship in her identity, standing out in drastic contrast with the work of him who works only with his hands, we say it is he alone who will be able to approximate that degree of perfection only attainable through the medium of mathematical demonstrations.

Griffiths also displayed this new experimental nature in the use of half-hull lift-models in figures throughout his *Treatise*. Figures 2.3 and 2.4 for example, illustrate how Griffiths suggested that the half-hull lift-model could be used to determine the ship's displacement by using either hydrostatic balance or comparative weight.

Figures 2.5 and 2.6 also show how Griffiths suggested that creators of half-hull lift-models could carve into the usually smooth wooden sides some diagonal lines that could help designers better "exhibit the manner and direction in which resistance to motion is met on vessels."[60]

By 1840, Griffiths had replicated Beaufoy's studies and added his own, more practical approaches. Griffiths then offered his findings in lectures to fellow shipwrights at the American Institute, the headquarters of marine architecture in New York.[61] Unfortunately, he received a great deal of criticism. Griffiths responded by denouncing his shipwright detractors in language quite reminiscent to Evans's in the 1790s when he denounced his neighboring Brandywine millwrights for failing to support his mill design:[62]

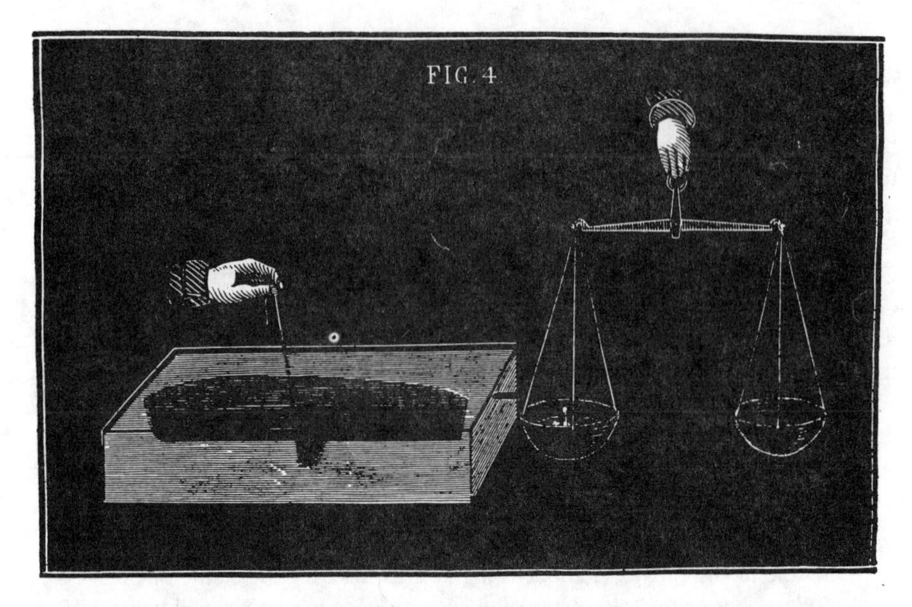

Figure 2.3. Figure 4 from Griffiths's 1851 *Treatise*, demonstrating the use of the half-hull lift-model to determine vessel displacement by hydrostatic balance

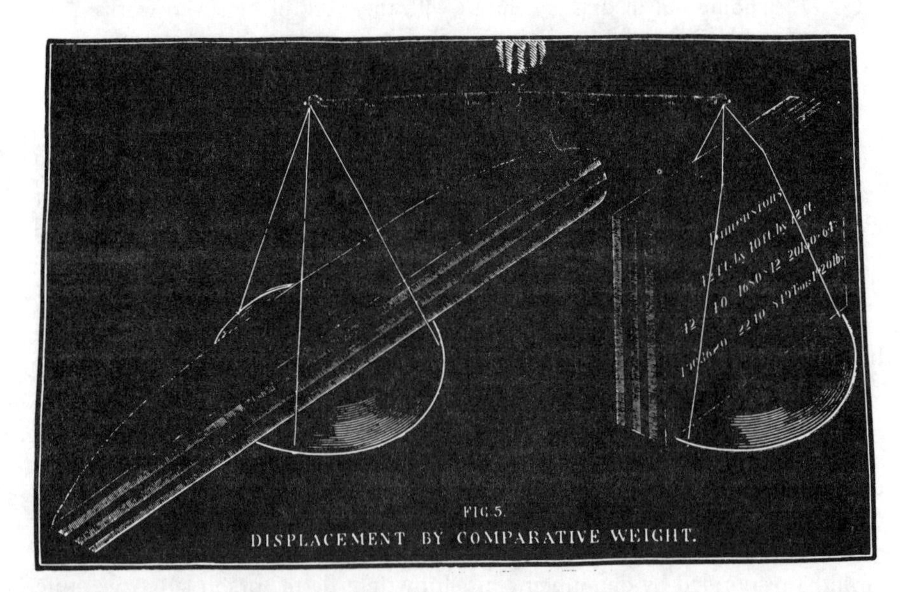

Figure 2.4. Figure 5 from Griffiths's 1851 *Treatise*, demonstrating the use of the half-hull lift-model to determine vessel displacement by comparative weight

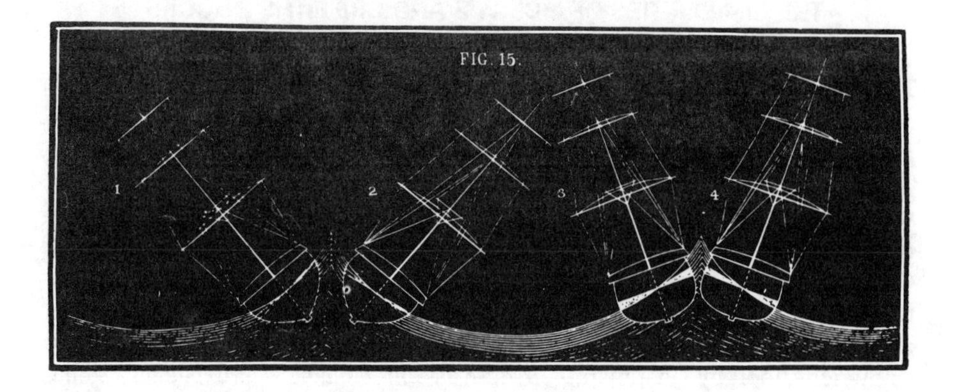

Figure 2.5. Figure 15 from Griffiths's 1851 *Treatise*

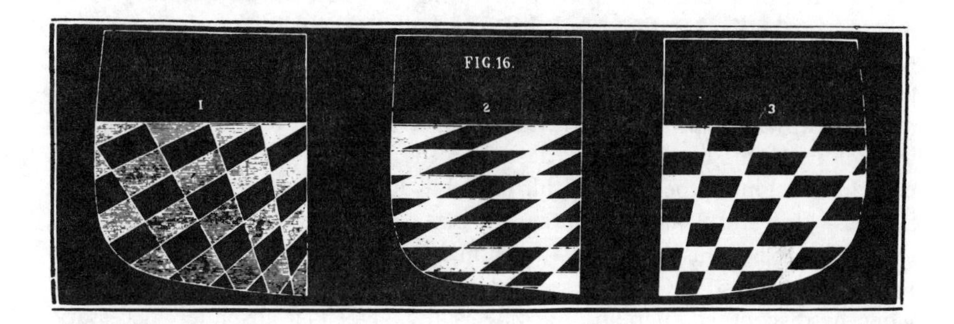

Figure 2.6. Figure 16 from Griffiths's 1851 *Treatise*, demonstrating the use of the half-hull lift-model to elucidate the right angled pressure principle

> How often have the results of years of patient investigation and midnight toil become the theme of denunciation, or the subject of brazen plunder, on the first glance of the skeptic, or the wary business man!

Yet Griffiths persevered, and again like Evans, he built a model [63] and exhibited it at a fair at the American Institute in February 1843.[64] Two years later, having gained a reputation through his lectures and model, Griffiths received the commission for his first "radically designed" clipper ship, the *Rainbow*.[65]

THE SIMILARITIES OF MCKAY'S AND GRIFFITHS'S BOOKS

There are many similarities between McKay's and Griffiths's books. Both addressed audiences who were novice shipwrights without scientific backgrounds: McKay's title page announced that his book was intended for the "instruction of the inexperienced," and Griffiths's "Preface" stated that his book was "elementary instruction for those who have not previously studied the principles of science in modeling and building ships." Both McKay and Griffiths called their works "treatises"—Griffiths, on his cover; McKay, on the top of each page. Calling their books "treatises" could have been a way to suggest similarity in content and design to a previously published, popular British shipwright book by Chapman, *A Treatise on Ship-building with Explanations and Demonstrations* (1820), or simply a way to indicate a methodical discussion of the shipwright design process.[66] Both authors also employed styles that they claimed were plain and intelligible: McKay noted in his Introduction that "Writing to exhibit learning, by the use of scientific terms, has not been my purpose."[67]

It can also be surmised that Griffiths's book self-consciously commented on McKay's earlier book because both contain identical asides. For example, each author commented on the forces acting on a ship's rudder even though such comments were not absolutely necessary to understanding the overall shipwright process.[68] Furthermore, Griffiths addressed a point made by McKay. In his book, McKay described the types of wood that should be used in developing the half-hull lift-model as follows: "and then prepare your boards—pine and Spanish cedar are the best, the colors being contrasted, and the stuff soft, and easy to work—if you make them all pine you cannot see the form of the lines."[69] Griffiths seemed to have this line specifically in mind when he discounted such absolutes by saying, "Nor is it absolutely necessary that the sheer-pieces should be alternately of cedar and pine."[70]

Perhaps more interesting than the similarities are the ways in which McKay's and Griffiths's books differ. For example, McKay presented his description of how to create the half-hull lift-model in only six pages (13-19) very early in his text. This placement probably made sense to him because McKay had stated in his "Introduction" that he intended to lead "the builder through his course of instruction as if he were in the ship-yard." McKay probably reasoned that if he had discussed a half-hull lift-model early in his own shipwright training with his master, Isaac Webb, the discussion of a model should also occur early in his own book. Positioning his discussion of the half-hull lift-model in this fashion suggests that McKay believed it was his job as an author to replicate the process taught to him in his apprenticeship. Such a belief may have arisen because McKay wrote his book a mere decade after completing his apprenticeship, when he may not have achieved sufficient shipwright experience to appraise critically the training he had received from Webb. Or, McKay may have wanted to replicate the process taught to him in his apprenticeship because he believed that what he learned from Isaac Webb was indeed the best way to design ships.

However, McKay's link between his book and his earlier training with Webb went far beyond an idea of where to put half-hull lift models in his book. The link was, in fact, essential to his book's rhetoric. In his Introduction, McKay described what he had desired to do in his book: "I have endeavored to afford necessary information, in conformity with common practice and without ostentation" and "having in the body of this book given instructions according to the plans most common in this country." McKay's book was very much like a cookbook: tried-and-true instructions in sequential order without much authorial intrusion. McKay did not believe that his role as an author was to move his audience beyond the standard practices of shipwrights. *Conformity with common practice* became McKay's essential means of persuasion—he believed the more his book conformed with common practice, the more persuasive it became. In this spirit, McKay initiated his discussion of the creation of half-hull lift-models with the following sentence that grounded his discussion in *common practice*: "As vessels are almost universally built from models in the United States, and as it is much the most accurate and preferable method, I shall commence by showing that mode of construction."[71]

McKay ended his discussion of the half-hull lift-model with an allusion to years of common practice: "and we, like our forefathers, conclude . . . "[72] McKay believed that his work would be followed by his audience so long as it conformed to common practice—the practice he himself had learned from Isaac Webb.

GRIFFITHS WANTED TO PRESENT IMPROVEMENTS IN ADVANCE OF THE AGE

In contrast to McKay, Griffiths needed nearly 30 pages—or 300% more pages than McKay—to present his description of creating the half-hull lift-model. Moreover, Griffiths's description of the development and use of the half-hull lift-models did not occur until Chapter 3, 90 pages into his book. Before even introducing the half-hull lift-model, Griffiths, like Oliver Evans, created his "first principles"[73] by describing the history of shipbuilding up to its revival in the Middle Ages; the laws of buoyancy, stability, resistance, and propulsion; the importance of ascertaining a ship's center of effort; and the "deleterious effects" of the Tonnage Laws—a topic McKay broached in only the last 10 pages of his book. Thus, in contrast to McKay's rhetorical organization on a sequential basis or as in the *common practice*—a rhetoric that does not reveal much of the presence of an author—Griffiths created a new rhetorical organization founded on his personal credibility (i.e., by relating the history of shipbuilding first, Griffiths hoped to establish his expertise with the audience), the science or experimental knowledge he understood science to be, and graphics. Certainly he could not depend on *common practice* as McKay had because, as noted in his own Introduction, Griffiths wanted to convey: "a few original improvements, and it is confidently believed that it will not only be found to embody in all departments, the latest improvements, but to be in advance of the age, in this complicated art."

Griffiths's writing was based on what he learned from Beaufoy, his own hull experiments, his work at a number of different shipyards, and two decades of experience. From such a self-confident vantage point, Griffiths was able to critically appraise the *common practice* he had learned from Isaac Webb. Accordingly, he found that plodding in the tried-and-true fashion was not to his liking. In fact, Griffiths's inability to proceed in this way may hint at a less than happy termination between master and apprentice 20 years earlier as suggested in the following two passages:[74,75]

> It will be found by a strict inquiry into the various opinions of those who model vessels, that there is very little originality of opinion with regard to shape [the key point of originality in Griffiths's design of clipper ships]; what is often termed experience, is rarely more than hereditary notions, handed down from father to son; and if the young man dare to form opinions from his own observation, which conflict with those of his sire, he is but too often branded as an addle-pated enthusiast, or a reckless adventurer upon the ocean of fame.
>
> The day is coming, nay, we have already seen the day, when the conceited architect, who folded his arms in mysterious wisdom, and hove-to, scorning to learn one more principle from the page of science, awoke upon a fine morning to observe his industrious apprentice passing him to windward on the path to fame. We are of those who believe in making all sail in fair weather, always lay our course when we can, head our craft to the gale, and never "lie-to" under a free wind.

In *Marine and Naval Architecture...September 20th, 1844*, Griffiths had also alluded to "an addle-pated enthusiast." However, in this earlier work, a work closer by some seven years to his apprenticeship with Isaac Webb, his fury against precedent, as well as perhaps against his former master's constraints, seemed to have no bounds, as can be observed in the following passage:[76]

> Not that noble public opinion that gives hardihood and unshrinking fearlessness, but that wayward offspring of fashion and fear, that intolerant public opinion which weighs actions by their popularity, and not

> by their effects, which bids us to do that which is customary, rather than that which is right; a dishonest time-serving sycophancy, that is like an incubus upon our boldest thoughts, and best exertions, making cowards of the brave, and checking the spread of knowledge, that darkens both land and sea with its prejudices, and chills the heart with its cold decrees, deeming him an addle-pated enthusiast who dares to assert by precept or practice, that the fields of science are open alike to all. The prejudices interwoven with the present system of ship building, cling to the builder like poisonous ivy to the monarch of the wood, throwing fetters of bondage around his intellect, which, if not rent asunder, will cause him like the oak, to perish in its palsying embrace.

Griffiths knew of McKay's book and had, in fact, criticized McKay in the matter of wood to use for the half-hull lift-model. Perhaps Griffiths was equating McKay with the mighty oak that perished in the palsying embrace of the prejudices of "common practice." Eschewing common practice and precedent, hoping to rend asunder the fetters of precedent, and desiring to offer readers "a few original improvements," Griffiths needed some new rhetorical tactics.

Illustrations of the tactics Griffiths used can be seen in the 24 pages of his *Marine and Naval Architecture...September 20th, 1844*. At one point Griffiths appealed to established authority to buttress his case by using a letter of support from a shipwright with 30 years experience.[77] Griffiths, like Evans before him, also used an appeal to nationalism when he proclaimed the half-hull lift-model a "proud emblem of American skill."[78] Griffiths also repeatedly projected in his text an image of readers who are "able to approximate that degree of perfection;"[79] "the thinking-man;"[80] and the man who wants to avoid a "dishonest time-serving sycophancy."[81] Griffiths rhetorically pulled out all the stops in order to bring his audience to his point of view. Most of all, however, he used the rhetorical power of graphics in a way totally alien to McKay.

THE QUALITATIVE WEAVE OF TEXT AND GRAPHICS REVEALS DIVERGENT RHETORICAL STRATEGIES

In comparing Evans's two works, a quantitative analysis of graphical references could describe Evans's varied weaves of text and graphics and the implications of such weaves. Yet in comparing McKay's book to Griffiths's, such a quantitative analysis is more difficult to employ because of the disparate lengths of the passages. For example, McKay used two plates in his six-page description of the half-hull lift-model: He referred specifically to one of the four figures on one plate (only alluding to the others) and one figure on another plate. Three pages, or 50% of McKay's text describing the half-hull lift-model, had 14 specific references to graphics. Griffiths used 11 graphics in 27 separate, specific references on seven pages, or about 23% of his text describing the half-hull lift-model. Such a quantitative comparison does not seem to yield any clear message when the bases for comparison are so dissimilar.

However, a qualitative comparison of the graphics can yield some very interesting insights. For example, Figures 2.7 and 2.8 show the two specific graphics referred to by McKay in his passage on developing half-hull lift-models. Each of these graphics was carefully drawn to scale and presented in the context of the entire vessel, which was also drawn to scale. In this manner, McKay's graphics were lifelike and appropriately persuasive to his audience to whom he was trying to convey the "common practice." In a typology of graphics,[82] McKay's drawn-to-scale graphics (much like Evans's graphics[83]) could be categorized as "pictorial symbols and drawings."

"Pictorial symbols and drawings" attempted to "reconstruct the object in a highly realistic symbol" and "the viewer should be able to transfer a pictorial symbol to reality with ease."[84] It was this ease of transference to reality that allowed illustrations to offer "virtual witnessing" like that used by Robert Boyle in the Royal Society. Boyle's, Evans's, and McKay's use of lifelike graphics and their intended effects on an audience were well described by Mitchell when he noted:[85]

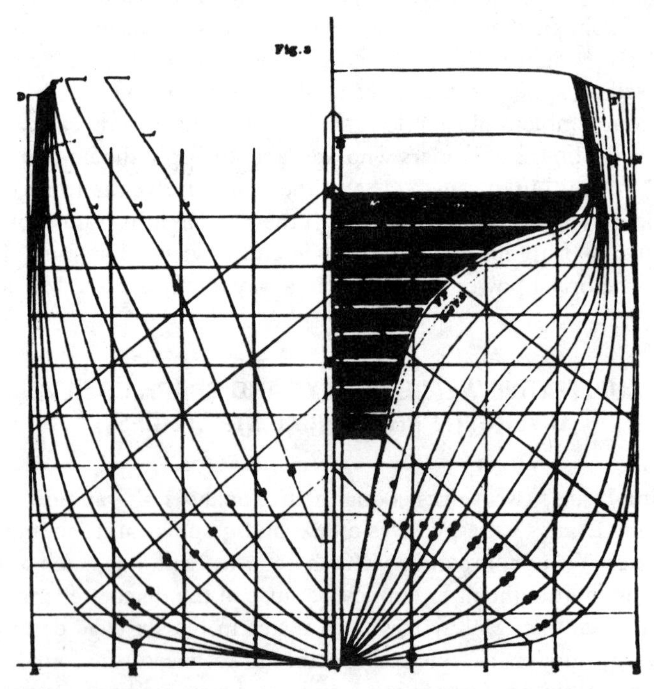

Figure 2.7. Figure 3 from McKay's 1839 *The Practical Ship-builder* demonstrating the positioning of the sternpost

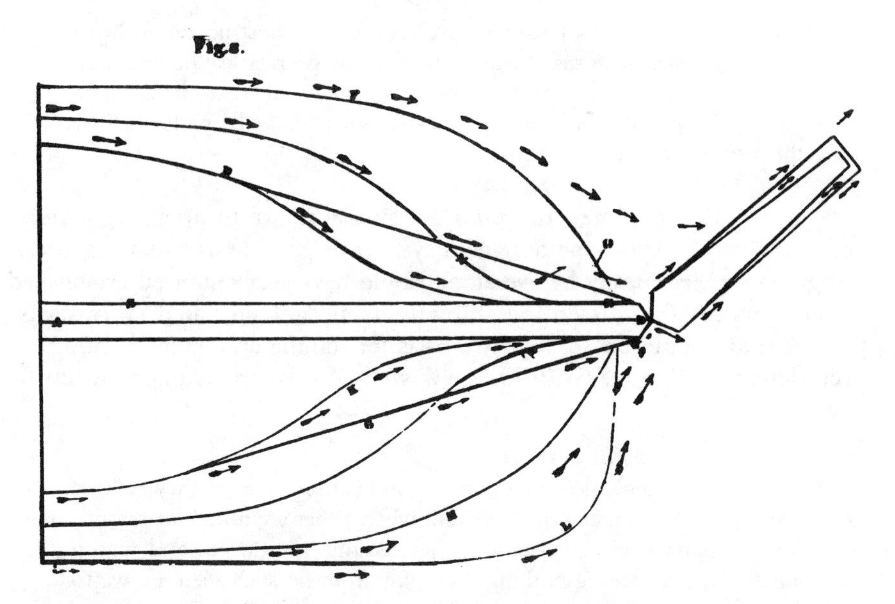

2.8. Figure 8 from McKay's 1839 *The Practical Ship-builder* demonstrating the forces acting on the rudder.

> The image is the sign that pretends not to be a sign, masquerading as (or, for the believer, actually achieving) natural immediacy and presence. The word is its "other," the artificial production of human will that disrupts natural presence by introducing unnatural elements into the world—time, consciousness, history, and the alienating intervention of symbolic mediation.

McKay justified the organization of his text as a reflection of the sequence of his actual training with Webb. His argument would, he hoped, prove persuasive to his audience because it reflected the common and accepted, real-world shipwright process of design. His graphics would be persuasive because they looked real. Consequently, remaining lifelike and adhering to precedent channeled McKay's rhetorical choices in graphics.

On the other hand, Griffiths's goal in discussing the creation of the use of half-hull lift-models was to transform their meaning and use so that the shipwright process would itself change. Griffiths could only present his "improvements . . . in advance of the age" by taking the half-hull lift-model and transforming it into an experimental scientific apparatus and taking the shipwright and transforming him or her into a scientist. Unfortunately, this goal proved very tricky. As a pure scientist, Beaufoy had proved troublesome for Griffiths's specific mechanician application, as earlier Smeaton had proved troublesome to Evans:[86]

> To whom are we to look for improvements in the construction of ships? Is it to the man who may bring forward some geometrical or mechanical series of curved lines for a ship's body, deduced from one or more curves? for this has been many times done, and may be performed by the mere dabbler in the art.

Yet, Griffiths knew he could not simply return to designing a vessel unscientifically by eye—one designer's eye was as good or as bad as another's. Designing unscientifically by eye alone would have easily allowed established shipwrights to label dissenting shipwrights as "addled-pated enthusiasts" because there would be no objective basis for settling an argument, and thus precedent would rule. Griffiths knew well of this predicament when he wrote:[87]

> to those who, regardless of any rules, build ships by what they call the eye! for there are many of these; and when either are asked for reasons for any particular construction, they assume mysticism, and would appear wise by saying nothing. Certainly from no such men are we to hope for improvements in a science pregnant with difficulties, to surmount which seems to exceed the force of human understanding.

However, Griffiths's work, as noted in the first page of his Preface (and echoing Evans's "Preface" 56 years earlier) "is designed to form the connecting link between science and practice." Later in the *Treatise*, Griffiths exclaimed: "How essential then it is that our ship-builders and ship-masters should be scientific as well as practical men."[88] To link the two, he had to turn shipwrights into scientists:[89]

> An idea has prevailed in the mechanical world, that scientific knowledge was reserved for the comprehension of minds of more than ordinary caliber. We apprehend this to be a great mistake; the term *science* may be applied to any branch of knowledge that may be made the subject of investigation, with a view to discover its first principles, as distinguished from *art*. . . . In its most restricted sense, it is but the ability to give a why and a wherefore for our daily practice.

In this process of making scientists out of shipwrights, objects already familiar to shipwrights were presented with a twist: "We are presenting the subject under a different aspect."[90]

For example, Figure 2.9 depicted exactly the same information as McKay did in Figure 2.7, but Figure 2.9 illustrates the placement of the sternpost in a dramatically different way. In Griffiths's figure, there is no scale, nor are there any carefully crosshatched lines to insure a lifelike quality. The absence of such

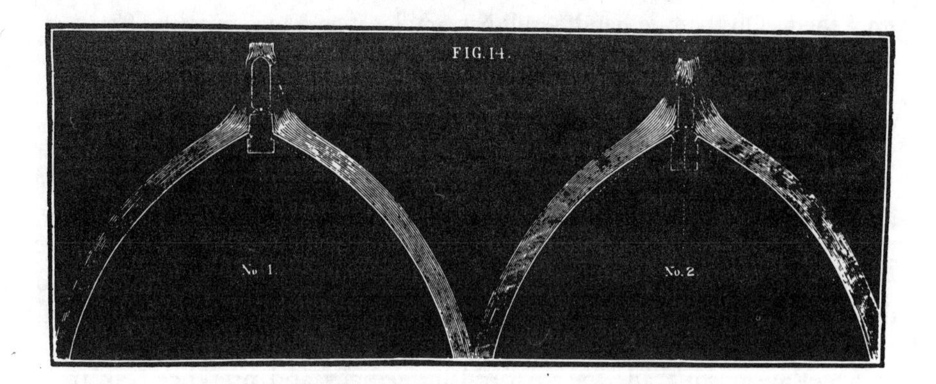

Figure 2.9. Figure 14 from Griffiths's 1851 *Treatise* demonstrating correct and incorrect positioning of the sternpost of a vessel

items could not have been an act of carelessness or an inability on Griffiths's part; after all, he had worked as a full-time draughtsman for years. Furthermore, he had presented a carefully drawn-to-scale model of a steam ship from three perspectives in another plate. Those lifelike graphics were just not the rhetorically correct graphic he needed for his text.

Instead, Griffiths's graphic evolved from a theoretical or conceptual nature. His was a "concept-related graphic symbol" that showed the reader a more stylized rendering of the essence of the object.[91] Such a symbol had "less detail" than the pictorial symbols made by McKay and Evans. Such concept-related graphics presented general information and drew attention to general concepts rather than specific dimensions. In short, they were acting as visual metaphors of ideas rather than as representations of actual objects.

For example, the placement of the sternpost, as McKay presented it in Figure 2.7, was almost hidden beneath all the lines needed to make the drawing lifelike. On the other hand, Griffiths's placement of the sternpost in Figure 2.9 was easy to see because he omitted details that were less important to his point even if they made the graphic less lifelike. Figure 2.4 showed how Griffiths intended the half-hull lift-model to be used to determine vessel displacement. With its disembodied hands floating in space, the figure had less value as a record of specific dimensions than of the general theory of how to use the half-hull lift-model in a fluid box to ascertain scientifically the vessel displacement. Griffiths's figures 2.5 and 2.6 continued in this metaphorical style.

Griffiths plowed the middle course between Beaufoy's scientific graphics (Figure 2.2), which were so abstract that they had no practical application, and McKay's specific realistic graphics (Figures 2.7 and 2.8), which presented difficulties to readers who endeavored to develop broad principles from them. On the one hand, Griffiths's graphics were recognizable. His readers could see how the graphics applied to the half-hull lift-models they could be holding in their hands. On the other hand, the graphics were abstract enough to present new general principles of shipbuilding.

Griffiths's graphics were also sufficiently abstract so readers could not substitute precedent for Griffiths's own innovations. Griffiths wanted his experimental, scientific method to lead others to "discover new principles in modeling . . . enabling them to take the helm and think and act for themselves."[92] His graphics, in fact, could be said to alienate readers from reality so they could manipulate objects and examine them in an objective, impersonal, scientific manner. In other words, whereas the figures created by Boyle, Evans, and McKay masqueraded as "natural immediacy and presence," abstract graphics such as Griffiths's were intentionally: "the artificial production of human will that disrupts natural presence by introducing unnatural elements into the world—time, consciousness, history, and the alienating intervention of symbolic mediation."[93]

THE MULTIPLE PURPOSES OF ANTEBELLUM SHIPWRIGHT GRAPHICS

Graphics and visual thinking played key roles in the evolution of antebellum shipwright manuals. Both McKay and Griffiths showcased the use of a new visual shipwright tool, the half-hull lift-model. Both McKay and Griffiths used graphics frequently in their text. Perhaps they did not employ graphics as frequently as Evans did in Part III of *The Guide*, but certainly McKay and Griffiths used graphics more frequently than did the British scientists Smeaton and Beaufoy. However, graphics played crucially varied rhetorical roles for McKay and Griffiths. McKay's graphics stressed lifelikeness because his text stressed precedent and common practice. Griffiths's, on the other hand, were specific enough to be recognizable to his readers but abstract enough to support new independent, scientific, experimental discoveries. As Griffiths wrote in 1855:[94]

> The day is coming, nay, we have already seen the day, when the conceited architect, who folded his arms in mysterious wisdom, and hove-to, scorning to learn *one* more principle from the page of science, awoke upon a fine morning to observe his industrious apprentice passing him to windward on the path to fame. We are of those who believe in making all sail in fair weather, always lay our course when we can, head our craft to the gale, and never "lie-to" under a *free wind*.

ENDNOTES

1. Griffiths, John W.: "Inland Navigation. Modeling Sail Vessels for the Lakes." *The Monthly Nautical Magazine and Quarterly Commercial Review* 1(1) October 1854: 10.
2. McKay, Lauchlan: *The Practical Ship-Builder*. New York: Collins, Keese and Co., 1839 (privately reprinted in 1940, and again in 1970. New York: Library Editions, Ltd.).
3. Griffiths, John W.: *Treatise on Marine and Naval Architecture or Theory and Practice Blended in Ship Building*. New York: D. Appleton & Co. 1851.
4. Apprenticeshop of the Maine Maritime Museum: *Half-Hull Modeling*. Bath, MA: Maine Maritime Museum (3rd ed.) 1986: 7–8.
5. Cutter, Carl C.: "Important Types of Merchant Sailing Craft." *The Marine Historical Association* 1(1) (December 15, 1930) (reprinted in *The Log of Mystic Seaport* 31(3) (Fall 1979): 71–8. See also Cutter, Carl C. *Greyhounds of the Sea*. Annapolis, MD: Naval Institute Press, 1984.
6. Apprenticeshop, op. cit., 1986: 6.
7. Apprenticeshop, op. cit., 1986: 6.
8. Steel, David: *The Elements and Practice of Naval Architecture; or, A Treatise on Ship-building Theoretical and Practical, On the Best Principles Established in Great Britain*. London: W. Simpkin and R. Marchall (3rd ed.) 1822: 258—" . . . then set the compasses to the thickness of the plank, and fix one leg where the half thickness of the post intersects the line squared down; and, with the other, describe an *arc*, from the back of which the water lines may pass through their respective spots, and end at the fore part of the half breadth plan, proceeding in the same manner was with the after part".
9. $$\frac{(L-3/5\ B) \times B \times 1/2\ B}{95}$$

   Fairburn, William Armstrong: *Merchant Sail. Vol. 2*. Center Lovell, MA: Fairburn Marine Educational Foundation, Inc. 1945–55: 1505. This formula was criticized by both McKay and Griffiths because, according to Griffiths (1844), "under this pernicious system [of design for the purposes of tariff tax cheating,] vessels were built narrow and deep, and thus not only inefficient but highly dangerous." Griffiths, John W.: *Marine and Naval Architecture, or The Science of Ship Building, Condensed into a Single Lecture, and Delivered Before the Shipwrights of the City of New-York, in the Rooms of the American Institute, September 20th, 1844*. New York: J. M. Marsh, 1844: 11.
10. Laing, Alexander: *Clippership Men*. New York: Duell, Sloan and Pearce, 1944: 179–85. See also Chapelle, Howard I.: *The Search for Speed Under Sail: 1700–1855*. New York: Bonanza Books, 1967: 150.

11. McKay, op. cit., 1839: Introduction. "The Baltimore clippers, as they are called from their fleetness, astonished and puzzled the British seamen in the last war. The writings of their naval officers are full of expressions of admiration and wonder at the feats of this species of craft upon the ocean." See also De Roos, Lt. the Hon. Fred. Fitzgerald: *Personal Narrative of Travels in the United States and Canada in 1826, Illustrated by plates, with remarks on the Present State of the American Navy*. London: William Harrison Amsworth, 1827 (qtd. in Chapelle, op. cit., 1967: 150). "I met with a builder who had a book of draughts of all the fastest-sailing schooners built at Baltimore, which so much puzzled our cruisers during the war." See also Chapelle, Howard I.: *The Baltimore Clipper*. Salem, MA: Marine Research Society, 1930; and *U. S. Nautical Magazine and Naval Journal* 4(3), June 1856: 184.
12. Apprenticeshop, op. cit., 1986: 7.
13. Griffiths, op. cit., 1851: 90–1.
14. Apprenticeshop, op. cit., 1986: 10.
15. Apprenticeshop, op. cit., 1986: 9.
16. Calhoun, op. cit., 1973: 244.
17. Chapelle, op. cit., 1967: 284.
18. Griffiths, John W.: "The Value of Calculations in Modeling." *Monthly Nautical Magazine and Quarterly Review* 2(1) April 1855: 9–11.
19. Chapelle, op. cit., 1967: 150.
20. Chapelle, Howard I.: *The American Fishing Schooners: 1825–1935*. New York: W. W. Norton & Co., Inc., 1973: 78.
21. Griffiths, op. cit., 1851: 92.
22. Hutchins, John G. B.: *The American Maritime Industries and Public Policy, 1789–1914*. Cambridge, MA: Harvard University Press, 1941: 115.
23. Chapelle, op. cit., 1973: 79.
24. Lane, Bruce M. & Lane, G. Gardener: "New Information on Ships Built by Donald McKay." *American Neptune* 42(2) (April 1982): 131.
25. *American Ships and Ship-building*. New York: Charles W. Baker, 1860: 10.
26. Reprinted in facsimile in 1969 by Stephen Austin & Sons Ltd., Caxton Hill, Hertford, England.
27. Reprinted in facsimile in 1977; see Pook, Samuel M.: "Laying Out in the Mould-Loft." *Nautical Research Journal* 35(3) (September 1990): 147.
28. Of this problem, McKay wrote in 1839: "There is now no access in our country to the rules and principles of the construction and building of ships within reach of those whose wealth will not allow them to obtain expensive books, or whose time will not permit them to go through a course of deep and intricate study. My purpose has been to supply this deficiency." Similarly, Griffiths observed in 1875 that: "Both Chapman and Stalkaart had come to the front, and subsequently Steel had appeared in print, but there were but two, if, indeed, there was more than a single

copy of these works in the United States; and these, not being adapted to the American need, were of little use to ship-builders." (*The Progressive Ship Builder*. 2 vols. New York: John W. Griffiths, 1875: 171).

29. Stillman, Charles K.: "The Development of the Builder's Half-Hull Model in America." *The Marine Historical Association* 1(1) (December 15, 1930) (reprinted in *The Log of Mystic Seaport* 31(3) (Fall 1979): 92). See also Chapelle, Howard I.: *The American Fishing Schooners 1825–1935*. New York: W. W. Norton & Co. Inc., 1973: 78.
30. Chapelle, op. cit., 1967: 151. See also: Laing, op. cit., 1944: 178–9; Chapelle, op. cit., 1960: 8; and Morrison, John N.: *History of New York Ship Yards*. New York: Wm F. Sametz & Co., 1909: 57.
31. McKay, Richard C.: *A Maritime History of New York*. New York: G. P. Putnam's Sons's, 1934: 353
32. Griffiths apprentice endenturement in 1827 is part of the Griffiths Collection held by the Transportation Division of the Smithsonian's National Museum of American History.
33. Calhoun, op. cit., 1973: 242, 252–3. Interestingly, the US Navy has just proceeded through the same revolution: "The previous paradigm [for designing nuclear attack submarines] was wooden mockups and engineering drawings, now it's virtual reality. In the earlier structure, the key to control was the paper, the drawings, now it's electrical charges moving around, to which everybody has access." Ashley, Steven: "Designing a Nuclear Attack Submarine." *Mechanical Engineering* 117(4) (April 1995): 67.
34. Chapelle, op. cit., 1967: 364.
35. 1(1) (October 1854): 60. See also Griffiths, op. cit., 1875: 173: "If we do not know what we want in the shape of a vessel we propose to build, and can only approximate our needs by comparison, that is to say, by comparing the design of one vessel with that of another, drawings should be compared with drawings and models with models, but neither will suffice alone, nor yet both without the aid of the calculus, if we would maintain a progressive state of ship-building."
36. Chapelle, op. cit., 1967: 364.
37. "Inland Navigation." *Monthly Nautical Magazine and Quarterly Commercial Review* 1(2) (November 1854): 69. See also Griffiths, op. cit., 1875: 175: "In the vessel, model or draft we find ourselves assuming the waterlines in the solid or diagonal lines in the frame model, to be resisting lines of direction to the molecules of the fluid in which the vessel is immersed, and through which she is to pass, when propulsory power is applied to move her in the line of her axis. Here is the mistake, it is at this point the eye is deceived..."
38. At one point, Lauchlan McKay worked at the competing New York firm of Smith and Dimon where John W. Griffiths had also moved and was working as a draughtsman. Laing, op. cit., 1944: 177.

39. McKay, Richard C. (ed.): "Lauchlan McKay" In McKay, Lauchlan. *The Practical Ship-Builder*. New York: Collins, Keese and Co., 1839 (privately reprinted in 1940, and reprinted again in 1970, New York: Library Editions, Ltd.). It was with some justice that Newburyport which gave birth to the half-hull model would also give birth to the first book for shipwrights using the half-hull model as a design tool.
40. Lane & Lane, op. cit., 1982: 132.
41. *Prospectus of an Intended Work on the Theory & Practice of Naval Architecture Particularly Illustrating American Ship Building* was developed by Donald in 1859.
42. Lane & Lane, op. cit., 1982: 134.
43. McKay, R., op. cit., 1934: 353.
44. New York: J. M. Marsh, 1844, 24 pages. Held in the Griffiths Collection, Transportation Division, National Museum of American History, Washington DC, and at the United States Naval Academy and Dartmouth College.
45. 2 vols. New York: John W. Griffiths. See also *U.S. Nautical and Naval Journal* 6(6), September 1857: 438.
46. In 1847 and 1849 he also "presided over a marine architectural institute" in New York (Chapelle, op. cit., 1967: 324).
47. La Grange, Helen: *Clipper Ships of America and Great Britain: 1833–1869*. New York: G. P. Putnam's Sons, 1932: 49.
48. Laing, op. cit., 1944: 176–7.
49. Wright, Thomas: "Mark Beaufoy's Nautical and Hydraulic Experiments." *Mariner's Mirror* 75(4) (November 1989): 313–327.
50. *Nautical and Hydraulic Experiments*. London, 1834.
51. Wright, op. cit., 1989: 316.
52. Wright, op. cit., 1989: 320.
53. Griffiths, op. cit., 1851: 204.
54. Wright, op. cit., 1989: 322.
55. Griffiths, op. cit., 1851: 204.
56. Wright, op. cit., 1989: 317.
57. Robert Fulton had earlier used Beaufoy's data when he was attempting to design the best proportions of length, width, and draft of his steamboat. (Letter of Fulton to Fulmer Skipwith, December 12, 1802, LeBoeuf Collection, New York Historical Society, qtd. in Brooke Hindle: *Emulation and Invention*. New York: New York University Press, 1981: 52.)
58. "Inland Navigation. Modeling Sail Vessels for the Lakes." *The Monthly Nautical Magazine and Quarterly Commercial Review* 1(1) October 1854: 10.
59. Griffiths, op. cit., 1851: 116.
60. Griffiths, op. cit., 1851: 112.

61. The American Institute in New York in the 1840s appears to have provided Griffiths with an intellectual environment of criticism and support very similar to that provided in Philadelphia for Evans in the 1790s by the Pennsylvania Society for the Encouragement of Manufactures and the Useful Arts, and the *American Museum* with Matthew Carey as its editor.
62. Griffiths, John W.: "Marine and Naval Architecture of the Crystal Palace." *The Monthly Nautical Magazine and Quarterly Commercial Review* 1(1) October 1854: 19.
63. It is interesting to note the role of the scale model at turning points in Evans's and Griffiths's debates with their audiences. Such use of scale models was so widespread that the Patent Act of 1790 called for a scale model to accompany written specifications and drawings whenever possible. For information on such 19th-century models, see Ferguson, Eugene S. and Baer, Christopher: *Little Machines: Patent Models in the Nineteenth Century*. Greenville, DE: Hagley Museum, 1979; Hindle, op. cit., 1981: 20–1; and Lawless, Benjamin: "Working with Working Models." *American Heritage of Invention and Technology* 1(2) Fall 1985: 10–7. It is also interesting to note that Evans and Griffiths had their first successes with very concrete graphics—the concreteness of graphic presentation never changed in Evans's work.
64. This was same institution that a year earlier had awarded Samuel Morse a gold medal for the result of his visual thinking—the telegraph.
65. Griffiths was also like Evans in that he, too, was a developer of patents. Griffiths secured one patent in 1862 (#35,232) for an improvement in keels, and two in 1875 (#167,036 and #171, 376) for machinery for bending wood. Copies of the patents are held in the Griffiths Collection, Transportation Division, National Museum of American History.
66. Compare the impact of the word "treatise" in McKay's and Griffiths's books to another contemporary shipwright design book, George W. Rogers's *The Shipwrights Own Book Being a Key to Most of the Different Kinds of Lines Made Use of by Ship Builders*. Pittsburgh: J. M'Millin, 1845. This book has not been included in this chapter because it is much less sophisticated than the other two (Calhoun, op. cit., 1973: 245), and, more importantly, the record of Rogers's apprenticeship training is unknown, whereas those of Griffiths and McKay are known and are identical. Also, both Griffiths and McKay continued learning in New York shipyards, whereas it appears that Rogers did most of his work in Pennsylvania. Thus, careful comparison and contrast of his work with the other two is much less plausible. Also, it is quite difficult to closely observe a copy of this book, which was printed in only one edition, because the only existing copy is in the Cleveland Public Library's rare book room.

67. One might note that McKay—and not Griffiths—included an "Explanation of Terms" in the back of his manual.
68. McKay, op. cit., 1839 17–9; Griffiths, op. cit., 1851: 102–111.
69. McKay, op. cit., 1839: 15.
70. Griffiths, op. cit., 1851: 98.
71. McKay, op. cit., 1839: 13.
72. McKay, op. cit., 1839: 18.
73. Remember that to begin with "first principles" was a widely used practice in the eighteenth century.
74. Griffiths, op. cit., 1851: 112. See also *U.S. Nautical Magazine and Naval Journal* 3(5) February 1856: 323. "How often has the reformer in marine construction been regarded as an addle-pated enthusiast..." This acrimony between former apprentice and master cannot be absolutely established because Griffiths may, in fact, have just been talking about the conservative members of the American Institute who had criticized his model and his ideas. In an 1856 article in his *U.S. Nautical Magazine and Naval Journal*, "The Ship-builder and the Apprentice," Griffiths gives voice to a kind of nostalgia for his apprenticeship days: "It was in those days that such men as . . . Isaac Webb . . . and a host of other builders of the first ranks, were inducted into the ranks of nautical mechanics. It was then usual for an apprentice to spend from five to six of the last years of his minority in the acquisition of mechanical knowledge: then they were boys until they were twenty-one years of age—but alas for this progressive age!...With a view to furnishing the means of remedying this manifest defect, we have taken up the subject in this *Magazine* and shall give such lessons . . . " 4 (1) April 1856: 15–6. Also, Griffiths may not have been referring so much as to a personal experience of authority reining in his innovative ideas as much as to the general problem of the dialectical relationship between authority and innovation. "At times, the effects of scientific authority can be stultifying: collective intellectual inertia blocked the reception of heliocentric astronomy for more than a century; Newton's posthumous authority retarded the reemergence of the wave theory of light. At other times, perhaps more frequently, authority and innovation interact beneficially; consider heliocentric astronomy between Copernicus and Kepler, the theory of light between Descartes and Newton, the concept of evolution in Darwin's early thought: in each of these cases we can see the positive results of the dialectical contest between authority and innovation." (Gross, Alan G.: *The Rhetoric of Science*. Cambridge, MA: Harvard University Press, 1990: 13.)
75. Griffiths, John W.: "The Value of Calculations in Modeling." *Monthly Nautical Magazine and Quarterly Review* 2(1) April 1855: 9–11.
76. Griffiths, op. cit., 1844: 6.

77. David Brown probably was part of the New York shipbuilding firm of Brown and Bell that had launched the first true clipper ship, the *Houqua*, a year before Griffiths's *Rainbow*.
78. Griffiths, op. cit., 1844: 14.
79. Griffiths, op. cit., 1844: 13.
80. Griffiths, op. cit., 1844: 22.
81. Griffiths, op. cit., 1844: 6.
82. Wileman, John: *Exercises in Visual Thinking*. New York: Hastings House, 1980: 25.
83. It is interesting to take a second look at Evans's graphics along this dichotomy of specificity and abstraction because Evans employed both in his book. Figure 1.17, the crucial Plate VIII, was drawn to scale like McKay's graphics so that Evans could convey the reality of his invention and not have it be dismissed as a mere projector's notions. But Figure 1.6 from the *Universal Asylum and Columbian Magazine* broadside had no scale and seemed more akin to Griffiths's more abstract graphics.
84. Wileman, ibid., 1980: 23. In his description of scientific illustrations, Alan Gross (op. cit., 1990) suggests much about the types of illustrations Evans and McKay employed—"Whatever their form, it is crucial that, through every alteration in their scale, every change in their angle of view, these [scientific] illustrations maintain the identity of the physical objects they purport to represent" (p. 74).
85. Mitchell, W. J. T.: *Iconology: Image, Text, Ideology*. Chicago, IL: University of Chicago Press, 1986: 43.
86. Griffiths, op. cit., 1851: 189.
87. Griffiths, op. cit., 1851: 189.
88. Griffiths, op. cit., 1851: 202.
89. Griffiths, op. cit., 1851: 185.
90. Griffiths, op. cit., 1851: 104.
91. Wileman, op. cit., 1980: 25.
92. Griffiths, op. cit., 1851: 112, 114.
93. Mitchell, op. cit., 1986: 43.
94. Griffiths, John W.: "The Value of Calculations in Modeling." *Monthly Nautical Magazine and Quarterly Review* 2(1) April 1855: 9–11.

Chapter 3

19th Century Oral Technical Communication: NCR's Silver Dollar Demonstration

Analyze! Visualize! Dramatize!
The "Golden Words" of Salesmanship
John H. Patterson, President of the National Cash Register Co.[1]

Chapters 1 and 2 explored how in antebellum America Oliver Evans, Lauchlan McKay, and John W. Griffiths succeeded in involving their millwright and shipwright readers by using pictures and three-dimensional models. Evans, McKay, and Griffiths engaged their readers through the use of such visual communication because it enabled them to reach the mind's eye of their mechanician readers. Yet, evidence for this crucial role of visual communication in 19th communication has come, thus far, only from printed books, pamphlets, magazines, and journals. Did visual communication play a similarly crucial role in the oral communication of technology in the 19th century? For example, what role did a speaker's hand gestures or the audiences' handling of objects in active participation with speakers play in this oral technical communication? The answer to this question and a glimpse of oral technical communication at the close of the 19th century can be found in a unique document, the National Cash Register Co.'s *Sales Primer*.[2] *The Sales Primer* recorded the words and "stage directions" of National Cash Register's top salesman in 1887. *The Sales Primer* was used for 30 years as a training tool for all other NCR salesman on how to communicate cash register technology to users: NCR's president at the time thought so highly of *The Sales Primer* that he would quiz salesman on its contents whenever he would visit a branch office. Thus, *The Sales Primer* was one of the key elements that allowed NCR to control over 95% of the domestic market and 50% of the foreign market by 1920, a short 40 years after the birth of the cash register itself.[3]

THE BIRTH OF THE CASH REGISTER

In February 1878, *Scientific American* published "The Cash Recording Machine," which described the birth of cash register technology and its moral purposes:[4]

> . . . a new machine for making people honest—a consummation to which (if it can be attained by machinery) no small amount of inventive genius is just now being brought to bear.
>
> Hitherto most efforts have been directed to the mechanical shoring up of the consciences of car conductors and stage drivers; but the present inventors have advanced higher, and propose to apply the same salutary influence to the moral sense of every class of employee within whose duties the handling and disbursing of cash is included.
>
> It must have occurred to any one who has noticed the Babel of confusion which exists in any large city retail dry goods store, for example, when crowded with shoppers, and when a constant stream of cash boys circulates between clerks and cashiers, that scarcely any system of checks and records depending upon the memory and fidelity of the employee can exist which does not leave loopholes for fraud. We are not prepared to assert that the present machine will at once substitute a system in which it is impossible to swindle, because it is a lamentable fact that there is perhaps nearly as much ingenuity enlisted in the service of sin as in that of virtue, and somebody may discover how to "beat" even the most thoughtfully contrived mechanism; but the new "cash recorders" certainly offer a very simple mode of keeping forcibly accurate records and, for our part, we fail to see where the chance to defraud it exists.

If "making people honest" and "keeping forcibly accurate records" were the general purposes for all cash registers, John H. Patterson, owner of the National Cash Register Company in Dayton, OH, made such claims specifically for NCR cash registers through his publicity. For example, in the August 1886 edition of the *American Store Keeper*, Patterson defined an NCR cash register as:

> An automatic cashier which records mechanically every sale made in a store. It never tires. It never does one thing while thinking of another, and never makes a mistake. It is a mathematical prodigy in brass and steel, all of whose computations are infallibly correct. It is a machine which will save the money you make and thus pay for itself over and over again.

A little more than a century after their invention, cash registers have become ubiquitous. Consequently, it is hard to imagine that NCR and Patterson would have had any difficulty selling them. It would seem that when cash registers were first introduced, they probably exemplified what Emerson is supposed to have said about mousetraps: "If a man can write a better book, preach a better sermon, or make a better mousetrap than his neighbor, though he builds his house in the woods, the world will make a beaten path to his door."[5] Nevertheless, many obstacles prevented the world from immediately making a beaten path to NCR's door.

First, the public was unsure of the cash register's place within all the technical innovations of this era of invention. Useful inventions—such as Sholes's typewriter in 1868 and Bell's telephone in 1876—were all mixed up in the public's mind with useless novelties, such as 1896's dimple maker or automatic hat that tipped itself to salute others. Furthermore, like all first-generation technologies, the cash registers of the 1880s were crude and cumbersome.[6]

A second obstacle to the cash register's acceptance was the fact that the prime potential locations for cash registers—stores and saloons—were operated by the very people cash registers were designed to control: pilfering

Figure 3.1. John H. Patterson said: "Analyze! Visualize! Dramatize!

clerks and bartenders. Many local "Protective Associations" of clerks and bartenders intercepted cash register advertising by destroying every envelope bearing an NCR headquarters postmark (Dayton, OH).[7] Hence, great resistance often greeted the arrival of a cash register salesperson and his or her sales message.[8]

Third, competition was fierce. Eighty-four companies competed with NCR for a market share of cash registers between 1888 and the mid-1890s.[9] Some of these competitors went so far as to offer free cash registers—"premiums"—that were simply given away as incentives by wholesalers and distributors of other products. For all competitors, as the historians of the cash register politely put it, competition was a no-holds-barred affair:[10]

> Every manner of legal and competitive tactic was used, many of which would not hold up under any reasonable standard of ethics, but such was the nature of competition during that day.

The confusion, resistance, and competition was so great in the cash register industry when Patterson bought the National Manufacturing Company in 1883 (later renamed the National Cash Register Company in 1894), that he tried immediately to sell it back to its original owner. The reply he received was: "I would not have it back as a gift."[11] In 1886 barely 1000 cash registers were sold. Yet by 1896 nearly 100,000 registers were sold.[12] What happened in 10 short years to raise sales 1000%?

Surely good communication about this new technology was essential, and some impressive printed materials were produced. The table of contents of the 1897 edition of the *Detail-Adding National Cash Registers* manual had headings that emphasized customers' actions rather than parts of the machine (see Chapter 9, on task-orientation): "To Operate the Cash Register," "To Ascertain the Amount of the Day's Sales," "To Regulate the Opening of the Cash-Drawer," and "Making Combination Sales." The manual also ended with a polite admonition: "Important. If the directions or the register are not clearly understood, or should there be *any* trouble, do not try to solve the problem yourself, but notify the agent."

Sometimes such manuals were accompanied by small instruction cards attached to the register drawers.[13] Figure 3.2 shows the text of such a register drawer card produced by an NCR competitor, the Boston Cash Indicator and Recorder Co. of Northhampton, MA.

The author of this register drawer card made it easy for readers to scan the information to find what they needed to perform their duties. For example, the format of the card's information was visually supported by its layout—the conceptually parallel information on the three sets of wheels was printed in a parallel manner. Also, each new "paragraph" (signified by white space before and after and by the initial indentation) described a single activi-

DIRECTIONS TO READ A NUMBER 3 BANK MACHINE
Each Wheel is Provided with 100 Teeth

On the RIGHT HAND set of wheels	Each tooth on the FRONT wheel represents 1c. Each tooth on the BACK wheel represents $1.00
On the CENTER set of wheels	Each tooth on the FRONT wheel represents 10c. Each tooth on the BACK wheel represents $10.00
On the Left Hand set of wheels	Each tooth on the FRONT wheel represents $1.00. Each tooth on the BACK wheel represents $100.00

To set the Machine: —Turn all the front wheels to the left until their pointers stand on division line next before 0; then turn the back wheels in the same way until their pointers also stand on division line next before 0. Now push down the 1c key in the right hand bank, the 10 key in the middle bank and the $1.00 key in the left hand bank and the machine is set with all the pointers on 0. The push button A controls the operation of the keys in three ways, viz.: Push back the button A and turn the lock key to the right and all of the keys are free to operate; turn the lock key to the left and you will be obliged to press back the button A each time, before a key can be operated. To lock all keys and machine, depress the blank key and turn the lock key to the right, and all the keys and machine are completely locked.

To operate the keys when drawer is open: — Push the lever located inside Cabinet center of machine in front to right; to lock keys when drawer is open, to left; necessitating shutting drawer before each depression of key.

To lock the blank key independent of the others, pull forward the lever which is located over it inside of Cabinet; to muffle the bell, push the short lever over the dials to the left.

To show the tags for dollars and fractions of a dollar; for example $8.72, press down the 8 dollar key and 70 cent key, hold the 8 dollar key down until you have put down the 2 cent key, then let them return to place.

Boston Cash Register Co. of Northhampton, Mass.

N.B. In registering *be sure to push down each key used to its fullest extent,* then it will return to its normal position by its own weight.

Figure 3.2. Example of a cash register drawer card

ty. Finally, each new paragraph had its own heading that was task-based rather than hardware-based like the 1896 cash register manual: "To set the Machine," "To operate the keys when drawer is open," "To lock the blank key independent of the others," and "To show the tags for dollars and fractions of a dollar."

Even though good, effective, printed communication may partly explain the dramatic growth of the entire cash register industry, nevertheless, something unique must have happened at NCR because NCR very quickly dominated the cash register market. One explanation for NCR's singular success was the use of an effective oral sales pitch with distinctive visual elements, the NCR *Sales Primer*.

THE EMERGENCE OF THE NCR *SALES PRIMER*

In 1887, John H. Patterson, the President of NCR, had a problem: He could not explain how his brother-in-law, Joseph H. Crane, consistently outsold all other NCR salesmen. Of course, Crane had had great success as a wallpaper salesman before joining Patterson at NCR. Moreover, in the mid-1880s, Crane was NCR's "general traveling agent," that is, he gathered "information from other experienced agents, and from close observation." Yet, perhaps most importantly, Crane learned from his mistakes and tried innovative solutions:[14]

> After the first man went out I went to the machine and said to myself: 'There is a point I forgot to explain, and there is another and another.' I meant not to forget these points again, so I got the back of a statement slip and made a note on the top of that slip of the things I had forgotten to tell "A" and the second and third things I had forgotten to tell "B." I asked myself the same question when another man went out without signing—I had forgotten to tell him certain things about the register. When those sixteen men had gone out, I found that there were still things I had forgotten, and I know the reason those people in Findlay, Ohio did not buy was because I had not done the machine justice. . . .
>
> When I used the memorandum I had prepared, I revised it from time to time, as I realized that to get a man's order it was necessary to explain the whole machine to him. I knew more at the end of the week than I did when I started. I knew it was a good thing, and I was being educated to the fact. I realized that it was necessary to transfer my information to a prospective purchaser. That was all that was necessary, but it had to be done in a systematic way. The first thing I mentioned was the sign, the next was the indicator, then I demonstrated the machine and used some money to illustrate it, and finally I got so I did not have to refer to the memorandum every time, and later it became unnecessary to refer to it at all. I tore it up, as I found I could explain the points in the same order, and almost in the same words. At the end

> of ninety days I realized fully that I was saying the same thing to every man and I had been successful—rather more so than anyone else.

Notice what Crane was being "educated to." In addition to writing down a memorandum of all that he should discuss about the machine, he realized that he should first point out the physical characteristics of the register ("the sign" and "the indicator"). Next, Crane discovered that he should demonstrate the machine and illustrate its use with actual money. What Crane was being "educated to" by his mistakes was what Evans and Griffiths had been "educated to" by their own early mistakes at communicating technology: the effectiveness of visual communication in transferring technical information. Visual communication in *The Guide's* illustrations had worked for millwrights; half-hull lift-models and visual metaphors had worked for shipwrights; and now, by using visual communication with potential purchasers of cash register, Crane claimed "I had been successful—rather more so than anyone else."

Crane consistently outsold other agents with his pitch having "committed it [to memory], just as an actor does his part."[15] Accordingly, NCR legend has it that one day NCR President Patterson asked Crane to come into his office and present the sales pitch just as if Crane were giving it to a prospective customer. Patterson was so impressed by the effectiveness of Crane's pitch that he had a secretary take down every one of Crane's words and describe every one of his gestures. Consequently, the resulting 16-page pamphlet, "How I Sell a National Cash Register by Joseph H. Crane," noted in its 1887 "Introduction" that these were the "exact words used by Mr. Jos. H. Crane in making sales," and Patterson evidently added, "We believe it is the best system for selling National Cash Registers to owners of stores."

Later, Patterson required all the other NCR salespeople to memorize Crane's pitch, soon renamed *The Sales Primer*, and use it: "Crane sells more machines than any of you fellows, and he sells them this way. I suggest that you all learn this."[16] Patterson began enforcing this edict five years later when he would drop in on NCR sales agents unannounced and quiz them on *The Sales Primer*. Although the sales and communication techniques of *The Sales Primer* are now common practice, Johnson and Lynch in *The Sales Strategy of John H. Patterson* (1932) claimed that in 1887 *The Sales Primer* "was the first appearance on any stage of the exhibition demonstration and the standardized sales presentation."[17] This novel exhibition demonstration in Crane's sales pitch soon became known in NCR as the *Silver Dollar Demonstration*.

The *Silver Dollar Demonstration* was only part of Crane's entire sales pitch, yet it was the part that displayed Crane's and Patterson's essential communication innovation both in its systematic coverage of the information and in its visual elements. The demonstration also comprised most of the printed sales pitch in 1887: five pages were devoted to it in comparison with two and two-and-a-half, respectively, for the opening and closing parts of the sales pitch.[18]

THE SYSTEMATIC COVERAGE OF INFORMATION AND THE NONVERBAL ELEMENTS OF THE SILVER DOLLAR DEMONSTRATION

The *Silver Dollar Demonstration's* effectiveness can be tied to four systematic organizational elements and an all important nonverbal element:

- organizing information from the user's point of view;
- building new information on established, known information;
- parallelism and repetition;
- ensuring that the information is task-oriented; and, most importantly,
- making the information vivid through participatory demonstrations and visual communication.

The following example is taken from the 1903 edition of *The Sales Primer.* (Stage directions are printed in bold; "P.P." is short for Prospective Purchaser)[19]

> Mr. Blank, we make over two hundred and fifty different styles and sizes of cash registers. It is impossible for me to carry a sample of everything we make, but I can demonstrate to you our other registers from these.
>
> I am going to show you how this register would stop the leaks in your store and save you money.
>
> The daily transactions with your customers of which you should keep a record, may be arranged in five classes, thus: (**Count them on your fingers**.)
>
> 1. You sell goods for cash.
> 2. You sell goods on credit.
> 3. You receive cash on account.
> 4. You pay out cash.
> 5. You change a coin or bill as an accommodation.
>
> Am I right? (**Make him admit this.**)
>
> Now, then, Mr. Blank, this register makes the proper entries. The sign reads "Amount Purchased," and this amount is exhibited by metal indicators showing through the glass below. The indicators showing the last recorded transaction are always visible, and the records are made by pressing the keys and turning the handle.
>
> When you sell for cash, these amount keys in connection with the clerk's initial make the record; when you sell on credit, this red "Charge" key in connection with the amount keys and the clerk's initial makes the record; when you receive cash on account, this yellow "Received on Account" key in connection with the amount keys and the

clerk's initial makes the record; when you pay out cash, this blue "Paid Out" key in connection with the amount keys and the clerk's initial makes the record; when you change a coin or bill, the initial key makes the record.

When one or more keys are pressed and the handle turned, nine results are accomplished at the same time: 1st, the indicators showing the last recorded transaction disappear; 2nd, the indicators appear, showing the new transaction; 3d, the cash-drawer is unlocked; 4th, a spring throws it open; 5th, a bell is rung announcing that a registration has been made; 6th, the proper entry is made inside; 7th, a check is printed; 8th, the check is cut off and thrown into the receiver; 9th, a detailed transaction is printed by the register.

I am going to give you a practical demonstration, and will use a little money, just as you would in your daily transactions. Suppose you sell something for $1. (**Show the P.P. the dollar.**) Press your initial key, the "$1" key, turn the handle, put the dollar into the cash-drawer, and shut the drawer. You have locked up the money in a safe place (**show him that the cash-drawer is locked when you shut it**), and have made a record of the transaction at the same time, while the indicators show everybody in the store what has been done. In addition to this, the register issues this check showing the number of the sale, the date, your business card, the amount and kind of transaction, and the initial of the clerk who made it. This check is also a receipt to the customer for cash paid.

The demonstration then proceeds to show the P.P. how to register different amounts, make change, handle credit sales, deal with money received on account, pay out cash, and change money. The demonstration concludes with the following:

You remember that we divided the daily transactions with your customers into five classes.

Now, I have shown you that this register takes care of each and all of them in the same way. Whether you sell for cash, sell on credit, receive cash on account, pay out cash or change a coin or bill, you go straight to the register and record what you have done by pressing the keys and turning the handle. A clerk won't forget this, for in each case he has something in his hand which he must deposit in the drawer to complete the transaction. If it is a cash sale, he has the cash in hand; a credit sale, the charge slip; cash on account, the 'Received on Account' slip and the cash; cash paid out, the 'Paid Out' check; a coin or bill to be changed, the coin or bill. (**Illustrate by showing each slip or coin in your hand**).

> Every clerk must go to one and the same place to complete each transaction, and so he forms a habit of doing this without fail.
>
> Our system of recording transactions and taking care of receipts in a store like yours, Mr. Blank, is easy to learn, quick to operation, safe from mistakes, and is always ready for use.

First, after spending time with customers as NCR's "general traveling agent," Crane knew the customer's manner of classifying activities. Therefore, he began his demonstration with a review of the five classes of a customer's daily transactions ("sell goods for cash;" "sell goods on credit;" "receive cash on account;" "pay out cash;" and "change a coin or bill"). Such a review helped Crane establish a rapport with his customers because it implied: "I took the time to understand your problem from your point of view before I ever approached you with this communication." In a 1932 book on NCR's sales techniques, the authors pointed out the novelty of this technique at the time:[20]

> The extraordinary effectiveness of *The Sales Primer* lay in the fact that, for probably the first time in history, it built up the selling argument consistently from the standpoint of the prospect's own desires and interests, and not from that of the company's desire to make a sale.

Thus Crane's first systematic element was that he organized his information from the customer's point of view. Crane also intuitively knew that an effective communicator must compel an audience to become involved with co-creating the information and understanding the organization being used. Thus after the classification of daily transactions, Crane compelled his audience to become involved by immediately using a rhetorical question, "Am I right?" and then would "Make him admit this" to ensure the customer's further involvement and understanding of the organization.

Second, in the *Silver Dollar Demonstration*, Crane described the physical aspects of the cash register (e.g., the signs such as "Amount Purchased," "Received on Account," or "Paid Out"), using the now verified five-part classification of activities ("sell goods for cash;" "sell goods on credit;" "receive cash on account;" "pay out cash;" and "change a coin or bill"). This transition from the classification of transaction types to the signs in the register's window was a brilliant example of taking what the audience already knows (the classification) and then building new information on it (the signs in the glass). This building of new information on established information—the so-called "Given-New Principle"[21]—allowed audiences quickly to understand and use information. It must have allowed Crane's customers to understand how the familiar actions of their store- or saloon-keeping were physically captured in the features (the signs) of the cash register.

Third, Crane used repetition as part of his systematic organization. He presented and received the customer's verification of the five classes of activities. Next, he repeated the identical five classes of activities in exactly the same order and linked them with the physical features of the signs in the window. Crane repeated these same five classes of transactions in detail and in the same order of presentation while performing each one of the five actions (these sections were deleted from the earlier excerpt). Finally, Crane repeated the five classes of actions at his conclusion as illustrated in the third from the last paragraph. This repetition helped to overcome the limitations of short-term memory. By the conclusion of the *Silver Dollar Demonstration* the customer may have gotten lost in the details of the signs or the actions, but the repetition of the five classes of transactions ensured that the customer saw the overall order and purpose of the *Demonstration*. Maintaining the parallelism of each of the five classes of transactions was also helpful for the NCR salesmen because such parallelism and repetition probably soon became a mnemonic aid for them just as parallel and repeated rhyme and rhythms worked for the bards and storytellers of preliterate days. Such parallelism and repetition certainly seemed to work for Crane himself as he noted earlier:[22]

> Finally I got so I did not have to refer to the memorandum every time, and later it became unnecessary to refer to it at all. I tore it up, as I found I could explain the points in the same order, and almost in the same words.

Fourth, the *Silver Dollar Demonstration* was systematically organized to be task-oriented; it explained how to use the machine rather than why or how the machine worked. Patterson repeated over and over again to his salesmen, "Don't talk machines; don't talk cash register; talk the prospect's business:"[23]

> It is worthy of special notice that this exposition of register functions in *The Manual* [subsequent name of *The Sales Primer*] was not a technical, mechanical exposition, and at no time did Mr. Patterson equip the salesmen with knowledge of the mechanical construction and operation of the machine. They were taught what the machine did, and why it did it; but how the machinery worked was outside their province. The salesman was required to know how to operate the register, how to reset it to zero for purposes of demonstration, and in general how to do with it anything that the user would be expected to do. But beyond that, the mechanical side of the register might remain a profound mystery to the salesmen, and Mr. Patterson rather preferred that it should.[24]

Crane in his *Demonstration* and Patterson in all his later sales practice directives understood the benefits of such a task orientation. Such an orientation focused on the customers and their needs rather than on the machine and its mechanics, and it also allowed Crane and all those who followed him to understand what was extraneous information. If the information presented was relevant to the task of the customer using the machine, then they would include it; if not, they deleted it. Surely such brevity in demonstrating the cash register was as helpful to the salesmen as it was to the busy customer audience.

All these four elements were essential parts of Crane's systematically organized demonstration of the cash register. Furthermore, like Evans, McKay, and Griffiths, Crane also had to make his presentation vivid in the audience's mind's eye by using visual means of communication. Patterson himself constantly used charts, posters, and illustrative diagrams to make abstract concepts concrete. Furthermore, the three "golden words" of salesmenship he drilled into his salesman were "Analyze! Visualize! Dramatize!:"[25]

> The effectiveness of an argument, according to Mr. Patterson, depended upon the amount of the dramatic element it contained. Mere words, whether written or spoken, he held to be ineffective without something definite in the way of a situation or a graphic portrayal to hold the attention. Comparatively few people understood words, he said, but anybody could grasp the significance of a picture of a situation.[26]

Thus, the visual element of the *Demonstration* began with the salesman's preparation of a special kit and scenery before the customer's arrival:[27]

> Mr. Patterson's strong belief in visualization was in evidence, not merely in the use of charts and illustrative diagrams, but throughout the whole demonstration wherever it could be applied. He insisted, for example, that the salesman should use real money in illustrating the function of the register, and real packages of merchandise in conducting the imaginary transactions in the demonstration room. A part of the salesman's regular equipment was a special purse or wallet containing exactly the right amount in bills and change for visualizing any transaction. Most of the agency demonstrating rooms were supplied draw-curtains, painted to represent store interiors, and many of the men, when they were away from headquarters and demonstrating in hotel rooms and the like, carried portable equipment of the same character. Nothing was overlooked that might contribute to the desired object of keeping the prospect's attention from being distracted, and to make him feel at ease in a familiar atmosphere.

After the briefest of introductions, and as Crane verbally listed the five classes of transactions, salesmen were directed to count them on their fingers. Thus the visual element of finger counting supported the words listing the transaction classes (see Figure 3.3).

Next the physical parts of the register were pointed out to the customer (see Figure 3.3). Moreover, not only were the register parts labeled with recognizable words for the customer, but the keys were also color-coded: red for "Charge," yellow for "Received on Account," and blue for "Paid Out." Again, a visual element—color—supported the words on the register labels. In a quite similar fashion—50 years later, when the first computer, ENIAC, was being used—scientists noticed a large number of errors arose because the rows and rows of switches that had to be set to process information in ENIAC were all identical and aligned. The number of processing errors plummeted dramatically when the rows of switches were color-coded, differentiating one from another. So, too, in the *Demonstration*, the keys could be remembered by label as well as by color.

Another visual element of the demonstration occurred when the salesman actually gave the "practical demonstration." In addition to the salesman requiring the customer to use actual money and press the key, a nonverbal "sentence" of sights and sounds occurred. The visual sentence began for the customer when he or she pressed a key and turned a handle. In response, the customer saw a sign in the register window, the cash drawer sprang out, a check was printed out, and a bell rang. The final punctuation of this visual sentence came when the customer closed the cash drawer.

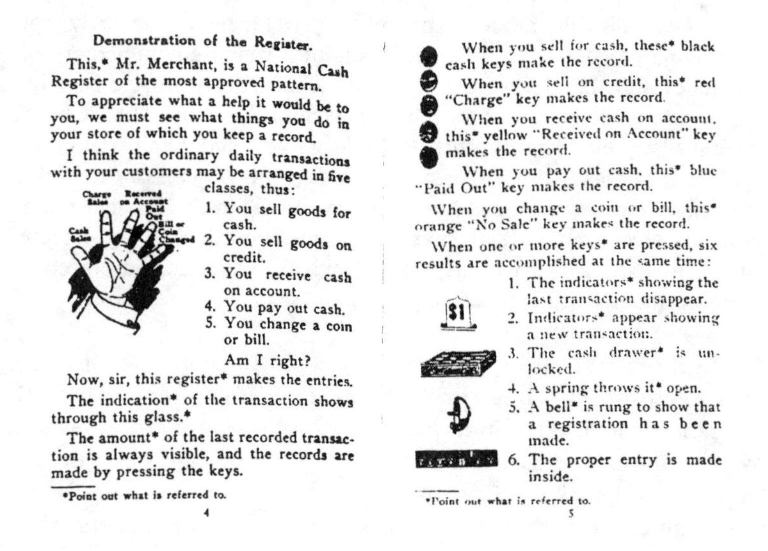

Demonstration of the Register.

This,* Mr. Merchant, is a National Cash Register of the most approved pattern.

To appreciate what a help it would be to you, we must see what things you do in your store of which you keep a record.

I think the ordinary daily transactions with your customers may be arranged in five classes, thus:

1. You sell goods for cash.
2. You sell goods on credit.
3. You receive cash on account.
4. You pay out cash.
5. You change a coin or bill.

Am I right?

Now, sir, this register* makes the entries.

The indication* of the transaction shows through this glass.*

The amount* of the last recorded transaction is always visible, and the records are made by pressing the keys.

*Point out what is referred to.

4

When you sell for cash, these* black cash keys make the record.

When you sell on credit, this* red "Charge" key makes the record.

When you receive cash on account, this* yellow "Received on Account" key makes the record.

When you pay out cash, this* blue "Paid Out" key makes the record.

When you change a coin or bill, this* orange "No Sale" key makes the record.

When one or more keys* are pressed, six results are accomplished at the same time:

1. The indicators* showing the last transaction disappear.
2. Indicators* appear showing a new transaction.
3. The cash drawer* is unlocked.
4. A spring throws it* open.
5. A bell* is rung to show that a registration has been made.
6. The proper entry is made inside.

*Point out what is referred to.

5

Figure 3.3. First sales manual, *The Primer*

Thus, the words on the key labels and in the register window were reinforced by many kinesthetic visual actions—the pressing of keys, the turning of handles, and the pushing in of drawers. These visual kinesthetic activities were repeated by the customer for each class of transaction and for various amounts of money: The salesperson had the customer repeat the visual sentence seven times. In almost the same fashion as a drill sergeant's teaching, the repetition of these activities transformed the problem of linguistic memory into a series of rote physical responses.

Thus, like McKay's and Griffiths's shipwrights who transferred an abstract, logarithmic problem into a kinesthetic carving activity with the half-hull models, Crane and Patterson at NCR made abstract cash register concepts concrete and easy to remember by using real money, opening and closing cash drawers, turning handles, ringing bells, and displaying register indicators. In the same way Evans knew his son should draw machines to understand them,[28] and in the same way that Griffiths knew that he needed to help his shipwrights "produce the exact counterpart of the living shapes which genius has configured in the 'lifts' of the mind,"[29] Patterson knew that he and his salesforce had to teach through the eye:[30]

> "Teaching through the eye" is the term most frequently applied to the process [of business]. Mr. Patterson himself often used that phrase in describing it. The term in its literal meaning, however, is hardly broad enough. "Teaching through the eye" in the sense of utilizing pictures and diagrams and graphic symbolism generally to focus the attention and emphasize the point of an argument is only one phase of the system. It was a highly important part of the system, and it was rarely that a meeting or conference was held without a running commentary of sketches and diagrams on the pedestal chart. But in his use of the phrase, Mr. Patterson was not referring exclusively to that. "I have been trying all my life," he said, "first to see for myself, and then to get other people to see with me. To succeed in business it is necessary to make the other man see things as you see them." *Seeing* in this broader sense was the objective. Mr. Patterson "taught through the eye" not merely through charts and diagrams, but by dramatic analogies and objective demonstrations. In the broadest possible sense he was a visualizer.

At the close of the 19th century, Joseph Crane initiated a new era of effective technology sales through the creation of the Silver Dollar Demonstration. His brother-in-law and boss, John H. Patterson, ensured the effectiveness of the *Silver Dollar Demonstration* by applying Crane's innovations to all teaching and sales processes at NCR and reinforcing such innovations by his own communication practices as president. Crane's and Patterson's secrets were systematic organization from the audience's point of view and visual, dramatic, kinesthetic reinforcement. Such effective oral technical communication was certainly one element that allowed NCR to hold a virtual monopoly on cash register sales in the early 20th century. The monopoly was so large—a 96% share of the market in 1907—that the Department of Justice charged NCR under the Sherman Antitrust Law.[31]

NCR's methods of oral technical communication spread throughout the U.S. economy because, as one magazine of the time noted, "between 1910 and 1930 one-sixth of the nation's top executives had been trained—and fired—by Patterson."[32] In particular, the communication techniques of the *Silver Dollar Demonstration* were taken by NCR's head of sales and second-in-command to a new company in 1914; the salesman was Thomas Watson, Sr. and the company was CTR, whose name was later changed to IBM in 1924. Thomas Watson, Sr. often said, "Nearly everything I know about building a business comes from Mr. Patterson."[33] Part of what he brought from NCR to IBM was the famous IBM slogan—"THINK"—which was originated at NCR in 1911.[34] Perhaps what NCR and IBM both meant by that slogan was not just a concerted, thoughtful process of business and planning, but John H. Patterson's continuous attempts to see and to get others to see with him:[35]

> First to see for myself, and then to get other people to see with me. To succeed in business it is necessary to make the other man see things as you see them.

ENDNOTES

1. Johnson, Roy W. and Lynch, Russell W.: *The Sales Strategy of John H. Patterson. (Sales Leaders Series)*. Chicago and New York: The Dartnell Corp., 1932: Foreword.
2. For further information see: Johnson and Lynch, op cit., 1932: "Chapter 29. The Silver Dollar Demonstration"; Bernstein, Mark: "John Patterson Rang Up Success with the Incorruptible Cashier." *Smithsonian* 20(3) June 1989: 150–66; Crandall, Richard L. and Robbins, Sam: *The Incorruptible Cashier, Vol. 1. The Formation of an Industry, 1876—1890*. Vestal, NY: Vestal Press, 1988; *1884–1922 The Cash Register Era*. Dayton, OH: NCR, 1984 (one of a 5-part company centennial celebration publication): 7, and Cortada, James W. R.: *Before the Computer: IBM, NCR, Burroughs, and Remington Rand and the Industry They Created, 1865–1956*. Princeton, NJ: Princeton University Press, 1993: 68–9.
3. Cortada, op. cit., 1993: 70.
4. February 16, 1878: 95.

5. Attributed to Emerson (in a lecture) by Sarah S. B. Yule and Mary S. Keene in *Borrowings* (1989).
6. Marcossen, Isaac F.: *Wherever Men Trade: The Romance of the Cash Register*. New York: Dodd, Mead & Co., 1947: 93.
7. Johnson and Lynch, op cit., 1932: 85.
8. In fact, miniature cash registers called "three key samples" were created to be discreetly carried by NCR salesman past the clerks and to the boss's office. See Bernstein, op cit., 1989: 150; see also Johnson Lynch, op. cit., 1932: 86.
9. Crandall and Robbins, op cit., 1988: 31.
10. Ibid., 1988: 32.
11. Bernstein, op. cit., 1989: 152.
12. Ibid., 1989: 156. See also NCR, op cit., 1984: 15.
13. Crandall and Robbins, op. cit., 1988: 81, 119: These pages show pictures of two such register drawer quick reference cards.
14. Johnson and Lynch, op cit., 1932: 146–8—"A Wall Paper Salesman and the Original Primer." John Carroll in *The Nurnberg Funnel: Designing Minimalist Instruction for Practical Computer Skill*. Cambridge, MA: MIT Press, 1990: 90. Carroll noted the role of trial and error in experience in the work of Henry Dreyfus. "Henry Dreyfus, the industrial designer responsible for the classic type-500 Bell telephone handset, routinely went out into the world to experience the situations he was trying to alter through design. Reading his account of committed design work makes it all too tangible how merely being systematic is inadequate: 'I have washed clothes, cooked, driven a tractor, run a Diesel locomotive, spread manure, vacuumed rugs, and ridden in an armored tank. I have operated a sewing machine, a telephone switchboard, a turret lathe, and a Linotype machine. When designing the rooms in a Statler hotel, I stayed in accommodations of all prices. I wore a hearing aid for a day and almost went deaf.'"
15. Johnson and Lynch, op cit., 1932: 148.
16. Johnson and Lynch, op cit., 1932: 150—"A Wall Paper Salesman and the Original *Primer*." This claim is somewhat suspect if one considers the sales demonstrations being offered by the sewing machine and mower/reaper industries in the 1860s and 1870s as detailed in Chapter 5.
17. Ibid., 1932: 150.
18. Both the initial and final parts of the sales pitch were in the 1880s—and even today—rather standard elements of any sales pitch. The initial part, the "Approach," was the part of the pitch in which the salesman would introduce the concept of a cash register and make an appointment to demonstrate it. The final part, the "Closing Arguments," was the part in which the salesman would get the prospective purchaser to make a financial commitment to purchase the machine.

19. Johnson and Lynch, op cit., 1932: 185—"A Wall Paper Salesman and the Original Primer." In 1892, the *Primer* was rewritten and checked for correctness by a Harvard English professor. The work existed as *The Primer* at least until 1916 when it was expanded into the *Book of Arguments* and then later into the *Manual*.
20. Ibid., 1932: 157.
21. See Selzer, Jack: "What Constitutes a 'Readable' Technical Style." In Anderson, Paul; Brockmann, R. John; and Miller, Carolyn R. (eds.) *New Essays in Technical and Scientific Communication: Research, Theory, Practice* Farmingdale, NY: Baywood Publishing Company, Inc., 1983: 82.
22. Ibid., 1932: 146–8.
23. Ibid., 1932: 199—"His main interest was always in the application of the register to the specific needs of the market, and the main objective of the *Manual* was to encourage the same attitude on the part of the salesman."
24. Ibid., 1932: 197–8.
25. Ibid., 1932: Foreword.
26. Ibid., 1932: 129.
27. Ibid., 1932: 169–70.
28. Oliver Evans letter, James L. Woods's collection of Evans's papers in the Franklin Institute Library. Also in Bathe, Grenville and Dorothy: Oliver Evans: *A Chronicle of Early American Engineering*. NY: Arno Press, 1972: 248. "I therefore request Mr. Cadwalader to come immediately up and stay two months and make accurate drawings of every thing he may see from actual measurements and get the whole fixed in his mind."
29. Griffiths, John W.: "Inland Navigation. Modeling Sail Vessels for the Lakes." *The Monthly Nautical Magazine and Quarterly Commercial Review* 1(1) October 1854: 10.
30. Ibid., 1932: 125–6.
31. For the sentencing of Watson and Patterson in this suit and their subsequent falling out, see Sampson, Anthony: *Company Man: The Rise and Fall of Corporate Life* New York: Random House, 1995: 48.
32. Bernstein, op. cit., 1989: 158–9.
33. Watson, Thomas Jr.: *Father Son & Co.* New York: Bantam Books, 1990: 13. Carroll, Paul: *Big Blues: The Unmaking of IBM*. NY: Crown Publishers, Inc.: 62. Watson, Sr. brought with him the penchant for jotting things down and drawing—even keeping butcher paper by his desk to jot down ideas. This eventually got out of hand in the mid-1980s. "People learned that the way to get ahead wasn't necessarily to have good ideas. That took too long to become apparent. The best way to get ahead was to make good presentations. People would say of comers: 'He's good with foils,' referring to the overhead transparencies that began to dominate IBM meetings. People began spending days or weeks preparing foils

for routine meetings. They not only made the few foils they actually planned to use but made a huge library of backup foils, just in case someone had a question. They became such a part of the culture that senior executives began having projectors built into their beautiful rosewood desks."

34. According to Frederick Fuller (*My Half-Century as an Inventor*. NY, 1938: 42), Watson himself invented the sign and slogan in 1911 while at NCR working as second-in-command to Patterson.
35. Johnson and Lynch, op cit., 1932: 125–6.

THEME 2

The Power of Genre in American Technical Communication

> So tellers of stories and listeners have an implicit agreement. Tellers will only tell standard stories, stories that are easy to understand. When tellers find some events to relay that are incomprehensible, they will not relay them, or they will force them into a format that makes them look comprehensible. In fact, tellers have no other choice.
>
> —Roger C. Schank[1]

Researchers in memory and storytelling suggest that there are story skeletons—high-level outlines of narrative action—into which storytellers fit their stories:[2]

> A storyteller might be more accurately described as a story-fitter. Telling stories about our own lives, especially ones with high emotional impact, means attempting to fit events to a story that has already been told, a well-known story that others will easily understand.

Consequently, to communicate effectively, storytellers must understand the story skeleton their audience knows and expects and fit their story to the appropriate skeleton. Sometimes this action of fitting events into a story skeleton is quite unconscious because the storytellers cannot understand or make sense of the original events unless and until the events conform to story skeleton expectations they themselves have. Thus the fitting-events-to-a-story-skeleton is both an element of perception as well as an element of memory and communication.

Technical communicators probably do not imagine that they are telling stories,[3] and yet, like storytellers all over the world, technical communicators fit information into forms that are familiar and easily understood by their audiences.[4] The difference is that technical communicators might call these forms "genres" or "subgenres" rather than story-skeletons.[5] A genre or subgenre implicitly includes many set elements such as:

- tone—for example, the instruction manual genre usually employs the you-attitude tone in which audiences are directly addressed in second person pronouns;
- length—for example, the instruction manual genre tends to be a text under 100 pages rather than 21 volumes; and
- weave of text and graphics—for example, the patent genre includes illustrations and their words refer to them frequently.

Not only do these genre expectations aid readers, but they also aid the creators of the texts. Technical communicators have actually felt fear and loss when they did not follow what they understand to be standard genre or subgenre requirements when developing texts. For example, in 1988, computer user documentation writers at Hewlett-Packard were asked to follow some new design doctrines that recommended they leave information purposefully incomplete ("minimal manual" concepts; see Chapter 9 for a complete discussion). When these technical communicators reported their experiences at a convention of their peers, the writers described the "fears" they experienced when they violated genre expectations of computer user documentation: [6]

> Fears of "letting go" of standard procedural techniques when developing GE+P [minimal manual] prototypes were our toughest problems to overcome. As technical writers, we were trained to write specific, clear and complete procedures for end-users. The GE [minimal manual] approach ran contrary to our training in that it required us to write truncated 'hints' which often left information purposefully incomplete.

Likewise, when readers fail to find their genre expectations being fulfilled in what they read, confusion and inability to comprehend surely follow.

In Theme One, the genres used by Evans, McKay, and Griffiths to communicate their technical information were books such as treatises or guides. Because treatises or guides had such a long-standing set of genre expectations it was probably not by chance that they were the first genres of technical communication to be successfully used in America. Yet, by the 1790s, science had already developed some unique genres for communicating its particular information. For example, scientists had established genres such as the "erudite letter"[7] and the philosophical transactions. Moreover, it was in these early philosophical transactions that Evans read of John Smeaton 's research (in the London Royal Society's *Philosophical Transactions*) as well as William Waring's (in Philadelphia's *Transactions of the American Philosophical Society*). In a similar way in the antebellum period, just as technology was beginning to be conceptually separated from legal, business, or scientific endeavors, mechanician authors began to create their own genres and subgenres to communicate about technology.[8]

What were some of these genres and subgenres of technical communication, and how did they develop and change in response to audiences?[9] To answer these questions, Chapters 4, 5, and 6 offer a bird's eye view of a large number of technical documents that together describe the genres into which individual texts were fit. For this second theme, I examined hundreds of patents and dozens of instruction manuals over some 200 years in order to discern some of the genres into which technical communicators have been fitting information about new technology, and the way these genres and their elements have evolved. Specifically, I examined the elements of two genres, patents and instruction manuals. Patents were defined, in the modern sense, for the first time in 1791 in the U.S. Constitution.[10] Instruction manuals, on the other hand, made their American debut 60 years later in the late 1850s when they accompanied the first factory-made consumer goods, sewing machines and mower-reapers.

ENDNOTES

1. Schank, Roger C.: *Tell Me A Story.* New York: Charles Scribner's Sons, 1990: 168–9
2. Schank, op. cit., 1990: 169. See also Donald Norman: *Things that Make Us Smart: Defending Human Attributes in the Age of the Machine*. Reading, MA: Addison-Wesley Publishing Co., 1993: 128–30.
3. Denning, Lynn H.: "The Nature of the Narrator in Technical Writing" In *The 1993 Proceedings of the 40th Annual Conference, June 6–9, 1993.* Arlington, VA: Society for Technical Communication, 1993: 154— "Writers of technical information rarely, if ever, intrude and directly address their readers; nor do they acknowledge themselves as authors or narrators, as participators who are in charge and are telling what happened or what they did."

4. Killingsworth, M. Jimmie, and Gilbertson, Michael K.: *Signs, Genres, and Communities in Technical Communication*. Amityville, NY: Baywood Publishing Company, Inc., 1992: 94.
5. The definition of genre I have used throughout is the recurrent use of situationally appropriate responses in form and substance to recurrent rhetorical situations; see Berkenkotter, Carol, and Huckin, Thomas N.: *Genre Knowledge in Disciplinary Communication: Cognition/ Culture/Power*. Hillsdale, NJ: Lawrence Erlbaum Associates, Publishers, 1995.
6. Vanderlinden, Gay., Cocklin, Thomas G., and McKita, Martha: "Testing and Developing Minimalist Tutorials" In *The Proceedings of the 34th International Technical Communication Conference, May 1987*. Washington, DC: Society for Technical Communication, 1987: RET-198.
7. Garfield, Eugene: "Has Scientific Communication Changed in 300 Years?" *Current Contents* 8(2) (1980): 6.
8. Michael E. Connaughton: "Technical Writing in America: A Historical Perspective." In Mathes, J. C., and Thomas E. Pinelli (comps.): *Technical Communication: Perspectives for the Eighties*. Washington, DC.: NASA Conference Publication 2203, 1981: 15–24; see also "Notes towards a History of American Scientific Writing." Paper presented at the 1981 Modern Language Association convention, Houston, Texas, December 28, 1981. Although Connaughton is one of the few researchers in technical communication doing the type of historical research described in this book, he confuses scientific writing and technical writing. "Technical Writing in America: A Historical Perspective" needs to be retitled to reflect more accurately the subject matter he examined, the publications of the American Philosophical Society, which he himself noted represent the "first substantial scientific publications in America" (32). Later, he corrected this confusion by correctly titling his second article "Notes towards a History of American Scientific Writing." For a good discussion of the differentiation between technology and science, see Carolyn R. Miller: "Technology as a Form of Consciousness: A Study of Contemporary Ethos." *Central States Speech Journal* 29 (Winter 1978): 228–236.
9. Miller, Carolyn R.: "Genre as Social Action." *Quarterly Journal of Speech* 70 (1984): 151—"I will be arguing that a rhetorically sound definition of genre must be centered not on the substance or the form of discourse but on the action it is used to accomplish."
10. Patents had existed earlier in Europe, notably in England. Generally they were used to grant monopolies to a particular trade or activity in exchange for money or political influence. For example, the king could grant one of his favorites the patent for travel on the upper reaches of the Thames, and that patentee could make a good deal of money by charging others for the right to travel on the Thames. The U. S. patent is

unique in history in that it was created to increase diffusion of technical information by recognizing the intellectual property of inventors by granting them a limited monopoly of some years in exchange for a full and complete disclosure of their invention or process. Thus full and complete disclosure of technology was the basis for U.S. patents, whereas political connection was the basis for previous patents. See Hindle, Brooke: *Emulation and Invention*. New York: New York University Press, 1981: 18–20, and Basalla, George: *The Evolution of Technology*. Cambridge: Cambridge University Press, 1988: 119–124.

Chapter 4

Technical Style from 1795-1980 in the Patent Genre[1]

[T]he expressions are very strong and seem intended to accommodate the description, which the patentee is required to give, to the comprehension of any practical mechanic, skilled in the art of which the machine is a branch, without taxing his genius or his inventive powers.
—Justice Washington on the Constitution's patent amendment (1793)[2]

Without a historical perspective, the unexamined past often becomes shadowy legend.[3] Like the golden age of the Greeks and Romans, technical communication legends often populate the past with superior communicators who surmounted the problems inherent in the ambiguity of language better than present-day, fallen, technical communicators. This repeated element in many technical communication legends reveals the influences of logical positivists who question language for its failure to be a flawless conveyor of information.[4] There are many examples of this mythic outlook on a past peopled with superior communicators; however, two examples suffice.

When Della Whittaker discussed Joseph Priestly's writing techniques in *Technical Communication*, she claimed: "In contrast to modern scientists, eighteenth-century scientists took responsibility for what they discovered and included themselves in their descriptions. He [Priestly] personalized his narrative with 'I.'"[5] Richard Ramsey seemed to concur when he wrote that: "when Archimedes discovered the law of floating bodies, he did not stolidly report, 'It has been observed that. . . . ' The grammatical structure which tends to prevail in century technical English is the third-person passive voice."[6] The message explicit in Whittaker and implicit in Ramsey was that writers in the past rarely used the objective-impersonal style characterized by passive voice and third person pronouns.[7]

One obvious problem in the leitmotif of a golden communication past is blindness to the great differences in the rhetorical situation of then and now. Carolyn Miller, in explaining the concept of "rhetorical situation," pointed out that one of the constituents in a rhetorical situation is the genre used to communicate information.[8] In the first comparison, Whittaker ignored the very special genre in the rhetorical situation of those 17th-century scientists. The "erudite letter" was the genre most used by 17th-century scientists, who worked alone and who described their experiments and transmitted their ideas by letter:[9]

> Normally, a letter was written to one particular scientist or, perhaps, copied by hand and sent to three or four people at a time. The recipient might show the letter to some other friends, but the number of people who actually saw a letter was never large. In addition, the recipients rarely criticized or debated the contents.

It is nearly impossible to compare the rhetorical situation of Priestley's observations contained in his genre of erudite letters addressed to the small close-knit Royal Society with that of a modern scientist addressing a large, generally unfamiliar audience, in the genre of a formal, refereed scientific article.[10] One cannot compare the situation of today's scientific rhetoric with its established genres and subgenres with the rhetorical situation of emerging genres in Priestley's time. For example, Bazerman noted that the second half of Priestley's *The History and Present State of Electricity* (1767) would today be found in such a disparate variety of genres as "children's activity books through advanced textbooks, equipment manuals, and research journals . . . [none to be found] . . . under the same cover."[11]

As difficult as it is to compare Priestley's rhetorical situation with the present, it is even more difficult to compare the rhetorical situation of what Archimedes is supposed to have said to himself in the bath with the genre a technical communicator uses, for example, a typeset pamphlet to describe objects or processes to a distant, large audience. These comparisons of the past and present by Whittaker and Ramsey simply do not comprehend the many varied rhetorical factors at work either then or now.

Yet perhaps Whittaker and Ramsey did not literally want to compare the writings of Priestly and Archimedes to those of modern technical communicators; perhaps they were just suggesting that there has been an increasing use of the objective-impersonal style over time.[12] However, is this more general observation any more accurate in its description of the past? To answer this question, I examined 111 patents written over 185 years from (1795 to 1980) to detect changes in the stylistic choices of impersonality or personality in the patent genre.

PATENTS AS EXPERIMENTAL ARTIFACTS

Despite an interest in developing an historical perspective, the lack of convenient artifacts to study may stymie such efforts. Theme One did show that there are artifacts of the technical communication past to study, yet some of these items—Evans's PORTFOLIO, Griffiths's nautical journals, and Crane's *Silver Dollar Demonstration*—are held in specialized archives or the rare book rooms of select libraries. Could it be that technical communication historical researchers will be required to sequester themselves in such archives or libraries just as historical researchers in business communications had to in their work at the British Museum?[13] David Dobrin observed:[14]

> The last kind of difficulty [in historically studying technical communication] is inherent. . . . It would have to do with the temporary quality of technical writing and the technology it projects . . . the practice will resist historical inquiry. Corporations do not conceive of themselves historically, nor do they suffer willingly the intrusion such a conception would warrant.

If Dobrin is correct, technical communication historians might be like paleontologists who are unable to discover when life began on earth because there are no rocks sufficiently old, accessible, and unchanged to contain the remnants of the earliest life forms. The genre of patents, however, may offer a group of artifacts to technical communication historians that the paleontologists do not have—a large number of sufficiently old, accessible, and unchanged items.

THE ORIGIN OF THE AMERICAN PATENT

The leaders of the United States at the Constitutional Convention knew that they had to stimulate technological growth in the new nation, and they knew also the impediment that "secret" technological development engendered.[15] Therefore, to increase the diffusion of technology, the United States Constitutional Convention invented the modern patent in 1790, and, ever since then, the United States Patent Office has dutifully catalogued and stored these documents. It costs less than a dollar to have a copy of a patent sent from the United States Patent Office. Thus, there are countless artifacts readily accessible for study in the government's patent collection.

Another major problem confronting historical researchers that Dobrin described could, however, still undermine historical research: "Even if we investigate [technical communication history], our conclusions are likely to be local, and our substantiating details will be private."[16] Dobrin suggested that even with sufficiently old artifacts, they would be so individualized by rhetorical particularities that no overall conclusions would be possible. Dobrin's suggestion may certainly hold true for some of the observations made in Theme One because the rhetorical situations of Evans, Griffiths, Crane, and Patterson were quite different than those of today. However, patents can allow the technical communication historian to coordinate many of the rhetorical variables. This coordination is possible because patents as a technical communication genre have had the same purpose and audience since 1790:[17]

> In return for a patent, the inventor gives a complete revelation or disclosure of the invention for which he seeks protection. . . . A Detail Description of the invention and drawings follow the general statement of the invention and a brief description of the drawings. This detailed description, required by Rule 71, section 608.01, must be written in such particularity as to enable anyone skilled in the pertinent art or science to make and use the invention without involving extensive experimentation.

Patent disclosures as a genre of technical communication have had the same media, pictures, and words, and each patent has the same subject matter, a new invention. Furthermore, although no rules have ever been given constraining the style of writing patents, the traditional guidelines have been of such a general nature that there has always been latitude for the expression of personal and period characteristic styles. The only traditional guidelines for the writing of patents have been that:[18]

- [the patent description] must be short, specific, and clear,
- unknown or complicated elements of the inventions must be completely explained,
- critical limits must be spelled out,
- if necessary, a clear mode of operation must be included,
- an applicant can use his or her own terminology so long as it can be understood,

- grammatical perfection is not examined, and
- technical words are not necessary.

In fact, in 1837, Phillips pointed out that:[19]

> [W]here superfluous processes or machinery are described for the purpose of misleading the person consulting the specification, and preventing him from getting knowledge of the process or machine patented, or of the best mode of using it, this will vitiate the patent.

Thus in the many artifacts to study in the genre of patents, the rhetorical variables of purpose, audience, media, subject matter, and occasion are relatively stable. As P. James Terrango noted: "[P]atents are different from journal literature in that they are standalone documents and have a uniformity of presentation."[20] Finally, patents as a genre reflect technical communication rather than legal, business, or science writing. Although some parts of patents may have been executed pro forma by lawyers, the part traditionally called the Detailed Description has customarily been written by the mechanicians or inventors themselves for other mechanicians to use.[21,22]

Despite all these positive attributes of patents as technical communication research artifacts, there are two cautions to observe in using them. First, the purposes of all patents may not be identical, i.e., they may not all be written to disclose information. Although the explicit legal goal of patents has always been to disclose technological secrets to the interested public in return for a number of years of exclusive control by the patentee, the implicit goal changed in some cases. In 1896, the Supreme Court ruled that patentees were bound neither to use a discovery nor permit others to use it.[23] Using this decision, some 20th century companies filed patents with generalized language early in the discovery process so that other companies will stay away from their new area of research, and thus some 20th century company's patents withhold rather than disclose information about a line of anticipated research, and development.[24] However, the impersonal-objective style examined here served both functions—to disclose and to withhold—equally well. Thus, even if researchers are unable to discriminate between patents designed to disclose information and those designed to withhold information, the same stylistic qualities function for both purposes.

Another caution to observe in using Patents has to do with the ability to differentiate different parts of the patent. The Patent Law of 1790 required only a "specification in writing." No model, drawings, or legal claims were specifically separated from or called for in the Act even though they all later became required in the early nineteenth century. Thus, the demarcation line between the various parts of patents—e.g., the Detailed Specification and the legal claims—in the 27 earliest patents is ambiguous. In the Patent Law of 1836, however, the parts were required to be more clearly separated.

SPECIFIC ANALYSIS USED IN SAMPLE PATENT EXPERIMENT

In the analysis of the patent genre, I chose a group of three patents to represent a span of five years—1795 to 1980; a total of 111 patents. Such a span of time seems sufficient for the prediction to be revealed—an increase in impersonality in the stylistic choices made by patent writers.[25] Among the group of three patents, I made sure that no two patents were written by the same author so that I would detect general, not individual, trends. Also, I made sure that the patents were all for devices, not procedures, so that there was a uniformity of content. With these sampling constraints, I was able to coordinate many of the variables of the patent's rhetorical situation: audience, purpose, content, subject matter, and medium.

Once I had the 111 patents that met these criteria, I examined three stylistic groups of choices in the patents that combine to create the objective-impersonal style: the erasure of human agency, the erasure of time, and the erasure of references inside the text to realities outside the text, exophoric references (see Table 4.1).[26] (Also, when I examined the 111 patent Detailed Descriptions I tried to be sure to examine stylistic choices made by writers and

Table 4.1. Predicted Changes in Frequency of Stylistic Choices Demonstrating An Increase in the Objective-Impersonal Stylistic Choices Over Time.

Element	Specific Stylistic Choice	Predicted Change of Frequency
(1) Erasure of human agency	Use of direct reference to reader or author	Decrease
	Use of personifications	Increase
	Use of agent-less passives rather than transitive active verbs	Increase
(2) Erasure of time	Use of nonfinite over finite verbs	Increase
(3) Erasure of exophoric references	Use of metaphors	Decrease
	Use of metaphors as adjectives	Increase
	Use of impressionistic adverbs or adjectives	Decrease

not their accommodations to the requirements of usage, syntax, or grammar.[27]) Table 4.1 lists these three areas of choice, the specifics that were examined, and the predicted frequency changes in the choices arising from Whittaker's and Ramsey's articles that I discussed earlier.[28]

Erasure of Human Agency

If writers can erase the evidence of human agency, they can make information appear more objective, and there are three specific stylistic choices writers can make to do this: (1) by not referring to either the reader or themselves, (2) by using personifications, and (3) by using agent-less passives when an agent-specified active voice verb could have been used. First, writers can erase human agency by choosing not to make any references to either the reader—such as in Fielding's asides to the "Dear Reader" in *Tom Jones*—or to themselves—such as in "I noticed" in Franklin's scientific journals. Writers may choose not to make any references to either the reader or themselves because, as Whitburn has described:[29]

> Science has attached to humanity certain connotations of frailty, and serious human activity suffers a loss of dignity from the injection of personality. One small manifestation of this trend is the banishment of "I" from exposition.

Writers wishing to erase the evidence of human agency would also avoid the use of second person pronouns, whether explicitly stated or understood as in the imperative mood.

Second, writers can choose to use personifications in which objects rather than humans take on agency and activity. This can be seen in the choice of writers to say "parts perform their duties," "edges operate," or "posts run." Carolyn Miller discussed this tactic when she noted: "Under the sway of positivism, scientists adopted as conventions the obvious stylistic conventions for staying out of the way of the subject matter—personifications and third person pronouns."[30]

Third, writers choose agent-less passives when presented with the ability to choose between agent-less passives or an agent-specified active voice verb.[31] This element would, for example, be reflected in the choice of a writer to use "the ball was kicked," rather than "John kicked the ball." In regard to this verb choice, Alan G. Gross explained:[32]

> Regardless of its surface features, at its deepest semantic and syntactic levels scientific prose requires an agent passive before the only real agent, nature itself. By means of its patterned and principled verbal

> choices, science begs the ontological question: through style its prose creates our sense that science is describing a reality independent of its linguistic formulations.

Walker Gibson also discussed this tactic under the heading of "stuffy rhetoric," his name for the objective-impersonal style: "A key characteristic of Stuffy rhetoric is just this refusal to assume personal responsibility. It is accomplished by at least two stylistic characteristics. . . . One is the use of the passive verb."[33]

Erasure of Time

If an author can erase the marks of time or tense, then the action in the text seems to have an aura of eternity and objectivity. A writer can erase the marks of time by choosing nonfinite verbs rather than finite verbs—finite verbs can be tense marked and thus time marked. For example, a writer's choice to use "booting the CPU" rather than "the CPU was booted" shows that he or she attempted to create text that is asituational and seemingly timeless.[34]

Erasure of Exophoric References

Three stylistic choices erase references in the text to things outside of the text, or exophoric references.[35] The fewer exophoric references, the more the text becomes asituational, timeless, and objective.[36] It is probably with the elimination of such references in mind that Sir Thomas Sprat mandated the style for the Royal Society "In all reports of experiments to be brought into the Society, the matter of fact shall be barely stated, without any prefaces, apologies, or rhetorical flourishes."[37]

To erase exophoric references, writers avoid the use of all metaphors. For example, writers would never write "that stations initially linking together through telecommunications *are like* two humans shaking hands." The words *are like* call attention to the human-craftedness of the text and diminish its aura of objectivity.

Yet, even if writers did use metaphors, they could undercut the metaphor's metaphorical qualities as well as any attention to its exophoric reference. Thus, a second way for a writer to erase outside references is to choose to use metaphors in the adjective position where their metaphorical quality is undercut and understated.[38] In using this approach, computer programmers would not say "that stations initially linking together through telecommunications are like two humans shaking hands;" rather they would, do, in fact, say, "the two stations went through a *handshaking* procedure."

The third way writers can erase exophoric references is to avoid the use of impressionistic adverbs, adjectives, adverb phrases, or adjectival phrases. For example, authors would choose not to use "more or less," "at the pleasure of the operator," or "in the well-known and usual manner," because to do so calls attention to a human crafting of the information.

WHAT THE PATENT EXPERIMENT REVEALED

My analysis of the patents reveals three conclusions:

- Symptoms of the increase of objective-impersonal style choices predicted by Whittaker and Ramsey were verified, however the change from a personal to an objective-impersonal style was neither large nor dramatic over the 185 years studied.
- Stylistic choices over time were often consistent and, in some cases, indicated no evidence of a shift from a personal to an objective-impersonal style.
- As much individual deviation in the stylistic choices occurred in the present as in the past.

The following graphs illustrate these general findings: Figures 4.1, 4.2, and 4.3 illustrate the frequency of stylistic choices that erase human agency. Figure 4.4 illustrates the frequency of stylistic choices that erase time specificity. Figure 4.5, 4.6, and 4.7 illustrate the frequency of stylistic choices that erase exophoric references. Table 4.2 also summarizes the inferences based on these graphs.

Table 4.2. Predicted Versus Observed Changes in Stylistic Choices Demonstrating An Increase in the Impersonal-Objective Style Over Time.

Element	Specific Stylistic Choice and Relevant Figure to Refer To	• Predicted Change • Observed Change • Prediction Verified?
(a) Erasure of human agency	Use of direct references to reader or author or second person pronouns (Figure 4.1)	• Decrease • Decrease • **Prediction verified**
	Use of personifications (Figure 4.2)	• Increase • Decrease—but overall not a stylistic technique often used • **Prediction not verified**

Table 4.2. Predicted Versus Observed Changes in Stylistic Choices Demonstrating An Increase in the Impersonal-Objective Style Over Time (con't).

	Use of agent-less passives rather than agent-specified active transitive verbs (Figure 4.3)	• Increase • Weak increase • **Prediction verified**
(b) Erasure of Time	Use of nonfinite over finite verbs (Figure 4.4)	• Increase • Nearly consistent • **Prediction not verified**
(c) Erasure of Exophoric References	Use of metaphors (Figure 4.5)	• Decrease • Nearly consistent • **Prediction verified**
	Use of metaphors as adjectives (Figure 4.6)	• Increase • Nearly consistent • **Prediction not verified**
	Use of impressionistic adjectives, adverbs, adjectival phrases, and adverbial phrases (Figure 4.7)	• Decrease • **Prediction not verified**

Erasure of Human Agency

Figure 4.1 displays the plot of raw total numbers of pronouns found in each sampling of patents. The straight line is that which best fits the data points.[39]

The frequency of personal pronouns decreased by 54% over the 185 years, from a mean frequency of 12 first- and second-person pronouns to a mean of 5. Data points falling on the X axis indicate that no first- or second-person pronouns were found in any of the patents sampled that year (i.e., none was observed in samples for 1795, 1815, 1975, and 1980). Figure 4.1, then, implies that in the decreasing use of first- and second-person pronouns the trend Whittaker and Ramsey predicted did exist.

Figure 4.2, however, appears to contradict the implications of Figure 4.1. Figure 4.2 displays the plot of number of personifications found in each individual patent in each sampling of three patents (37 separate samples). The straight line, the line which best fits the data points, rather than disclosing an increase in the use of personifications—as should happen when objects increasingly take over agency from humans as the style becomes more objective-impersonal—shows a decrease in the use of personifications. The decrease in the frequency of personifications is contrary to expectations based on a hypothesis of increasing use of the objective-impersonal style choices. However, even though the frequency of use decreases from a mean of 0.4 personifications per sample for 1795 to 0 personifications per sample for 1925 to 1980, 30 of the 37 separate samples studied (81%) had no personifications at

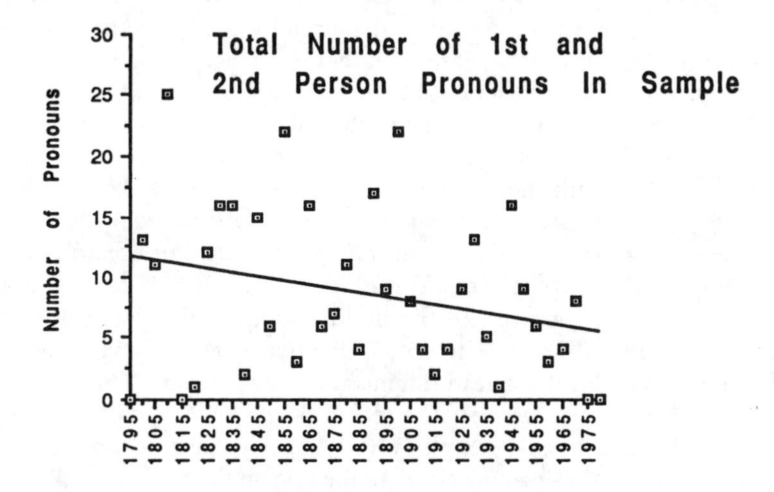

Figure 4.1. Decreasing use of first and second person pronouns

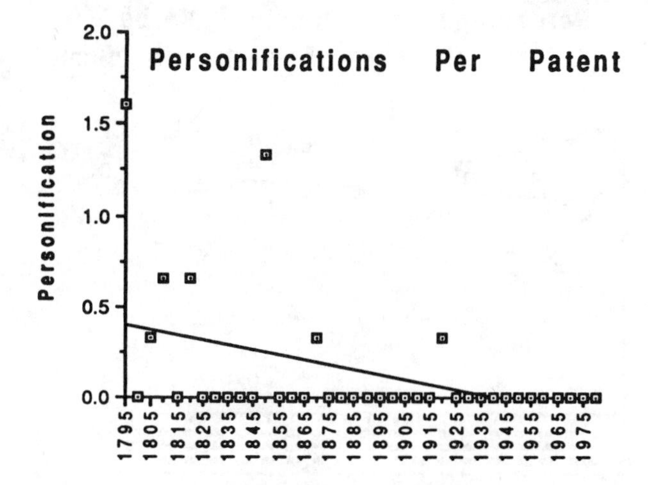

Figure 4.2. Quick disappearance of personifications

all. Thus, although the implications of Figure 4.2 are counter to the implications of Figure 4.1, the data base for Figure 4.2 is rather limited, and, in general, patent writers made little use of personifications at any time over the last 185 years examined.

Figure 4.3 plots the percentage of verbs that are passive rather than active, the final of the three stylistic choices that erase human agency.[40] Figure 4.3 displays the percentages of total verbs that are passive transitive in each sample of patents. From 1795 to the present, the passive verb form was chosen by writers—rather than the active verb—form by a rather constant ratio of roughly 3 to 1. There is a small increase in the percentage of passive verbs in a passage (beginning with a mean of 58% in 1795 and increasing to a mean of 65% in 1980). However, because the straight line is nearly level, there is only a very weak correlation between ratio of passive transitive verb choices to active verb transitive verb choices and time.

Yet, as weak as the correlation is, the variations from the line do reveal a general tightening of data points around the line. For example, in the first 10 samples, the range is over a spread of 40% (from 75% to 35% of the transitive verbs were passive), in the middle samples the data points are spread out only 28% (from 73% to 45% of the transitive verbs were passive), and in the last samples the spread diminishes to 22% (from 75% to 53% of the transitive verbs were passive). The reason why such a tightening around the line changed was because samples with low percentages of passive transitive verbs decreased, thus coming closer to the line on the bottom, while, concurrently, samples with high percentages of active transitive verbs also decreased. With the percentages of high passive transitive verb choices

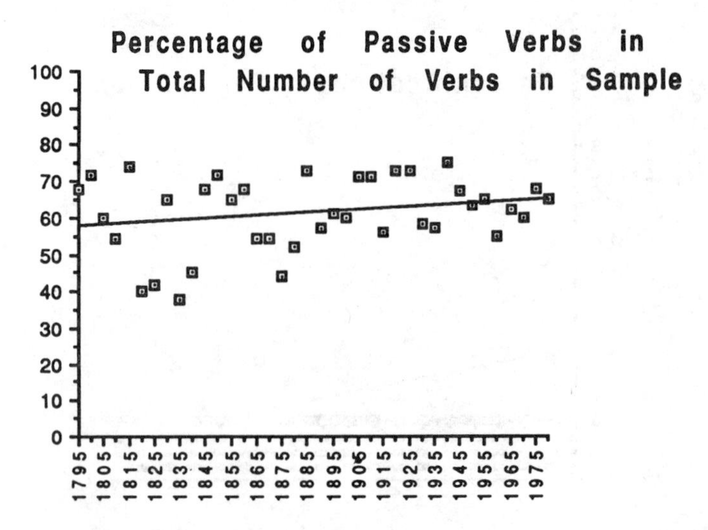

Figure 4.3. Large percentage of passive verbs consistently used by patent authors

increasing and the percentages of low active transitive verb choices decreasing, the spread of data points eventually diminishes.

Thus, in comparing the experimental finding about writers' choices related to the erasure of human agency with the predictions from Whittaker's and Ramsey's suggestions:

- one area of stylistic choice (pronoun choice) supports the assumption of a dramatic increase in the impersonal-objective style;
- a second area of choice (personifications) fails to reveal a significant trend because the technique was little used; and
- a third area of choice (passive versus active) shows only a very slight increase in the use of the passive voice rather than the active voice—7% over 185 years.

Nonetheless, it is interesting that since 1795 the passive has been the predominant verb form, with a rather consistent ratio of 3 to 1 of passive voice verbs to active voice verbs. In short, the findings reported in the first three figures do not support the hypothesis of a dramatic change from a personal to an objective-impersonal style.

Erasure of Time

Figure 4.4 displays stylistic choices made to erase tense: a choice indicated by using nonfinite verbs rather than finite verbs. If writers choose nonfinite verbs, they tend to make the writing asituational and impersonal. Here again, howev-

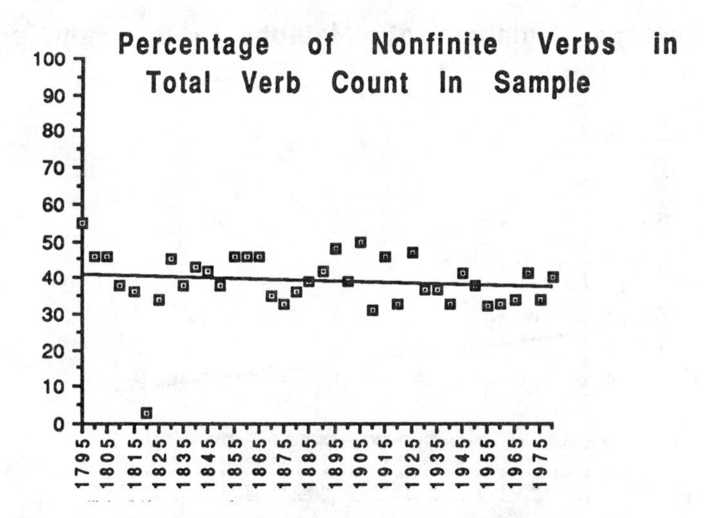

Figure 4.4. Consistent use of nonfinite verbs

er, the straight line is nearly level demonstrating a weak relationship between the percentage of nonfinite verbs chosen and the year from which the sample was taken. However, although weak, the line is a very slight downward one, suggesting that more objective-impersonal stylistic choices were made in 1795 (55% of all verbs were nonfinite) and more personalized stylistic choices were made in 1980 (when only 40% of all verbs were nonfinite). The analysis of this stylistic choice, then, appears counter to the predictions implicit in Whittaker's and Ramsey's observations. In fact, by observing the specific data points rather than the general line, one can see that, in 1795, 55% of all verbs were nonfinite, as were 46% both in 1800 and 1805, whereas from 1950 to 1980, most of the specific data points are much lower: 32% in 1955, 33% in 1960, 34% in 1965, and 34% in 1975. This stylistic aspect thus suggests more personalized stylistic choices were actually made by writers in later years.

Erasure of Exophoric References

The last three figures display information related to the third element of the objective impersonal style, the erasure of the exophoric references. Figures 4.5 and 4.7 appear to support the prediction of an increasing use of the objective-impersonal style. Figure 4.5 provides support because the line is downward. Figure 4.7 provides such support even more strongly. However, the data in Figure 4.6 proves to be counter to the predictions.

Figure 4.5 displays the average number of metaphors found in each patent sample group, and the straight line initially decreases just as predicted by the hypothesis—if the objective-impersonal style were to be used more

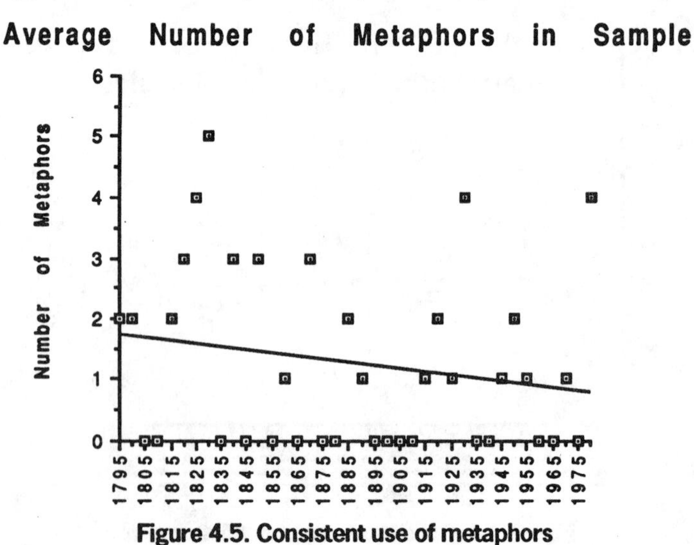

Figure 4.5. Consistent use of metaphors

often, the number of metaphors should decrease. However, nine of the samples prior to 1900 had no metaphors as did eight of the samples after 1900. Figure 4.5 suggests, therefore, that there were just as many metaphors used by patent authors in the late 20th century as there were in the early 19th century; however, at no time were metaphors used frequently. Thus the predicted decrease in the use of metaphors accompanying the hypothesized increase in the objective-impersonal style is only somewhat verified in this instance.

Figure 4.6 shows the percentage of the metaphors found in Figure 4.5 that were used in adjective positions. (Thus, if no metaphors were present in the five-year sample, it was removed as a data point in Figure 4.6—nearly half or 17 data points were removed.) If the metaphors were used in a adjectival position, their metaphorical power could be muted; the exophoric references could be diminished; and the prediction of an increasing impersonal-objective style verified. Conclusions to be drawn from Figure 4.6 are made difficult because the use of metaphors as adjectives cluster either around 0 (no metaphors used in the adjectival position in the patent sample) or 100 (all metaphors used in the adjectival position in the patent sample). Moreover the straight line itself is nearly level at 50% suggesting that half the time this technique of the the objective-impersonal style is used, half the time it is not. Even more to the point, by dividing the data points from 1795 to 1900 from those of 1900 to 1980, it appears that prior to 1900, when metaphors were used by patent writers, they were muted in their metaphorical qualities by being used in adjective positions by a ratio of 3:1. However, after 1900 when, according to the predictions one would expect such muting of metaphors to increase, the ratio of metaphors as adjectives to metaphors used as metaphors is 1:1.

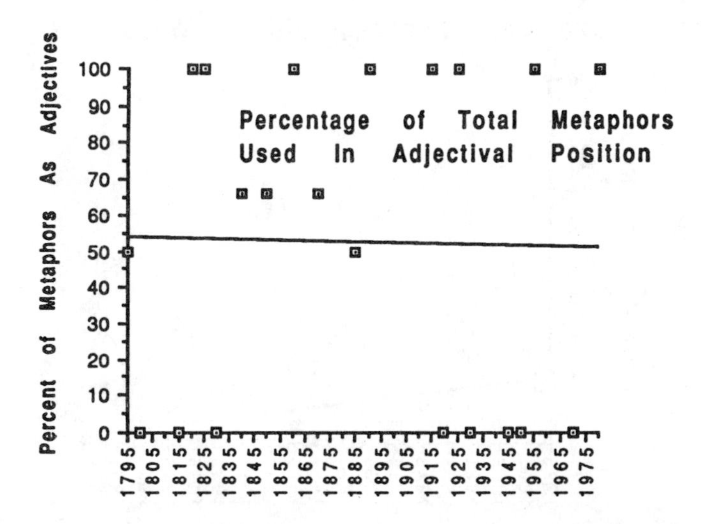

Figure 4.6. Consistently varied position of metaphors in adjective positions

The implications of Figures 4.5 and 4.6 are that both the overall presence of metaphors in the patent samples—few were ever used—and their use—as unmuted metaphors—was consistent rather than changing. Just as few metaphors were used in the early 19th century as in the late 20th century, yet, prior to 1900, the technique of metaphorical muting was used more often than later. Thus, these data on metaphors are counter to Whittaker's and Ramsey's predicted trends and suggest a consistent style.

In comparison with Figures 4.5 and 4.6, however, Figure 4.7 has a line falling much more than for most of the samples. Figure 4.7 displays the raw number of impressionistic adverbs, adjectives, adverbial phrases, and adjectival phrases found in each patent sampling. The graph shows a decrease in the mean number of impressionistic items per sample, from 4.8 items in 1795 to 0 between 1865 and 1950. In fact, from 1795 to 1865, only one sample failed to have impressionistic items, whereas, in 18 of the sample years analyzed between 1865 and 1950, only two had any impressionistic items. Up to the 1955 sample, such a strong decrease in the presence of impressionistic items strongly supports Whittaker's and Ramsey's predication of an increasing use of the impersonal-objective style. However, the samples from 1955 to 1980 undercut such dramatic support. In these six samples, five had impressionistic items. Perhaps these most recent patent samples did not have some of the large numbers present prior to 1865 (e.g., 18 in the 1810 sample), but this aspect of objective-impersonal writing style does indeed increase at the end of the samples.

Stylistic choices appear to be more complicated than Whittaker and Ramsey and other commentators on technical communication history have

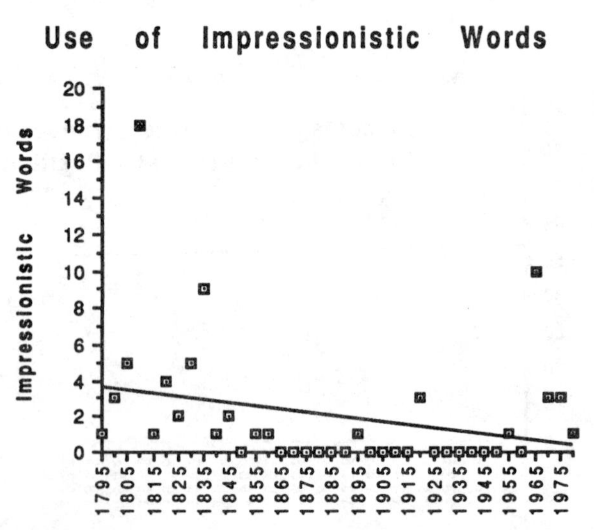

Figure 4.7. Decrease in the use of impressionistic words and phrases

described: There is scant evidence supporting the suggestion of a large-scale abandonment of a personal style in favor of an objective-impersonal style in the genre of patents for the 185-year period examined in this experiment. Although it is true that some of the evidence supports the abandonment hypothesis (Figure 4.1), others only weakly or partially support such predictions (Figures 4.3 and 4.7), and some appear to counter such predictions (Figure 4.2, 4.4, 4.5, and 4.6). Perhaps as important as finding or not finding overwhelming evidence to support or reject the specific predictions made by Whittaker and Ramsey is the conclusion that the evidence does not exhibit many dramatic changes in the specific stylistic choices examined. In fact, there is more evidence for consistency of stylistic choices over the 185 years than of change.

PATENT STYLISTIC CONSISTENCY

The lack of a clear-cut dramatic trend denoting a change in writing style to an increasingly objective-impersonal style within the patent genre suggests that many views of the technical communication past are based more on legend and philosophical prejudice than on data. Additionally, because consistency, not change, may be the most logical interpretation of the data, perhaps the objective-impersonal style is a key element in technological communication and not simply a 20th-century stylistic device. It is interesting to note in this regard that analysts of scientific texts have found one element of the objective-impersonal style, the passive voice, to be a consistent feature of such texts over many decades.[41] Perhaps, then, objective-impersonal stylistic choices are not a result of the obstinacy of poor contemporary writers as some commentators would suggest, but a stylistic expression of a necessary alienation of humans from their tools—a psychological state directly descended from 17th-century "Learned Latin."[42] Walter Ong (1977) suggested much in this regard:[43]

> Learned Latin was a scientific medium servicing the kind of noninvolvement that science demands. . . . Whatever the deeper psychological connections between the Learned Latin and the beginnings of modern science, their alliance has broken up. . . . The fact that the entire academic enterprise out of which modern science emerged had been conducted in an international no man's language would appear on the face of it to have implications which would be at the very least bizarre and possibly profound.

Moreover, why was Learned Latin such a necessity for science?[44] Ong suggested that its "no man's language" quality allows for a certain element of alienation of people from their world:[45]

> Alienation, cleavage, is not all bad. To understand things and themselves, to grow, human beings need not only proximity but also distance, even from themselves. Out of alienation, and only out of alienation, certain unities can come. Persons at ease with their origins and with their own unconscious welcome certain alienations, for they can put them to good use. The evolution of human consciousness would be impossible—would be unthinkable—without the alienation introduced by writing, print, and the electronic transformations of the word.

And perhaps by the objective-impersonal style.

Is it remarkable to consider that the demise of Learned Latin around the middle of the 17th century was soon followed by the initiation of the objective-impersonal style exhibited in Sprat's stylistic guidelines for using vernacular English?[46] Andrew Wanning (1936) traced this need for impersonality to "that desire of the scientific mentality to base its generalizations on objective procedure divorced from the variability of individual subjectivity."[47] Kathryn Riley (1991) concurred in regard to one element in the objective-impersonal style when she noted:[48]

> These findings suggest that passive voice plays an important role in reinforcing the relationship between authors and their subject matter. Scientific writers are expected to be more distanced from their subject matter in the expository sections, and thus they use the passive voice more frequently.

Thus, perhaps the cries of frustration at the intransigence of the objective-impersonal writing style should be tempered. Perhaps the intransigence of this style in this genre says less about the stubbornness of certain writers and, as Ong says, more about the incredibly old and consistent alienation that mechanicians and technologists intuitively knew they needed to perform their work.[49] Perhaps such a style is not a stylistic fad of the 20th century as many would believe, but is indeed a constant characteristic of technical communication.

APPENDIX 1: PATENTS EXAMINED

As noted, the patents were variously gathered from two government departments, the Patent Office and the National Archives (National Archives and Records Service, General Services Administration; Commissioner of Patents and Trademarks, Patent and Trademark Office, U.S. Department of Commerce).

Year*	Patentee	Name of Patent	Patent Number
1797	Dorr	Knifewheel	46
1797	Newbolds	Plough	177
1797	Bruff	Teeth Extractor	178
1801	W. Stillman	Veneering Plough	322
1802	Reed	Nailmaker	385
1803	Hopkins	Boat Building	449
1805	Pryor	Bark Mill	621
1805	Folsom & Hayden	Washing Machine	649
1805	Lester	Iron Molds	730
1810	Clarke	Marble and Pipe Stone Turner	1259
1810	Peabody	Grain Cleaner	1315
1810	Hall	Diving Dress	1405
1815	Phinney	Cornsheller	2267
1815	Cooper	Cradle	2281
1815	W. Stillman	Shears	2302
1820	Crain	Rope machine	3162
1820	Hatch	Watch chain maker	3169
1820	Eartman	Circular Saw	3183
1825	Wheeler	Lever making machine	4039
1825	Cad. Evans	Plough	4087
1825	Sargent	Single rail transport	4106
1830	Fisher	Distilling Machinery	5935
1830	Moore	Stove	5951
1830	Gilpin	Paper finish	6048
1835	Ross	Corn thrusher	8629
1835	Mott	Stove	8983
1835	Burr	Loom	8964

*Patents from 1790 to 1835 were obtained from the National Archives, Record Group 241; the numbers relate to the National Archives's "x" numbers used for filing and not to originally assigned patent numbers.

1840	Lamb	Mowing Machine	1645
1840	McCarthy	Cotton Gin	1675
1840	Keane, Keane, & Kane	Screw wrench	1720
1845	Ray	Railroad truck	3962
1845	Janes	Air heating furnace	4140
1845	Schneider	Tuning reeds	4142

*Patents from 1840 and 1845 were obtained from the National Archives, Record Group 241; "Records Relating to Numbered Patents 1836 to 1923."

1850	Clinton	Improvement in boiler lids	7099
1850	Fessenden	Pill box making machinery	7162
1850	Havey	Shutter and Sash Fastener	7289
1855	Aylsworth	Metal Milling Machinery	12,117
1855	Lee	Machine for Forging Wheels	12,135
1855	Wells	Burglar Alarm	12,177
1860	Messenger	Chair for invalids	30,149
1860	Heydrich	Fire escape	27,127
1860	Byrn	Corkscrew	27,615
1865	Finkle	Heddle	49,251
1865	Marqua	Hobby horse	46,258
1865	Munger	Lock	46,531
1870	Byland	Walking Planter	100,974
1870	Schevenell	Cartridge case	102,051
1870	Gumpel	Steering apparatus	106,358
1875	Libert	Refrigerating table	161,694
1875	Hull	Bed bottom	165,166
1875	McAllister	Fire escape	171,030
1880	Reeve	Blacksmith's shears	226,105
1880	Stewart	Sewing machine	227,187
1880	Murray and Baker	Folding Boat	228,470
1885	Owen	Bicycle	323,066
1885	Wolcott	Car dump	327,506
1885	Moorman and Moorman	Music-leaf turner	330,413
1890	Merrill	Door latch	419,966
1890	Menendez	Drill	423,620
1890	Glass	Disk harrow	427,854
1895	Fenimore	Combination feeder and mixer	550,912
1895	Nutt	Paper-folding machine	538,622
1895	Holland	Potato digger	545,229
1900	Sharp	Egg case	644,808
1900	Hanstein	Educational appliance	646,661
1900	Westcott	Knitting machine	649,378
1905	Bassell	Gun sight	783,092
1905	Ames	Cloth-splitting machine	802,545

1905	Burmeister	Clock	802,910
1910	Tucker	Measuring Instrument	953,933
1910	Weber	Dental chair	958,048
1910	Smith	Diving apparatus	975,727
1915	Evans	Locomotive	1,136,947
1915	Rose	Snare drum	1,142,843
1915	Steeley	Elevator	1,159,809
1920	McCormick	Toaster	1,329,421
1920	LaBatt	Package fastener	1,347,861
1920	Sandberg	Door lock and Burglar alarm	1,358,249
1925	Harris	Chiropractic table	1,530,719
1925	Cullinan	Tennis racquet	1,532,232
1925	Gehrke	Milk cooler and aerator	1,542,691
1930	Degener	Typewriter	1,773,098
1930	Carr	Valve-spring lifter	1,783,827
1930	Becker	Windshield clearing	1,786,932
1935	Ricks	Mechanism for shoe bottoms	2,018,808
1935	Hewgly	Fluid operated lift	2,021,997
1935	Beall	Track shunting device	2,023,301
1940	Davis	Window screen	2,195,361
1940	Canfield	Dual wheel vehicle	2,213,383
1940	Birdseye	Reflecting electric lamp	2,221,629
1945	Lambert	Disc brake	2, 368,417
1945	Lyon	Wheel cover structure	2,369,239
1945	Strother	Coffee maker	2,386,278
1950	McRae	Heat controller	2,499,964
1950	Scieri	Power can opener	2,518,190
1950	Feiertag	Card-sorting machine	2,525,405
1955	Thornburg	Apparatus for curing latex	2,707,804
1955	Herbst	Hydraulic rachet wrenches	2,708,383
1955	Palmer	Aircraft turret mechanism	2,709,394
1960	Coryell	Portable escalator	2,929,482
1960	Bonami	Poultry stuffing device	2,940,117
1960	Blodgett	Steam generating unit	2,962,006
1965	Pittman	Tube hold-down device	3,191,684
1965	Najimian	Anchor and bow light	3,192,376

1965	Parse	Landing nets	3,224,131
1970	Kirk	Mailing piece separator	3,507,211
1970	Krzyszczuk	Pump or compressor	3,521,981
1970	Levi	Book packing apparatus	3,530,643
1975	Heesemann	Grinding machine	3,859,757
1975	Honig	Teaching device	3,871,114
1975	Orndorff	Rubber bearing with nonmetalic support	3,932,004
1980	Dick	Pedal driven vehicle	4,227,712
1980	Bobbe	Half tone imaging system	4,227,795
1980	Thomas	Lens holder and sterilizer	4,228,136

*Patents from 1850 to 1980 were received directly from the Patent Office.

APPENDIX 2: PATENT ANALYSIS DATA

Year	Raw # of 1st and 2nd Person Pronouns in Sample	Personifications per Patent	% of Passive Verbs	% of Non-finite Verbs	# of Metaphors per Patent	% of Metaphors Used in Adjective Position	Total # of Impressionistic Items in Sample
1795	0	1.6	68	55	2	50	1
1800	13	0	72	46	2	0	3
1805	11	0.33	60	46	0	0	5
1810	25	0.66	54	38	0	0	18
1815	0	0	74	36	2	0	1
1820	1	0.66	40	3	3	100	4
1825	12	0	42	34	4	100	2
1830	16	0	65	45	5	0	5
1835	16	0	38	38	0	0	9
1840	2	0	45	43	3	66	1
1845	15	0	68	42	0	0	2
1850	6	1.33	72	38	3	66	0
1855	22	0	65	46	0	0	1
1860	3	0	68	46	1	100	1
1865	16	0	54	46	0	0	0
1870	6	0.33	54	35	3	66	0
1875	7	0	44	33	0	0	0
1880	11	0	52	36	0	0	0
1885	4	0	73	39	2	50	0

APPENDIX 2: PATENT ANALYSIS DATA (con't)

Year	Raw # of 1st and 2nd Person Pronouns in Sample	Personifications per Patent	% of Passive Verbs	% of Non-finite Verbs	# of Metaphors per Patent	% of Metaphors Used in Adjective Position	Total # of Impressionistic Items in Sample
1890	17	0	57	42	1	100	0
1895	9	0	61	48	0	0	1
1900	22	0	60	39	0	0	0
1905	8	0	71	50	0	0	0
1910	4	0	71	31	0	0	0
1915	2	0	56	46	1	100	0
1920	4	0.33	73	33	2	0	3
1925	9	0	73	47	1	100	0
1930	13	0	58	37	4	0	0
1935	5	0	57	37	0	0	0
1940	1	0	75	33	0	0	0
1945	16	0	67	41	1	0	0
1950	9	0	63	38	2	0	0
1955	6	0	65	32	1	100	1
1960	3	0	55	33	0	0	0
1965	4	0	62	34	0	0	10
1970	8	0	60	41	1	0	3
1975	0	0	68	34	0	0	3
1980	0	0	65	40	4	100	1

ENDNOTES

1. Portions of this chapter appeared in an earlier version: "Does Clio Have a Place in Technical Writing? Considering Patents in a History of Technical Communication." *Journal of Technical Writing and Communication* 18(4) (1988): 297-304. Publisher, Baywood Publishing Co., Inc.
2. Phillips, Willard: *The Law of Patents for New Inventions*. Boston: American Stationers Company, 1837: 241. See also Fessenden, Thomas Green: *An Essay on the Law of Patents for New Inventions*. Boston: Charles Ewer, 1822.
3. An interesting attempt to go beyond a mythic conception of the technical writing past can be found in Carol Lipson's deconstructionist reappraisal of Francis Bacon and his paradoxical contribution to the Royal Society's concept of plain style ("Francis Bacon and Plain Scientific Prose: A Reexamination." *Journal of Technical Writing and Communication* 15 (1985): 143-55).
4. Dobrin, David: "What's Technical About Technical Writing." In (Anderson, Paul V., Brockmann, R. John, and Miller, Carolyn R., eds.). *New Essays in Technical and Scientific Communication: Research, Theory, Practice*. Farmingdale, NY: Baywood, 1983: 235-7.
5. Whittaker, Della: "Priestley's Personal Style." *Technical Communication* 26(1) (1979): 28.
6. Ramsey, Richard David: "Grammatical Voice and Person in Technical Writing: Results of a Survey." *Journal of Technical Writing and Communication* 10(2) 1980: 109; see also Trammell, M. K.: "Enlightened Use of the Passive Voice in Technical Writing." In Mathes, J. C., and Thomas E. Pinelli (comps.): *Technical Communication: Perspectives for the Eighties*. Washington, DC: NASA Conference Publication 2203, (1981): 181-90.
7. Whitburn, Merrill D.: "Personality in Scientific and Technical Writing." *Journal of Technical Writing and Communication* 6(4) (1976): 203 Merrill Whitburn also seems to fall victim to the logical positivism leitmotif: "Communicators, then, could use models of scientific and technical information that involve audience adaptation and employ the human element to arouse interest. One source for such models that remains relatively unexplored is the scientific and technical writing of the past century, particularly the writing of the latter part of the seventeenth century when the rise of science was beginning and the public was eager to learn about the new field. The writing of this time has the additional advantage of not having been shaped with any standard of depersonalization in mind."
8. Miller, Carolyn: "The Rhetorical Genre: An Explanatory Concept for Technical Communication." Paper presented at the 1980 Modern

Language Association convention, Houston, Texas, December 29, 1980: 2; see also Miller, Carolyn: "A Humanistic Rationale for Technical Writing." *College English* 40 (1979): 610-7.

9. Garfield, Eugene: "Has Scientific Communication Changed in 300 Years?" *Current Contents* 8(2) (1980): 6.
10. In fact, one would have to specify exactly what phase of the formal, refereed scientific article was being discussed because the elements of this genre were not firmly in place until 1800. See Bazerman, Charles: *Shaping Written Knowledge: The Genre and Activity of the Experimental Article in Science*. Madison, WI: University of Wisconsin Press, 1988: 63.
11. Bazerman, Charles: "How Natural Philosophers Can Cooperate." In (Bazerman, Charles, and Paradis, James, (eds.)). *Textual Dynamics of the Professions: Historical and Contemporary Studies of Writing in Professional Communities*. Madison, WI: University of Wisconsin Press, 1991: 18.
12. Another descriptor of this objective-impersonal style is the "thingishness" used by Killingsworth and Gilbertson (*Signs, Genres, and Communities in Technical Communication*. Amityville, NY: Baywood Publishing Company, Inc., 1992: 131).
13. Locker, Kitty: "The Earliest Correspondence of the British East India Company (1600-19)." In (Douglas, George H., and Hilderbrandt Herbert W., (eds)). *Studies in The History of Business Writing*. Urbana, IL: American Business Communication, 1985: 69-86.
14. Dobrin, David: "What's Difficult About Technical Writing?" *College English* 44 (1982): 135-40.
15. Mitchell, Charles Eliot: "Birth and Growth of the American Patent System." In *Proceedings and Addresses, Celebration of the Beginning of the Second Century of the American Patent System*. Washington, DC: Gedney & Roberts, 1892: 47.
16. Dobrin, op. cit., 1983: 231.
17. Deller, Anthony William: *Deller's Walker on Patents* (Vol. 5.) Rochester, NY: Lawyer's Co-operative Publishing Co.,1972: 116-7.
18. *Manual of Patent Examining Procedures*. Washington, DC: United States Department of Commerce, Patent and Trademark Office: Section 608.
19. Phillips, op. cit., 1837: 246.
20. Terrango, P. James: "Patents as Technical Literature." *IEEE Transactions on Professional Communication* 22(2) (1979): 101.
21. Bitzer, Lloyd F.: "The Rhetorical Situation." *Philosophy and Rhetoric* 1 (1968): 1-14.
22. For further information about patent writing, see (Joenk, R. J., (ed.)). "Patents." Special Issue of *IEEE Transactions on Professional Communication* 22.2 (1979): 46-127 or Smith, Arthur M.: *The Art of Writing Readable Patents*. New York: Practicing Law Institute, 1958. A very interesting and useful bit of information Smith includes that could

be easily used in the classroom is a list of nine patents ranging from 1916 to 1956 (Appendix D, "Examples of Patents Criticized as Written Documents") that the courts critiqued as being difficult to understand, or, if understandable, the claims were not properly written.

23. *Heaton-Peninsula Button Fastener* vs *Eureka Specialty Co.*
24. See Nobel, David F.: *American by Design: Science, Technology, and the Rise of Corporate Capitalism*. New York: Alfred Knopf, 1977: Chapter 6; Reich, Leonard S.: "Science." In (Porter, Glenn, (ed.)). *Encyclopedia of American Economic History* (Vol 1) New York: Scribner, 1980: 298.
25. I was unable to select three patents every fifth year from 1790 to 1800 because most patents were destroyed in the Patent Office fire of 1836. Thus, one patent from 1792 and two from 1797 make up the 1795 patent group; one from 1801, one from 1802, and one from 1803 make up the 1800 patent group. This problem also explains the presence in the examined corpus of two William Stillman patents in the 1800 and 1815 samples. William Stillman wrote 6 of the first 23 patents.
26. See Frawley, William J., and Raoul N. Smith: "Patterns of Textual Cohesion in Genre-Specific Discourse." In (Stephen Williams, ed.) *Humans and Machines*. Norwood, NJ: Ablex Publishing Co., 1985: 145-53.
27. Enkevist, Nils Erik: "On Defining Style." In (Spencer, M. Gregory and Enkevist, Nils Erik, (eds.)). *Linguistics and Style* London: Oxford University Press, 1966: 1-56—"The objective style is a set of collective characteristics which reveal a choice between alternative expressions." This failure to discriminate between stylistic choice and linguistic necessity has skewed many past judgments comparing personal and impersonal writing styles. For example, many times the use of intransitive verbs is included in the impersonal style counts even when the writer does not have a choice to use a personal subject. Contrast Enkevist's definition with Croll's description of the opposite of the objective-impersonal style—the personal style—that Croll defined as a style "that portrays the process of acquiring the truth rather than the secure possession of it." (Croll, Morris William: "'Attic Prose' in the Seventeenth Century." In (Patrick, J. Max; Evans, R. O.; Wallace, J. M.; and Shoeck, R. J., (eds.)). *Style, Rhetoric and Rhythm: Essays by Morris W. Croll*. Princeton, NJ: Princeton University Press, 1966: 89.)
28. The patent analysis produced a questionnaire with 11 questions. Questions 1 through 6 examined the authorial choices made to erase human agency. Questions 1 and 2 looked for the use of references outside the text either to the reader or the author. (No such references were found in any patent.) Questions 3 and 4 focused on the use of implied reference to a reader as in the imperative voice or the direct use of first- or second-person pronouns. Question 5 narrowed the choice between

the use of agent-specified transitive verbs in the active voice and agent-unspecified transitive verbs known as truncated passives. Based on Jane Walpole's observations (Walpole, Jane R.: "Why Must the Passive Voice Be Damned?" *College Composition and Communication* 30(3) (1979): 251-4), copulatives, intransitive verbs, prepositional verbs, full passives, and a number of special verbs were excluded; the contrast was thus narrowed to a writer's choice of agent-specified (active) verbs or agent-unspecified (passive) verbs. Question 6 looked at the number of personifications, which is another way to disguise or reduce the designation of agency. Questions 7 through 9 go beyond the erasure of human agency to the erasure of time specificity—the conscious use of gerunds, infinitives, and participles rather than time-specified verbs. Finally, questions 10 and 11 examine expressions of the craftedness of language (or as Richard Lanham would say "the opacity of language" (Lanham, op. cit., 1974)). Question 10 looked for metaphors and, if found, whether they were in the adjective position (e.g., "like the common key instrument" or "the key instrument"). Question 11 looked for the presence of impressionistic adjectives and adverbs (e.g., "more or less," "at the pleasure of the operator," "in the well-known and usual manner").

29. Whitburn, op. cit., 1976: 206.
30. Miller, op. cit., 1979: 614.
31. Killingsworth and Gilbertson, op. cit., 1992: 133.
32. *The Rhetoric of Science*. Cambridge, MA: Harvard University Press, 1990: 17. Michael Halloran concurred when he noted that "passive-voice constructions are common and in some cases are demanded by style sheets; hence we read 'It was observed that....' rather than 'I (we) observed that....' Halloran refers to the strategy of manufacturing "abstract rhetors." "Authors do not show or argue, 'the *data* show that....' or 'This *paper* will argue that....' The cumulative effect of such rhetorical strategies is the assertion or implication that what is said and done is all that *could* be said or done within the research situation." Halloran, S. Michael: "The Birth of Molecular Biology: An Essay in the Rhetorical Criticism of Scientific Discourse." *Rhetorical Review* 3 (1984): 74–75, qtd. in Prelli, Lawrence J.: *A Rhetoric of Science: Inventing Scientific Discourse*. Columbia, SC: University of South Carolina Press, 1989: 103.
33. Gibson, William Walker: *Tough, Sweet, and Stuffy: An Essay on Modern American Prose Styles*. Bloomington, IN: Indiana University Press, 1966: 94.
34. Wells, Rulon: "Nominal and Verbal Style." In (Donald C. Freeman, ed.) *Linguistics and Literary Style*. New York: Holt, Rinehart & Winston, 1970: 303.
35. These stylistic choices could also be called the "scientific style." Richard Lanham (*Style: An Anti-Text*. New Haven: Yale University Press, 1974)

described the differences between opaque and scientific styles in the following way. The opaque style was to be looked "at" rather than "through" (53); it was "contrived, literary, and self-conscious" (62). The scientific style was more like "a mathematical equation, pure significance, all sign" (52); "a world where concepts come first" (62).

36. Frawley and Smith, op. cit., 1985: 145-53. With fewer exophoric references, a text also becomes more closed to alternative interpretations; see Bazerman, Charles: *Shaping Written Knowledge: The Genre and Activity of the Experimental Article in Science*. Madison, WI: University of Wisconsin Press, 1988: 124—"The compelling effect of Book 1 of the *Opticks* is rather evidence of how well, totally, and precisely Newton has gained control of the reader's reasoning and perception, so that he can make the reader go through turn by turn exactly as he wishes. In modern literary theory such a text is called a closed text as opposed to an open one that allows the reader greater freedom in providing alternative interpretive procedures and meanings, and projecting personal considerations in the text. In the closed text we read only the author...."
37. Sprat, Sir Thomas J. (J. Cope and H. Jones,(eds.)). *History of the Royal Society*. St. Louis, MO: Washington University Press, 1958: 112.
38. Brooke-Rose, Christine: *A Grammar of Metaphor*. London: Secker and Warburg, 1958: 238.
39. The simple straight lines describing the data points in Figures 4.1 through 4.7 were generated by Cricket Graph software (Version 1.32).
40. An interesting similar attempt to characterize the use of active and passive verb choices in contemporary procedural manuals has been reported from the Project on Writing Quality conducted at the Rensselaer Polytechnic Institute (1989–1991). See such articles as the following in *1994 Proceedings of the 41st Annual Conference, May 15–18, 1994*. Arlington, VA: Society for Technical Communication, 1994: Krull, Robert: "Comparative Assessment of Document Usability with Writing Quality Measures." (216–218), Hosier, William: "Documentation Quality Metrics within Total Quality Management Systems." (219–221), and Hunter, Claudia M.: "Quality Measurement: Examples from a Document Quality Formula." (222–224).
41. See, for example, Huddleston, R. D.: *The Sentence in Written English: A Syntactic Study Based on an Analysis of Scientific Texts*. Cambridge: Cambridge University Press, 1971; Gopnik, M.: *Linguistic Structures in Scientific Texts*. The Hague: Mouton, 1972; and Bazerman, C.: *Shaping Written Knowledge: The Genre and Activity of the Experimental Article in Science*. Madison, WI: University of Wisconsin Press, 1988.
42. For example, Copernicus's *De revolutionibus* (1543) and Newton's *Philosophiae naturalis principis mathematica* (1687) were both written in Latin, not English.

43. Ong,Walter J.: *Interface of the Word: Studies in the Evolution of Consciousness and Culture*. Ithaca, NY: Cornell University Press, 1977: 38-8.
44. In Oldenburg's day most English Fellows of the Royal Society still read Latin. Newton wrote Latin as well as English, but few were at home in any vernacular except their own. Robert Hooke was said not to believe anything written in French, and French scientists were generally ignorant of English. (Boorstein, Daniel J.: *The Discovers*. New York: Vintage, 1985: 392.)
45. Ong, op. cit., 1977: 47.
46. Sprat, op. cit., 1958: 112. See also Bazerman, 1988: 141—"With the failure of ethos as the primary means of validating results unwitnessed by others, the burden of persuasion fell on detailed accounts of each individual experiment—that is, on the representation. . . . Thus authority now comes not from one's sources, nor from one's good person, nor even from a publicly witnessed fact, but from a representation of events, hewing closely enough to events and defining events as to answer all critics, seem plausible to readers with extensive knowledge and experience with similar events, and to hold up against future attempts to create similar events."
47. Wanning, Andrews: "Some Changes in the Prose Style of the 17th Century." Dissertation, Cambridge University, 1936: 36.
48. Riley, Kathryn: "Passive Voice and Rhetorical Role in Scientific Writing." *Journal of Technical Writing and Communication* 21(3)(1991): 253.
49. Dobrin, op. cit., 1983: 231. For a counter argument, see Deming, Lynn H.: "The Nature of the Narrator in Technical Writing." In *The 1993 Proceedings of the 40th Annual Conference, June 6–9, 1993*. Arlington, VA: Society for Technical Communication, 1993: 154–7. Deming suggests that the objective-impersonal voice arises from the inability of technical communicators to see themselves as narrators who are in charge and who construct, not neutrally transmit, the information presented. Because of this inability, technical communicators are inflexible in their choice of narrative perspectives and thus overemploy the third person, passive, narrative perspective.

Chapter 5

The Emergence and Development of a Technical Communication Genre, The Instruction Manual

Part 1. Sewing Machines and Mower-reapers: 1950-1915[1]

[N]early everybody understands . . . the fact that a vast deal of trouble, expense and advertisement spared to the trade of the present day is wholly due to the wise foresight of the trade's leaders during the past twenty years who saw that the fate of the business depended upon the people's intimate knowledge of sewing machinery and took care to supply them with it.

— "The People's Knowledge of Sewing Machinery" (1885)

THE BIRTH OF THE INSTRUCTION MANUAL GENRE

Most items of domestic technology up to the 1850s, except for firearms, were manufactured in the home or on the farm, and most information describing their use was transmitted by word of mouth. However, in the 1850s, factory-produced goods such as sewing machines and mower-reapers were introduced, and such novel technologies needed something more than word-of-mouth instruction.[2]

Figure 5.1. Early sewing machine

Nearly everyone has seen sewing machines in action, however, as horse-drawn, mower-reaper machinery has long been replaced by larger, multipurpose machinery with internal-combustion engines, mower-reapers need a bit of explanation. Mower-reapers were among the first agricultural machines to appear on the 19th-century farm and were, in fact, the first machines with which farmers sought to exceed the limits of their human labor:[3]

> The harvesting of small grains and grasses was the most difficult as well as the most critical task known to northern farming. The ability to command labor at wheat harvest or at haying time determined the acreage of those crops that could be raised and the number of animals that could be supported, which effectively established limits on the total product of an enterprise.

Until the 1850s, the hand-swung scythe was the chief instrument of harvesting, and the scythe required one person for each tool. Yet, the mower-reaper—sometimes separate machines and sometimes different configurations of the same machine—could multiply the power of farm labor and thus increase the opportunity to harvest a crop before weather conditions affected it.

Figure 5.2. Early Mower-reaper

A mower-reaper was drawn by horses and had a seat above the wheels for the farmer. A cutting bar extended out from the side of the seat. This cutting bar had reciprocating triangular knives linked by gears to the wheels.[4] When pulled through a field by horses, the mower-reaper's turning wheels would power the reciprocating triangular knives through the gearing. These knives either cut an entire stalk if the cutting bar was positioned low—in the case of hay mowing, for example—or just the head of the grain if the cutting bar was positioned higher—in the case of wheat reaping. This ability to raise and lower the cutting arm to perform varied functions was why this particular piece of farm machinery was called a mower-reaper.

Sewing machines and mower-reapers were quite different technologies; one has continued to be used to this day, whereas the other has disappeared. Nevertheless, both were born in the 1850s, and both entered an America in which factory-produced technology had not previously existed. Thus, instruction manuals, along with oil cans and screw drivers, played a key role in allowing Americans to use these new machines:[5]

> Each machine is furnished, without charge, with Hemmer, Braider, Quilter, Plate-Gauge and Screw, Shuttle, Six Bobbins, Twelve Needles, One Wrench, Screw Driver, Oil Can, Extra Throat Plate, Bottle of Oil, and Instruction Book.

THE NEED TO EXPLORE THE BEGINNINGS OF THE INSTRUCTION MANUAL

An appreciation of 19th-century instruction manuals for sewing machines and mower-reapers can help correct the notion that the technical communication profession and the instruction manual[6] genre were born in the military manual mills of World War II.[7] This notion could have been belied with an awareness of the existence of Evans's, Griffiths's, McKay's, and Crane's books, pamphlets, models, periodicals, and sales pitches examined in Chapters 1, 2, and 3, as well as the entire genre of patents reaching back to 1795 examined in Chapter 4. Nevertheless, perhaps the genre of instruction manuals written for factory-produced products by professional writers in self-conscious ways was different. Perhaps this most recognizable artifact of contemporary technical communication is indeed of more recent lineage. Yet, like the guides and treatises for millwrights and shipwrights, the instruction manuals not only existed as early as the 1850s of Griffiths and McKay, but these instruction manuals self-consciously employed many of the design techniques now ubiquitous in the contemporary genre of instruction manuals including:

- dense interweaving of text and graphics
- packaging text into small paragraphs
- including indexes and tables of contents
- using task-oriented headings

Moreover, in contrast to some of the other -century technical communicators and many patent writers, the writers in this genre of instruction manuals were self-conscious about their writing techniques. For instance, within five years (1877–82), three nationally distributed journals appeared: *Sewing Machine Advance*,[8] *Sewing Machine News*,[9] and *Sewing Machine Times*,[10] and each not only contained general instruction manual standards, but even critiqued the relative effectiveness or ineffectiveness of individual instruction manuals.

Not only is it inaccurate to believe that technical communication and the genre of instruction manuals in particular arose from the military manual mills of World War II, but the notion itself betrays a latent sexism; it seems to imply that technology and men are inextricably linked. This notion was itself created in the 1960s by a man for consumption by a profession of technical communicators primarily made up of men.[11] However, the first sewing machine instruction manual collected in the Smithsonian's National Museum of American History—Grover and Baker's *A Home Scene; Mr. Aston's First Evening with Grover and Baker's Celebrated Family Sewing Machine Containing Directions for Using*—was published in 1859, and it was written primarily for women.[12,13] Grover and Baker's 1859 manual even faced the sexist link of technology and males head-on in the short story that initiated the manual. In the six short-story pages prefacing the instructions, the author tried very hard to show women were equal, if not superior, to men when using sewing machine technology.

The gist of the opening short story was that Mrs. Mary Aston having learned how to run the sewing machine, was asked by her husband to teach him how to run the machine:[14]

> Mr. A., after tea, jocosely remarked: "I think I will make a good operator, Mary;" and seating himself at the machine, said, "See me; I take hold of the lower part of the wheel with my right hand, and pull it in a downward direction. What now? It does not run in the same manner in which I started it."

Her husband-student turned out to be a bumbling klutz who kept breaking needles and thread and was only saved from dire consequences by his wife's superior mechanical know how:[15]

> "Thank you, Mary. I am glad you prove so patient a teacher. I must confess that I was incredulous that you had so good a knowledge of the machine, and to test it, felt desirous of seeing how far you could help me out of any difficulty I could get into. I am more than convinced of its simplicity, and think with a large majority of people, that the Grover & Baker is the best made, and most easily managed machine for family use, that can be purchased. I feel that I have made a capital investment for saving time and labor, and gained with the Sewing Machine, an assortment of sweet smiles, pleased and contented looks; and many pleasant evenings we will have on account of the freedom which the Sewing Machine will give you, Mary."

Thus, as early as the 1850s, communication about sewing machines that enabled women to have "so good a knowledge of the machine" set many of the precedents for this genre of technical communication.

Perhaps the link of technical communication's birth to the military manual mills of World War II also carried a latent animosity of early commentators towards technology itself. Early published commentators on technical communication came from such fields as English literature, in which in the 1960s—when this notion of World War II birth was first promulgated—books such as Leo Marx's *The Machine in the Garden: Technology and the Pastoral Ideal in America* lamented the effects of technology and focused on its negative effects.[16] Linking technical communication to the military manual mills instantly categorized it as negative, destructive, and, perhaps, even lethal.

However, when technical communication and the genre of instruction manuals can be linked to the concerns of women sewing for their families on 19th-century farms, such negative connotations disappear. It makes quite a difference to see the first published description of reading an instruction manual appear not in the history of Oppenheimer's "Manhattan Project" but in the pages of Laura Ingalls Wilder's *Little House on the Prairie* series:[17]

> Pa took the endgate out of the wagon, and he and Ma and Laura lifted the sewing machine carefully down and carried it into the sitting room, while Carrie and Grace hovered around it excitedly. Then Pa lifted the box-cover off the machine and they stood in silent admiration.
>
> "It is beautiful," Ma said at last, "and what a help it will be. I can hardly wait to use it." But this was late on Saturday afternoon. The sewing machine must stand still over Sunday.
>
> Next week Ma *studied the instruction book* and learned to run the machine, and the next Saturday she and Laura began to work on the lawn dress.

Technical communicators can pridefully point to their role in introducing an early technology that was acclaimed at the time—for example, in the *New York Tribune*—for its effects on female emancipation:[18]

> At last relief came the Sewing Machine. We saw in it another step in the emancipation of woman. We saw that she would be exonerated from much that was monotonous, wearisome, and belittling in her lot. . . . Her brain will soon do its office in the world, as she has now opportunities to use it.

New York's *Independent* newspaper also cited the technology for its ability to make life easier for all:[19]

> Hand-sewing has become a great tax on the industry of the human race, employing millions of laborers of both sexes. With the progress of refinement and elegance, this tax has steadily increased, until it has become well nigh intolerable. Any invention that can abridge human labor without diminishing its products, is a public blessing. The reaper and mower, and the threshing machine do this for the farmer...and the sewing machines propose to do it in the manufacture of almost every species of clothing.

THE NINETEENTH-CENTURY DEFINES THE GENRE OF INSTRUCTION MANUALS

Yet, what was the 19th-century concept of the instruction manual genre that introduced these labor saving devices? Four critiques of individual instruction manuals that appeared in the sewing machine industry journals of the 1880s suggest what the genre of instruction manuals meant in the 19th-century.[20]

In October 1882, the following was written in *Sewing Machine Advance* to describe the instruction manual for the *New Davis Sewing Machine*:[21]

> We have at last seen what we always thought was a long felt want, namely, an instruction book for the sewing machine sufficiently explicit and sufficiently illustrated to make the directions for doing all varieties of work, both with and without the attachments, so clear as to enable the operator to master the entire gamut of work from plain stitching to fancy embroidery without the aid of a teacher. This gem of sewing machine literature has been gotten up by Mr. J. B. Collins, manager of the Davis Co.'s Chicago office, to accompany the New Davis. It contains sixty-four pages and seventy-two illustrations printed in the best style on fine paper. The whole makeup is in excellent taste and reflects the highest credit upon the sagacity of Mr. Collins.

Three years later, in September 1885, the following appeared in *The Sewing Machine News* noting that the *New Home* sewing machine instruction manual was:[22]

> a perfect model of what such a pamphlet ought to be, complete, plainly and briefly explaining every point of the minutest importance, filled with fine and accurate illustrations referred to by the text, and, in brief, filling all the duty that an experienced, conscientious and patient teacher could perform. There has been no expense spared in its mechanical make up. The book is finely printed on the best heavy calendared tinted book paper, and has a cover of thick pearl gray cardboard, with a beautiful engraving of a rustic scene on the front cover.

In June 1886, *Sewing Machine Advance* made these observations concerning the instruction manual for the Standard sewing machine:[23]

> No one, however little their knowledge of sewing machinery may be, can be at a loss in any of the ordinary operations of sewing after a careful study of the neatly covered and profusely illustrated little pamphlet. . . . There is much improvement over most instruction books in the

> close and terse character of the language used. Although but a brief paragraph is given to each subject, all that is necessary is said right to the point. The whole book only contains twenty-four pages and a cover, ten of which are devoted to attachments. The typographical work, binding and printing are excellent. Whatever the "Standard" Company are doing they are doing well.

Finally, in August 1886, the *Sewing Machine News* described the sort of pamphlet that "would be of greatest benefit to new dealers and others engaging in the trade for the first time:"[24]

> It would be of great benefit for each company to furnish its agents and dealers the fullest information upon the repairing and adjusting of their own make of machines, and this instruction should be printed in convenient and compact form . . . and should be so clearly written as to be readily understood and appropriately illustrated . . . Such instructions should be conveyed in a very few words, as simply as possible, so that they can be read in a few moments. . . . An elaborate treatise of forty or fifty pages on the construction, functions, and complaints of the various parts of the machine is not what is wanted. Simplicity is everything and brevity.

From these articles critiquing various individual sewing machine instruction manuals in the 1880s, three general genre elements seem consistent:

1. The instructions should be complete, brief and plain.
2. The instructions should be profusely illustrated with a correlative relationship between text and graphics.
3. The instructions should be packaged well and look good.[25]

It is interesting is note how these genre expectations for sewing machine instruction manuals recapitulate many of the audience's expectations for earlier mechanician books and patents. In the rules regulating patent descriptions, for example, the government required that the descriptions be "short, specific, and clear."[26] Here, again, in the instruction manual genre, brevity was what the audience wanted. The stories in Theme One of Evans, Griffiths, Crane, and Patterson repeatedly suggested how American audiences wanted information presented in a visual way, and such a preference was stated in each one of the instruction manual descriptions. In particular, the type of illustrations desired in instruction manuals, "accurate illustrations referred to by the text," was exactly the type of illustration—correlative illustrations—used by Evans in linking together his text and the illustrations of his automated flour mill and by McKay for his text and his half-hull lift-model illustrations. This type of illustration is defined as "correlative" because the text and graphic closely correlate with each other.[27]

Thus, the genre of the instruction manual for the sewing machine industry in the 1880s used many of the elements of other established genres of technical communication: the guide, the treatise, and the patent. Yet how consistent were these elements in the instruction manual genre? Were they stable and consistent like the consistency of the style used in patents, or did the elements of the instruction manual evolve in the 19th century and only slowly take shape? After all, these 1880s critiques were not printed until some 30 years after the first sewing machine instruction manuals were published. Moreover, a large portion of the first manual in 1859, Grover and Baker's *A Home Scene; Mr. Aston's First Evening with Grover and Baker's Celebrated Family Sewing Machine Containing Directions for Using*, was composed of material written in a very different genre, a short story. Thus, perhaps the instruction manual in its early years was more like a short story than the instruction manual described in the 1880s.

Not only is it important to know whether this instruction manual genre was stable or evolving, but it is also important to know whether the instruction manuals differed by technology as well as by the sex of their audiences. Was the genre of instruction manual in the 19th century considered to have application across technologies, or was it relegated only to the industry in which it first appeared, sewing machines? Additionally, in regard to genre differences based on the sex of audiences, one might expect some kind of difference because technology is part and parcel of a culture, and, in the second half of the 19th century when women had yet to receive suffrage, they were treated quite differently than men. Steve Lubar (1993) made the connection between technology and culture when he noted that technology is not neutral but rather: "is the physical embodiment of social order, reflecting cultural traditions. It is part culture, and, as such, it mediates social relationships."[28] Moreover, if technology is part culture, and, as such, mediates social relationships, then communication about technology was and is a primary means to mediate social relationships. Bazerman and Paradis (1991) expressed this idea very well in the introduction to their book: "Writing is more than socially embedded: it is socially constructive. Writing structures our relations with others and organizes our perceptions of the world."[29] If the instruction manuals differed for each audience, how exactly were they different, and what did this difference suggest about 19th-century social relationships?

To answer these theoretical questions, I closely examined the actual texts of a number of sewing machine and mower-reaper manuals. I compared and contrasted them to a set of guidelines, the elements of the instruction manual genre proclaimed in the sewing machine journals in the 1880s. In this examination, I rephrased the qualitative descriptions of the sewing machine journals in a more quantifiable fashion. Thus, I examined both the sewing machine and agricultural manuals for brevity, plainness, profuseness of graphics, and correlative illustrations. I specifically measured these characteristics in 29 manuals in the following ways:

- Brevity. To measure a manual's brevity, I first measured the percentage of the instruction manual pages containing explicit instructions rather than sales pitches, lists of spare parts, testimonials, lists of sales offices, patent explanations, spare parts descriptions, and so on. The higher the percentage of pages containing only instructions, the more the document fit the genre element of brevity. Second, I measured brevity by computing the percentage of all illustrations that were used in support of the instructions rather than in support of such items as sales pitches for spare parts or accessories.
- Plainness. In examining the manuals for plainness I looked at the individual sentences and paragraphs of the instructions and measured them by using the Flesch Reading Ease scale and by looking at paragraph length.[30] The closer the score on the Reading Ease scale was to 100 ("Very Easy") and the shorter the paragraphs, the more the manual fit the genre element of plainness in its sentences and paragraphs. (The average sentence length and number of syllables for each of these manuals, which determined their Flesch Reading Ease score, is given in the footnotes for each table rather than in the text.) To measure the relative paragraph lengths, I chose three paragraphs from the explicit instruction pages—one from the beginning, the middle, and the end—and averaged the number of sentences contained in them to compute the average paragraph length.
- "Profuseness" of illustration. I examined the manuals for "profuseness" by counting the number of illustrations per page of explicit instruction. The higher the number of illustrations, the higher the "profuseness."
- Correlative quality of illustrations. I measured the correlative quality of illustrations by determining the percentage of the instruction illustrations that were correlative—illustrations explicitly referred to in the text—or ancillary—illustrations communicating instruction disconnected to text instructions. (For example, an ancillary illustration would include illustrations with captions describing how a piece of a reaper should be attached without any textual reference, parallelism, or numbered callouts.) Ancillary illustrations convey instructions in a more purely visual manner, whereas correlative illustrations communicate instructions in a mix of text and pictures with the pictures often in a subordinate role (i.e., the picture alone could not convey the information the way the text could without the picture).[31] The higher the percentage of instruction illustrations that are correlative, the more the manual fit the genre element of correlative illustration.

(I did not examine two other genre elements contained in the 1880 descriptions—completeness and good packaging and printing. I did not examine the manuals for completeness because it is difficult, if not impossible, to ascertain today all the actions that could have been performed with each of the different sewing machines. Thus, I could not satisfactorily judge a manual's completeness. Of course, judging the completeness of the mower-reaper manuals would have been even more difficult because few mower-reapers still exist in good working order, even fewer are available for actual work, and today hardly anyone knows how to operate all their applications. I also was unable to measure the element of good packaging and printing because in archives many of the manuals' covers were missing or damaged and the quality of good looking covers and printing was hard to quantify and examine objectively.)

The instruction manuals examined here are not the result of scientific sampling from a large number of instruction manuals as was somewhat possible in the patent experiment in Chapter 4. Rather, the 14 sewing machine manuals and 15 mower-reaper manuals, listed in Table 5.1, represent the largest number of manuals that could be collected from a number of archives: the Agricultural Division and the Textile Division, National Museum of American History (Smithsonian);[32] the Chicago Historical Society; the Hagley Archives and Library (Wilmington, DE); the Henry Francis Du Pont Winterthur Museum Library (Wilmington, DE); and the Agricultural History Center, University of California, Davis. Additionally, among all the instruction manuals available for study at these archives, only 29 met various consistency criteria. I only chose manuals published between: the time of the first mass production and delivery of sewing machines and mower-reapers in the late 1850s, and[33,34] the time of the dramatic increase in design complexity in the mid-1910s when sewing machines were electrified and mower-reapers began to be powered by internal combustion engines.

Also I only chose 29 manuals from the many available in order to ensure an equal distribution for the two technologies over the 60 years (1850–1910) examined. Finally, I also tried to pick manuals produced by as many different companies as possible to ensure that I was describing general characteristics of a genre and not just those of a particular company. Of course, this was difficult in the sewing machine market because the Singer Sewing Machine Company accounted for such a large percentage of the market and because they are one of the few companies to have survived to this day. Thus three of the 14 sewing machine manuals are from the Singer Sewing Machine Company. In a similar manner in the mower-reaper industry, the choice of manuals from different companies was also hard to maintain, and thus of the 15 mower-reaper manuals chosen, two come from Adriance, Platt and Company, two from the Walter A. Wood Mowing and Reaper Machine Co., and four from the various companies that combined to form International Harvester.

Table 5.1. The 29 Manuals Examined (With the Sewing Machine Manuals Listed First).

Company Name	Date Published	# of Total Pages	# of Explicit Instruction Pages	Number of Illustrations		
				Total #	# of Correlative	# of Ancillary
Sewing Machines						
Grover & Baker	1859	36	7	27	0	0
Singer Sewing Machines	1862	4	3	0	0	0
Howe Machine Co.	1867	20	16	23	22	0
Singer Sewing Machines	1868	10	8	8	8	0
Finkle & Lyon M'fg Co	1869	12	12	0	0	0
W. G. Wilson Sewing Machine Co.	1870	8	8	4	4	0
Davis Sewing Machines	1876	16	15	0	0	0
New American Sewing Machine	1878	32	32	13	13	0
Florence Sewing Machine Co.	1878	13	13	9	9	0
A. G. Mason Mfg. Co.	1880	4	4	1	1	0
Domestic Sewing Machine Company	1880	45	42	29	24	0
Perry Mason & Co.	1899	19	19	11	11	0
Wilcox & Gibbs Sewing Machine Co.	1907	34	32	34	28	0
Singer Sewing Machines	1914	11	9	12	12	0
Mower-reapers						
Boas & Spangler	1857	24	1	0	0	0
Alzirus Brown	1864	21	1	6	1	0
Adriance, Platt & Co.	1865	18	2	11	0	4
Williams, Wallace & Co.	1865	22	4	4	0	3
Nixon & Co.	1865	17	2	6	0	3
Adriance, Platt & Co.	1867	5	2	1	0	1

Table 5.1. The 29 Manuals Examined (With the Sewing Machine Manuals Listed First) (con't).

Company Name	Date Published	# of Total Pages	# of Explicit Instruction Pages	Number of Illustrations Total #	# of Correlative	# of Ancillary
Alzirus Brown	1867	7	1	5	0	0
Walter A. Wood Mowing & Reaping Machine Co.	1873	40	9	18	18	0
Eureka Mower Co.	1879	15	1.5	2	0	1
Walter A. Wood Mowing & Reaping Machine Co.	1890	15	3	4	4	0
Wm. Deering & Co.	1891	6	2.5	3	2	0
International Harvester Company of America-Champion	1909	12	3	8	5	0
International Harvester Company of America-Deering	1910	10	3	7	6	0
International Harvester Company of America-Deering	1911	10	3	6	5	0
International Harvester Company of America-McCormick	1915	12	7	16	6	6

To examine these 29 manuals for brevity, plainness, and illustrations, I performed four experiments. The first was to establish whether the journals' standards had any relevancy when they were published in the 1880s, I examined four sewing machine instruction manuals of that period. The findings in this first experiment revealed whether there was any correlation between what was publicly described in the journals and what was actually done by individual companies in creating manuals. If there was a fit between the published standard and the actual manuals, the applicability of the genre standards would be established.

The second experiment was to determine whether these genre standards could be applied to the contemporaneous mower-reaper manuals. I examined three mower-reaper manuals and compared them to the four sewing machine manuals in Experiment One. The findings in this experiment would reveal how much divergence from the published genre standards there was between industries. If there was a complete divergence between the sewing machine journal standards and the mower-reaper manuals, the applicability of the genre standards outside of the sewing machine industry could be questioned.

Third, I examined all the sewing machine manuals to observe whether there had been any change in adherence to the published standards over a 60-year time period. The findings in this experiment could either confirm a consistent adherence to genre standards as was revealed in the patent experiments in Chapter 4, or the findings could show some kind of change.

Fourth, I compared all of the sewing machine manuals over 60 years to all of the mower-reaper manuals over the same time period to see how they adhered to the published genre standards. Again, the findings in this experiment could either confirm a consistent adherence to genre standards in both industries as was revealed in the patent experiments, or the experiment could show some kind of change. This experiment could also reveal differences based on the different technologies explained in the manuals as well as the different sexes addressed as audiences.

THE FIRST EXPERIMENT: COMPARING THE PUBLISHED STANDARDS TO CONTEMPORANEOUS SEWING MACHINE MANUALS

The four sewing machine manuals contemporaneous with the published genre expectations of the journals included:

- New American Sewing Machine, *New American Sewing Machine* (1878),
- Florence Sewing Machine Co., *Directions for Using the Florence Sewing Machine* (1878),

- A. G. Mason Mfg. Co., *Improved Common Sense Family Sewing Machine* (1880), and
- Domestic Sewing Machine Company, *Domestic Sewing Machine Without a Teacher* (1880).

The results of measuring their adherence to the published genre standards is presented in Table 5.2.

Because most of these manuals were made up of explicit instructions (between 93% and 100%), they exhibited a high correlation to the published genre standard of brevity. They also exhibited a high correlation to brevity because graphics directly supporting the instructions were primarily included (between 82% and 100% of the illustrations directly supported instructions). Thus, these contemporaneous manuals kept a very tight textual and visual focus on instruction.

When I examined these manuals for adherence to the genre standard of plainness, a varied adherence to the genre standard became evident. Flesch Reading Ease scores ranged between 50 and 72, thus indicating that the Reading Ease in these manuals ranged from "Fairly Difficult" to "Fairly Easy." Paragraphs also ranged from two to eight sentences in length; thus some manuals had portions with somewhat long, dense text, whereas others were written in a plainer manner.

Additionally, when I examined these four manuals for their adherence to the genre standard of visual presentation of the information, they again exhibited a great range of adherence to this standard, with as few as 25% of the instruction pages having illustrations to as many as 69% having illustrations. However, the more specific criterion of correlative graphics—graphics referred to and explained in the text—was universally adhered to in the contemporaneous manuals; all the illustrations used in all the sewing machine manuals were discussed and referred to in the text.

Overall, the sewing machine manuals published at the time the standards appeared in the industry journals exhibited the genre element of brevity by not including anything other than instructions and illustrations only when they were in direct support of the instructions. When the manuals used illustrations, they were, as described in the published critiques, carefully explained in the text and correlative. Nevertheless, these manuals did not universally exhibit plainness ("close and terse character of the language"), for example, the A. G. Mason Mfg. Co. manual had a Flesch Reading Ease score defined as "Fairly Difficult" and long paragraphs with an average of eight sentences. Additionally, most of the manuals had illustrations on only about 48%of their pages with one of the four manuals having illustrations on only 25% of its instruction pages. Thus, the actual measurements of manuals written at the time the genre standards were published in the 1880s suggest that the standards were more often ideals than accomplished facts. However, the examina-

Table 5.2. Contemporaneous Sewing Machine Manuals' Adherence to the Genre Standards.

Company Name	Instruction Pages as Percentage of Total # of Pages	Percent of Total Graphics Used for Instruction	Flesch Reading Ease Score[35]	Average # of Sentences Per Paragraph	# of Instruction Illustrations (% of Illustra-tions Per Page of Instructions)	Percentage of Instruction Illustrations that are Correlative or Ancillary
New American Sewing Machine	100	100	70	3	13 (40)	100-0
Florence Sewing Machine Co.	100	100	50	3	9 (69)	100-0
A. G. Mason Mfg. Co.	100	100	72	8	1 (25)	100-0
Domestic Sewing Machine Company	93	82	55	2	29 (57)	100-0

tion of these four manuals did show that the published genre standards did have applicability to real manuals.

THE SECOND EXPERIMENT: CONTEMPORARY SEWING MACHINE MANUALS COMPARED TO CONTEMPORARY MOWER-REAPER MANUALS

I examined three contemporaneous mower-reaper manuals and compared their elements to the sewing machine manuals and published genre standards of Experiment One:

- Walter A. Wood Mowing and Reaping Machine Co., *Directions for Putting Together and Using Walter A. Wood's Harvesting Machines* (1873)
- Eureka Mower Co., *Wilber's Eureka Mower, Direct Draft* (1879)
- Walter A. Wood Mowing and Reaping Machine Co., *Directions and List of Parts for Walter A. Wood's Two-Horse Enclosed Gear Mower with Tilting Bar* (1890)

The findings for the areas of brevity, plainness, and illustrations are presented in Table 5.3 and in Figures 5.3 to 5.7 the mower-reaper manuals are presented side-by-side with the contemporaneous sewing machine machine manuals.

These mower-reaper instruction manuals exhibited a very low adherence to the 1880 genre element of textual brevity since many pages were filled with testimonials, sales pitches, patent descriptions, and so on. Only 10% to 23% of the manuals was devoted to instructions in contrast to the 90% to 100% of contemporaneous sewing machine manuals devoted to instructions (see Figure 5.3).

However, when the mower-reaper manuals were examined for graphical brevity—the percentage of all illustrations devoted to instructions—there was a closer similarity to the sewing machine manuals (Figure 5.4). Like the contemporaneous sewing machine manuals, two of the three mower-reaper manuals had all their graphics devoted to instructions.

However, when I examined these mower-reaper manuals for their adherence to the element of plainness as measured by the Flesch Reading Ease scale and by paragraph length, these mower-reaper manual actually adhered much more closely to the published standards than did the sewing machine manuals. With Reading Ease scores averaging 85 ("Easy") and paragraphs including only 1 or 2 sentences, the instructions in these mower-reaper manuals were much more "close and terse" in their language (see Figures 5.5 and 5.6).

Table 5.3. Contemporaneous Mower Reaper Manual Adherence to the Genre Standards.

Company Name	Instruction Pages as Percentage of Total # of Pages	Percent of Total Graphics Used for Instruction	Flesch Reading Ease Score[36]	Average # of Sentences Per Paragraph	# of Instruction Illustrations (% of Illustra- tions Per Page of Instructions)	Percentage of Instruction Illustrations that are Correlative or Ancillary
Walter A. Wood Mowing & Reaping Machine Co.	23	100	93	2	18 (200)	100-0
Eureka Mower Co.	10	50	78	1	1 (66)	0-100
Walter A. Wood Mowing & Reaping Machine Co.	20	100	84	2	4 (133)	100-0

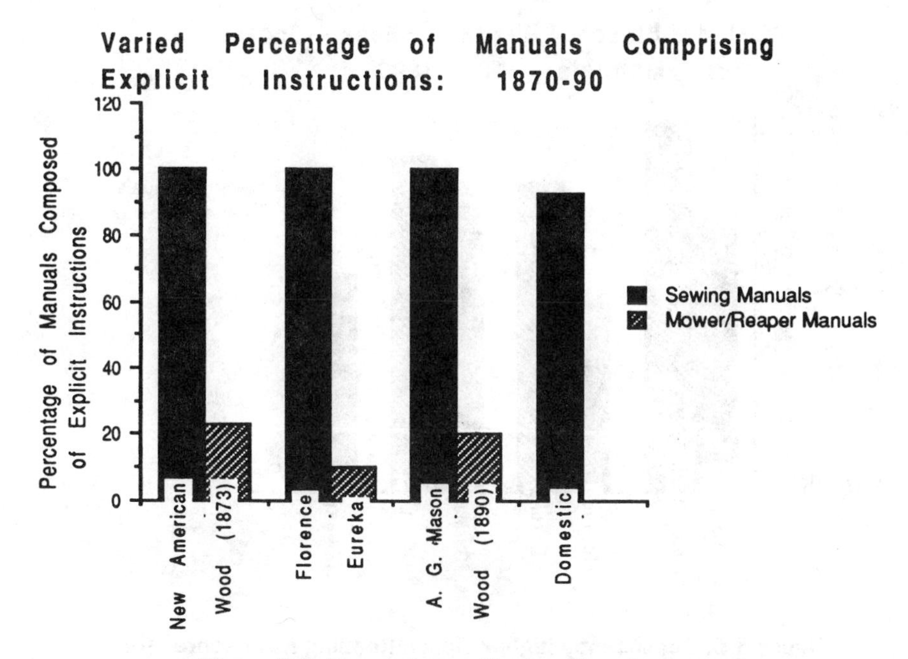

Figure 5.3. Percentage of instructions in manuals

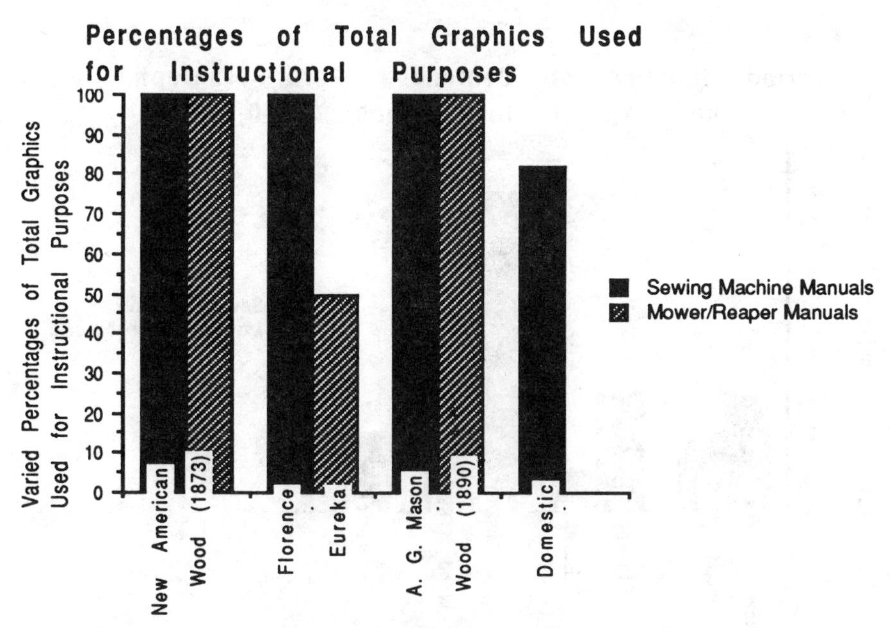

Figure 5.4. Similarity of percentages of total graphics devoted to instructional support

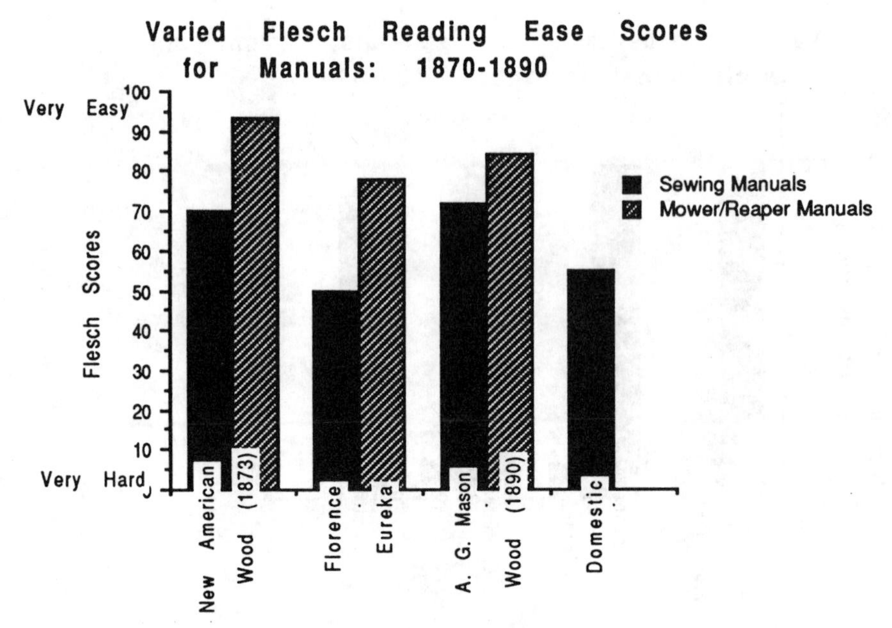

Figure 5.5. Consistently higher Flesch Reading Ease scores for mower-reaper manuals

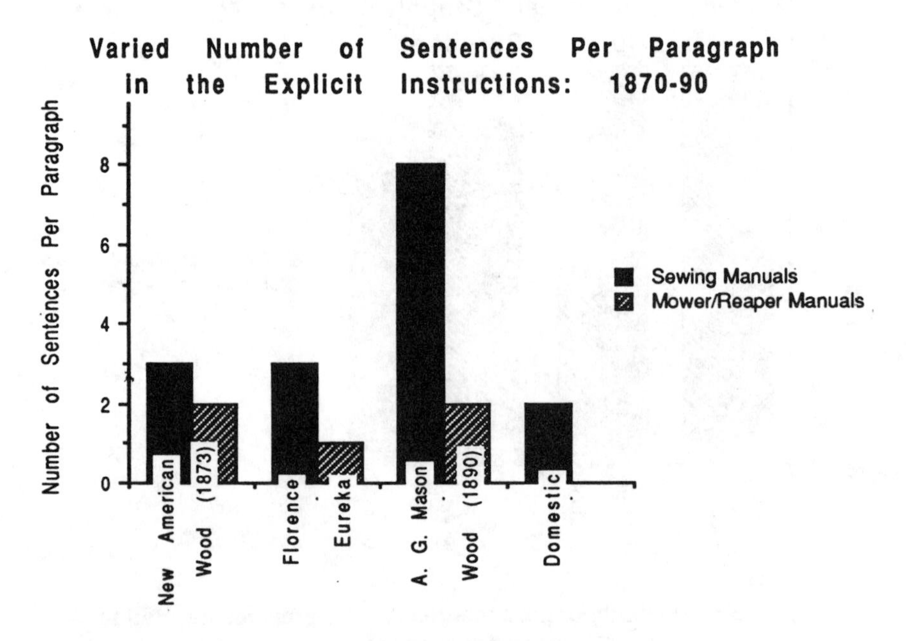

Figure 5.6. Consistently longer paragraphs in sewing machine manuals

Not only were the mower-reaper manuals written in a much plainer manner than the sewing machine manuals, but their instructions, when they were given, were better illustrated (see Figure 5.7). Two of the three mower-reaper manuals had many more illustrations per page of instructions than did the sewing machine manuals—the average was some 800% higher. Additionally, when the type of illustrations used in the mower-reaper instruction manuals was examined, most were found to be correlative. Only one manual, that of the Eureka Mower Co., had any ancillary illustrations.

Consequently, the published journal genre guidelines of the sewing machine industry had very real applicability to the mower-reaper manuals. In plainness and profuseness of illustrations, mower-reaper manuals were actually better at meeting the genre descriptions than were the sewing machine manuals themselves. The mower-reaper manuals also largely employed correlative graphics, and exhibited graphical brevity by devoting most of their illustrations to instructional purposes. However, the mower-reaper manuals greatly differed from the published genre descriptions in the area of verbal brevity. Only an average of 16% of the mower-reaper manual was used for purposes of instruction; in fact, most of the mower-reaper manual contained almost anything other than instructions and directions.

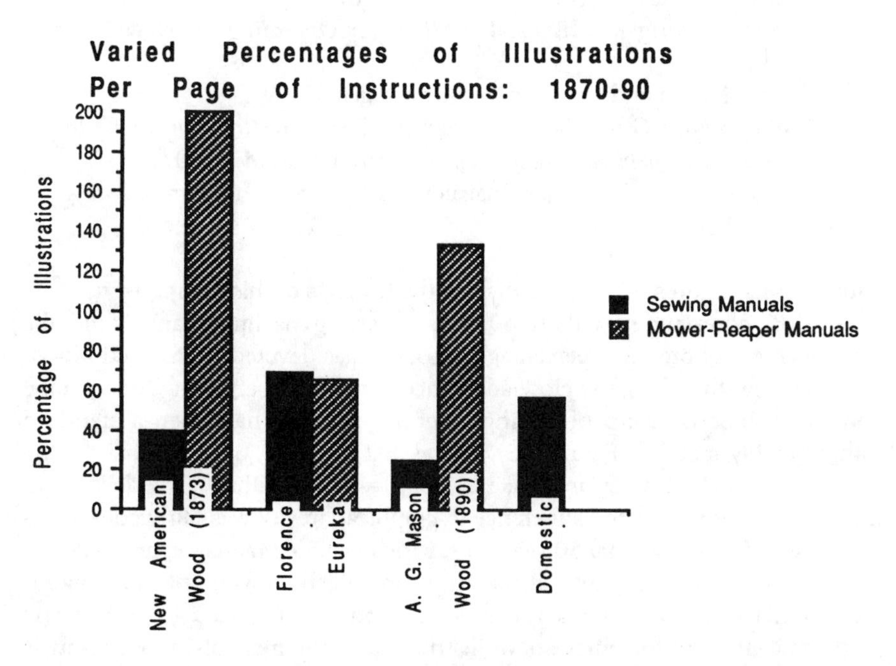

Figure 5.7. Varying percentages of illustrations per page of instructions

THE THIRD EXPERIMENT: COMPARING 60 YEARS OF SEWING MACHINE MANUALS TO THOSE WRITTEN AT THE TIME OF THE PUBLISHED GENRE STANDARDS

To determine the overall adherence of sewing machine instruction manuals over a period of 60 years to the genre descriptions published in the 1880s, I examined 10 additional sewing machine manuals:

- Grover and Baker, *A Home Scene; Mr. Aston's First Evening with Grover and Baker's Celebrated Family Sewing Machine Containing Directions for Using* (1859)
- Singer Sewing Machines, *Directions for Using Singer's Patent Straight-needle, Transverse Shuttle Sewing Machine* (1859-65)
- Howe Machine Co., *Instruction Book for the Howe Sewing Machine, Step Feed* (1867)
- Finkle and Lyon M'fg Co, *Directions for Using Finkle and Lyon M'fg Co's New Sewing Machine, Victor* (1867-72)
- Singer Sewing Machines, *Directions for Using New Family Sewing Machine* (1868)
- W. G. Wilson Sewing Machine Co., *Directions for Using the New Buckeye Under-feed Sewing Machine* (1870)
- Davis Sewing Machines, *Directions for Operating the New Davis* (1876)
- Perry Mason and Co., *The New Companion* (1899)
- Wilcox and Gibbs Sewing Machine Co., *Directions for Using the Wilcox and Gibbs Automatic Noiseless Sewing Machine* (1907)
- Singer Sewing Machines, Instructions for Using Attachments, *Style No. 14* (1914)

Table 5.4 and figures 5.8 to 5.12 display the findings of this comparison.

By observing how these 60 years of sewing machine manuals met the genre element of brevity—percentage of total pages devoted to instructions—it becomes clear that the fairly close adherence noted in Experiment One with the contemporaneous group of sewing machine manuals had been a standard rather quickly met (see Figure 5.8). In the first 20 years, only the first sewing machine manual—Grover and Baker's (1859)—had less than 75% devoted to instructions. Moreover, the adherence to graphical brevity was fairly consistently followed during the final 50 years of sewing machine manuals examined.

In the area of graphical brevity, nearly all the sewing machine manuals adhered to the published genre description (see Figure 5.9). When any graphics were used for purposes of instruction in the manuals, most manuals devoted between 82% and 100% to instructions. Only four of the manuals did not achieve this standard of graphical brevity during the 60 years examined,

Table 5.4. Adherence to the Genre Standards In Sewing Machine Manuals, 1859 –1914.

Company Name	Instruction Pages as Percentage of Total # of Pages	Percent of Total Graphics Used for Instruction	Flesch Reading Ease Score[37]	Average # of Sentences Per Paragraph	# of Instruction Illustrations (% of Illustrations Per Page of Instructions)	Percentage of Instruction Illustrations that are Correlative or Ancillary
Grover & Baker	19	0	53	6	0 (0)	0-0
Singer Sewing Machines	75	0	70	9	0 (0)	0-0
Howe Machine Co.	80	96	75	3	22 (138)	100-0
Singer Sewing Machines	80	100	51	5	8 (100)	100-0
Finkle & Lyon M'fg Co	100	0	45	4	0 (0)	0-0
W. G. Wilson Sewing Machine Co.	100	100	43	2	4 (100)	100-0
Davis Sewing Machines	94	0	74	6	0 (0)	0-0
New American Sewing Machine	100	100	70	3	13 (40)	100-0
Florence Sewing Machine Co.	100	100	50	3	9 (69)	100-0
A. G. Mason Mfg. Co.	100	100	72	8	1 (25)	100-0
Domestic Sewing Machine Company	93	82	55	2	29 (57)	100-0
Perry Mason & Co.	100	100	46	2	11 (58)	82-18
Wilcox & Gibbs Sewing Machine Co.	94	82	64	3	28 (88)	100-0
Singer Sewing Machines	81	100	74	3	12 (109)	100-0

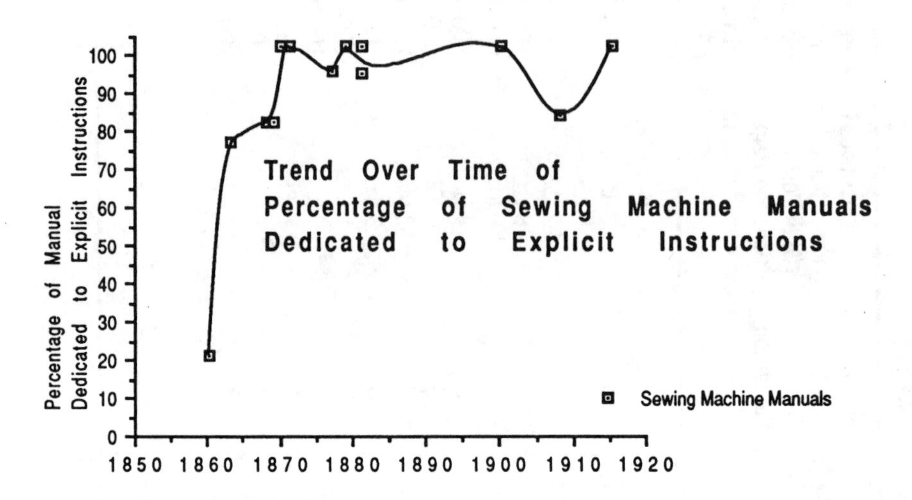

Figure 5.8. Increasing percentage of instruction in sewing machine manuals[38]

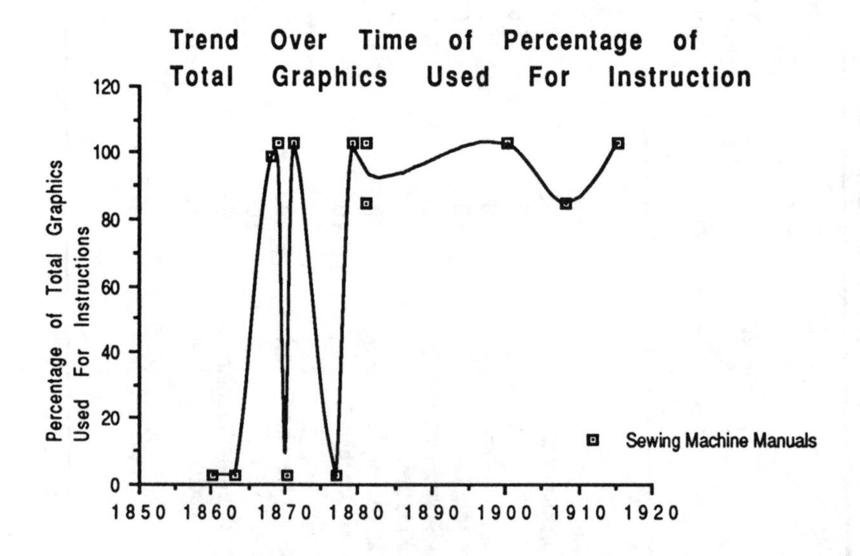

Figure 5.9. Increasing percentage of illustrations dedicated to supporting instructions

and three of those four had no graphics in the manual at all (e.g., Singer [1862], Finkle and Lyon [1869], and Davis [1876]). Thus in the area of graphical brevity, there was not much evidence of change except for the increasing general presence of illustrations. Moreover, such an increasing appearance of illustrations is better measured by the graph of profuseness of illustrations (Figure 5.12).

To examine the genre element of plainness, I analyzed the Reading Ease scores and the lengths of paragraphs (Figures 5.10 and 5.11). The line in Figure 5.10 does vary but only within a limited range because the Reading Ease scores have a plus-minus range of only 30 points over the 60 years; this suggests that the sewing machine instruction manual authors had some difficulty writing plainly sentence by sentence. There was no development or much of a change in this element of the genre; there are as many 70+ ratings before the 1880s as there are after the 1880s.

However, the line in Figure 5.11 shows a much more dramatic downward trend over time than does Figure 5.10, and the range plus or minus from the general trend gets much tighter after the 1880s. Thus, the sewing machine manuals were mixed in how they achieved the element of plainness. Sentence-by-sentence plainness, as measured by the Reading Ease scores, showed little change over the 60 years. Yet, paragraph-by-paragraph plainness, as measured by paragraph length, showed that the paragraphs were made easier to read by

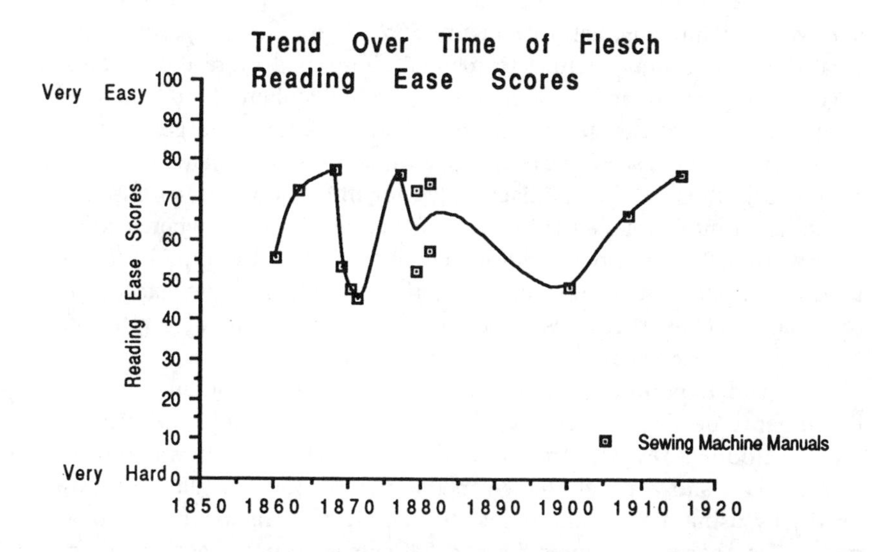

Figure 5.10. Lack of change in Flesch Reading Ease Scores of sewing machine manuals

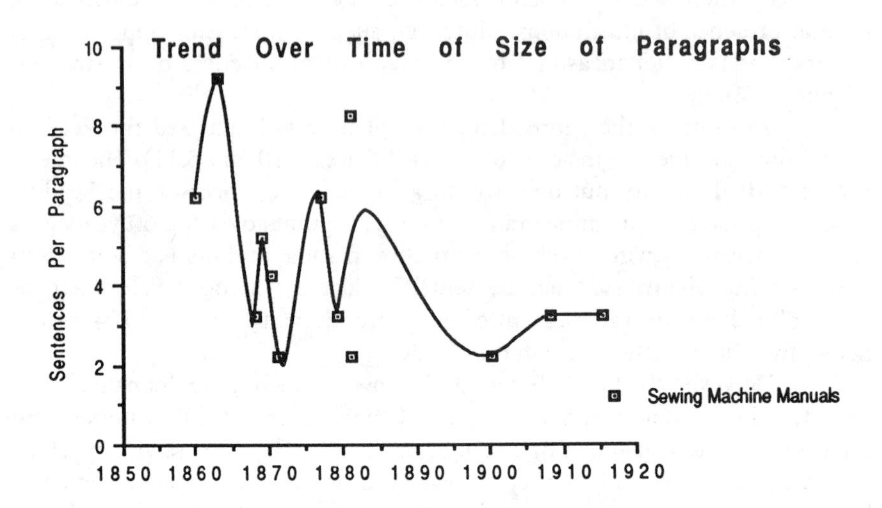

Figure 5.11. Decreasing size of paragraphs in sewing machine manuals

making them smaller. Thus, there was some change in the genre in this latter element.

Finally, I looked at the trend over 60 years in the use of illustrations in sewing machine manuals (See Figure 5.12). In the first 20 years, there was a great divergence from manual to manual: Some had more illustrations than pages of instruction, and some had no illustrations of any kind. Then, as with other elements of the genre, around 1880 the great divergence narrowed somewhat for the last 40 years. Thus, although the later manuals were never as profusely illustrated as perhaps called for in the journals, profuseness was much less a hit-or-miss situation than in the first 20 years where profuseness ranged from 0% to 138%. Additionally, in regard to the type of illustration used to support instructions, all illustrations were correlative except those in one manual. Thus, there was great consistency in the type of graphics used if graphics were used at all.

Did Experiment Three confirm a consistent adherence to the published genre descriptions as was revealed in the patent style experiments, or did the findings reveal some kind of change? The answer was mixed. The manuals did show a real change in the area of both textual and graphical brevity by using most of the pages and the illustrations for instructional purposes. The change in Figures 5.8 and 5.9 shows an initial accelerated change in the first 20 years, with a flattening out and maintenance of general consistency in the last 40 years. This same rapid initial change followed by consis-

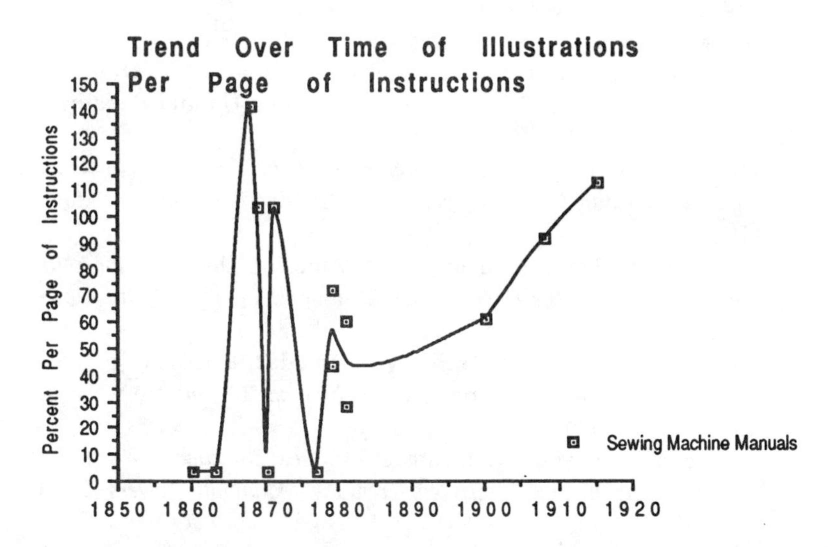

Figure 5.12. Increasing number of illustrations per page of instructions in sewing machine manuals

tency seemed also to be displayed in the size and the variations of paragraphs and in the range of profuseness of instruction illustration. Only in the plainness measured on the Reading Ease tests and in the use of correlative-type illustrations with instructions does there seem to be the kind of consistent adherence from the beginning as that found with the patents.

THE FOURTH EXPERIMENT: COMPARING THE ADHERENCE TO PUBLISHED GENRE STANDARDS OF SEWING MACHINE MANUALS AND MOWER-REAPER MANUALS WRITTEN OVER 60 YEARS

Along with the sewing machine manuals examined in Experiment Three, in Experiment Four, I looked at 12 mower-reaper manuals covering the same span of time from the 1850s to the 1920s:

- Boas and Spangler, *Ketchum's Celebrated Combined Mower and Reaper* (1857)
- Alzirus Brown, *Union Mowing Machine Circular for 1864-5* (1864-5)
- Adriance, Platt and Co., *Buckeye Mower and Reaper* (1865)
- Williams, Wallace and Co., *Hubbard's Combined Reaper and Mower* (1865)

- Nixon and Co., *Cayuga Chief and Cayuga Chief Number Two Mower and Reaper and Mower Combined* (1865)
- Adriance, Platt and Co., *Buckeye Mower No. 1* (1867)
- Alzirus Brown, *Union Mowing Machine Circular for 1867* (1867)
- Wm. Deering and Co., *Directions for Setting Up and Running the Deering Giant Mower* (Front Cut) (1891)
- International Harvester Company of America, *Directions for Setting Up and Operating Champion New Draw Cut and Big Draw Cut Mowers* (1909)
- International Harvester Company of America, *Directions for Setting Up and Operating the Deering New Ideal and New Ideal Giant Mowers* (1910)
- International Harvester Company of America, *Directions for Setting Up and Operating the Deering New Ideal Type "B" One-Horse Vertical Lift Mower* (1911)
- International Harvester Company of America, *Directions for Setting Up and Operating the McCormick No. 6 and Big 6 Plain Lift Mowers* (1915)

The specific numbers for the brevity, plainness, and illustration tests are presented in Table 5.5 and Figures 5.13 to 5.17.

Figures 5.13 and 5.14 show the results of examining the element of textual and visual brevity in the entire two sets of manuals. Figure 5.13 shows that the mower-reaper manuals never achieved as large a percentage of their pages dedicated to instructions as did the sewing machine manuals. Both sets of manuals did, however, progressively dedicate more of their pages to instructions, with the mower-reaper manuals increasing their percentage a little more dramatically, but the increase of instruction percentages in the mower-reaper manuals never came close to those in the sewing machine manuals until the very last one.

However, there was some change in the textual element of brevity in both industries, and this suggests that the manuals in both industries were not as consistent as the patents.

Figure 5.14, which shows the trend over time in the area of graphical brevity (percentage of overall graphics dedicated to instructional purposes), reveals dramatic increases in the trends. Both sets of manuals dramatically increased the percentage of graphics in instructions, but again the sewing machine increase was larger and faster than that of the mower-reaper. Thus, more sewing machine manuals more quickly incorporated graphics dedicated to instruction than did mower-reaper manuals. This also suggests that a large number of graphics in the mower-reaper manuals were used for purposes other than instruction on a continuing basis.

These results suggest that the sewing machine instruction manual authors quickly dropped secondary concerns and focused on instructions.

Table 5.5. Adherence to Genre Standards In Mower-reaper Manuals, 1857–1915.

Company Name	Instruction Pages as Percentage of Total # of Pages	Percent of Total Graphics Used for Instruction	Flesch Reading Ease Score[39]	Average # of Sentences Per Paragraph	# of Instruction Illustrations (% of Illustrations Per Page of Instructions)	Percentage of Instruction Illustrations that are Correlative or Ancillary
Boas & Spangler	4	0	85	12	0 (0)	0-0
Alzirus Brown	4	16	87	1	1 (100)	100-0
Adriance, Platt & Co.	11	36	88	3	4 (200)	0-100
Williams, Wallace & Co.	18	75	72	12	3 (75)	0-100
Nixon & Co.	11	50	22	4	3 (150)	0-100
Adriance, Platt & Co.	40	100	75	1	1 (50)	0-100
Alzirus Brown	14	0	85	1	0 (0)	0-0
Walter A. Wood Mowing & . Reaping Machine Co	23	100	93	2	18 (200)	100-0
Eureka Mower Co.	10	50	78	1	1 (66)	0-100
Walter A. Wood Mowing & Reaping Machine Co.	20	100	84	2	4 (133)	100-0
Wm. Deering & Co.	41	66	80	2	2 (125)	100-0
International Harvester Company of America-Champion	25	62	83	3	5 (167)	100-0
International Harvester Company of America-Deering	30	86	66	2	6 (200)	100-0
International Harvester Company of America-Deering	30	83	82	2	5 (166)	100-0
International Harvester Company of America-McCormick	75	75	66	2	12 (133)	50-50

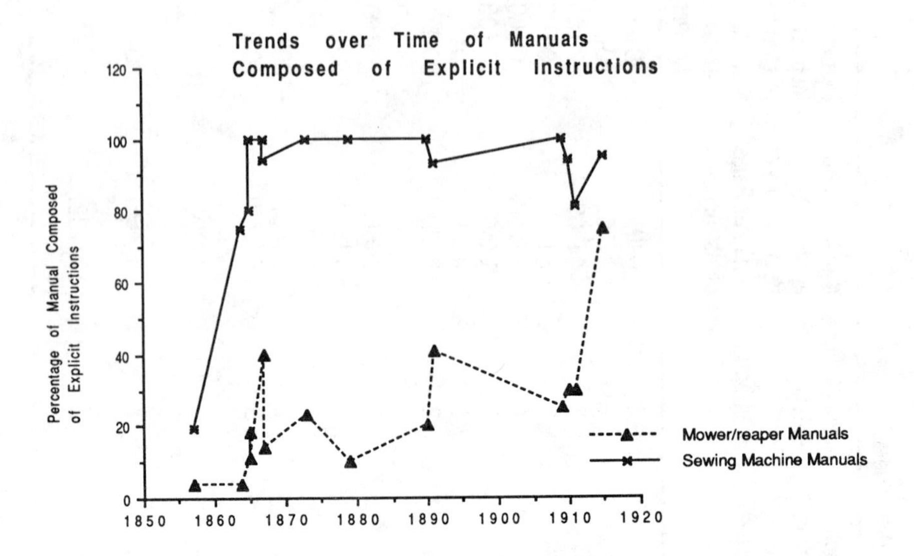

Figure 5.13. Similar increase in percent of manuals devoted to instructions

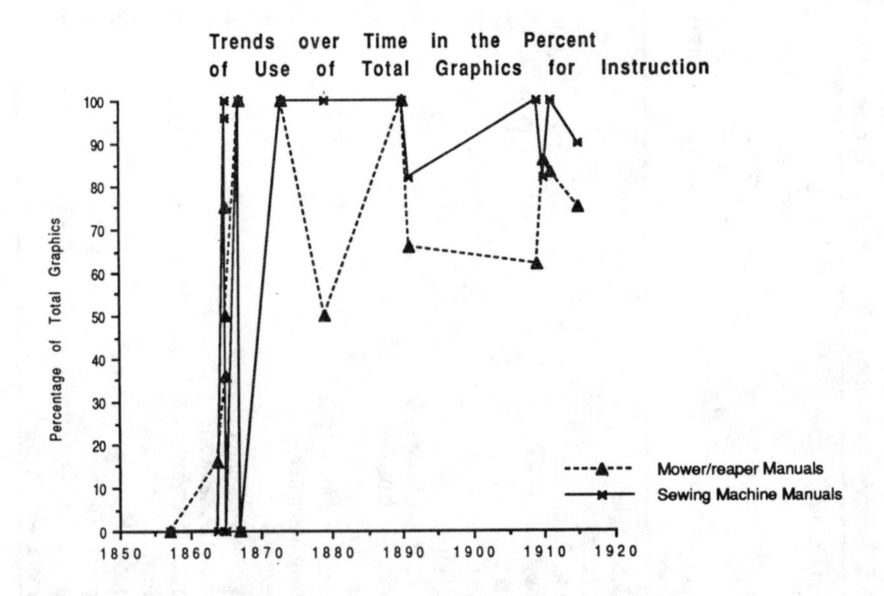

Figure 5.14. Rapid rise in graphics devoted to instructions

Thus, within 20 years, the sewing machine manual authors came to adhere fairly strictly to the quality standard of textual brevity. Mower-reaper manuals never achieved such an overall standard of textual brevity suggesting that, to the end of the period studied, they continued to have secondary concerns and a less strict focus on instructions.

Figures 5.15 and 5.16 display the trends in the two sets of manuals in achieving the genre element of plainness as measured by Reading Ease scores and paragraph size. As already noted in a preliminary way in Experiment Two, mower-reaper manual authors were more successful at achieving a plain style when measured sentence by sentence in the Reading Ease scores nearly consistently over the entire period. Both lines vary within a limited range, suggesting a fairly good consistency in the way the entire group of authors wrote in a "close and terse" fashion. Of all the genre elements studied in this chapter, this was perhaps the closest to the consistency of the objective-impersonal tone studied in the patents, and this element seemed to be the one most consistent over time, showing very little development with either set of manuals.

Figure 5.16, however, does show some very dramatic changes in both sets of manuals when the genre element of plainness was measured by looking at relative sizes of paragraphs. After an initial wide spread of paragraph sizes, the range of variations in the lines for both sets of manuals became less and nearly identical.

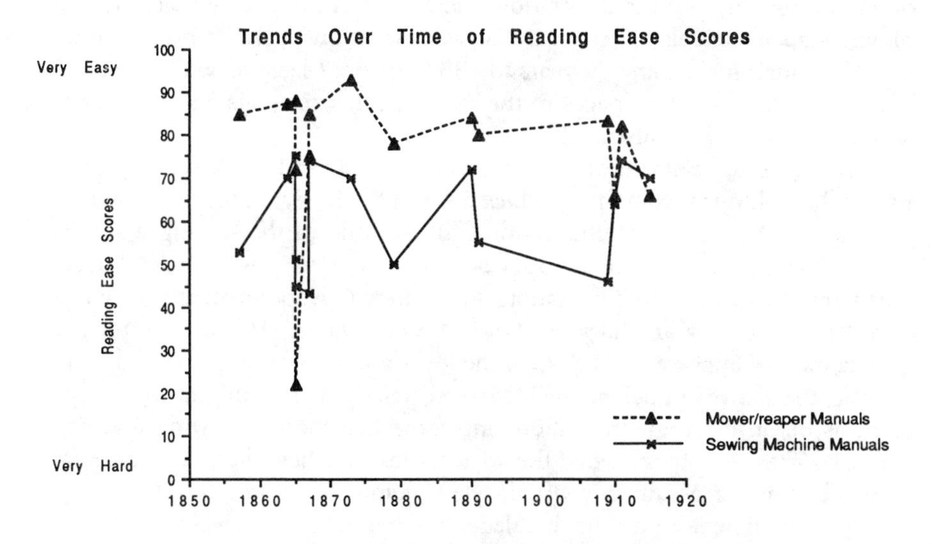

Figure 5.15. Convergence of Flesch Reading Ease scores in the two groups of manuals

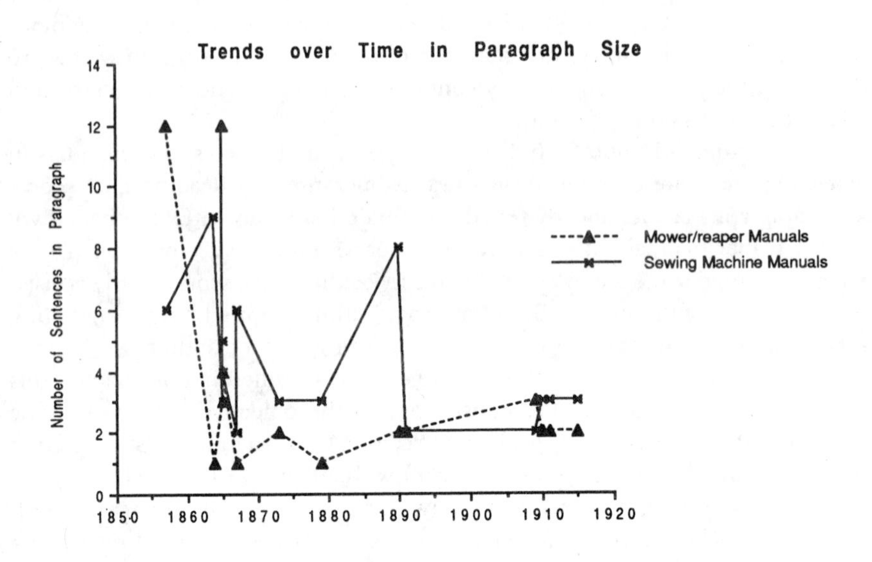

Figure 5.16. Similar trends in downsizing paragraphs

The relative measure of how well each set of manuals met the instruction manual genre element of illustration is given in Figure 5.17, which describes how profuse the illustrations were in both the sewing machine and mower-reaper manuals. After an initial wild variation in the trends for both sets of manuals for the first 30 years to 1880, the lines seem to level out somewhat with the instruction pages in the mower-reaper manuals becoming consistently more heavily illustrated.

The other element of illustrations measured can best be seen by comparing the right-most column in Tables 5.4 and 5.5 in regard to type of graphics used. If they had any illustrations, all but one of the sewing machine instruction manuals over all 60 years had ones that were correlative. Furthermore, up to 1880, illustrations in the mower-reaper manuals were just as apt to be wholly ancillary as they were correlative. However, after the Eureka manual appeared in 1879, all the illustrations used to support instructions in the mower-reaper manuals were correlative. Thus the mower-reaper manuals began using graphics more and more like those in sewing machine manuals. These graphics needed the adjacent text for their effective interpretation and became a less directly visual style of communication.

Experiment Four thus revealed a number of findings:

1. Other than the genre aspect of plainness as measured by the Reading Ease scores, and the unchanging predominant role of cor-

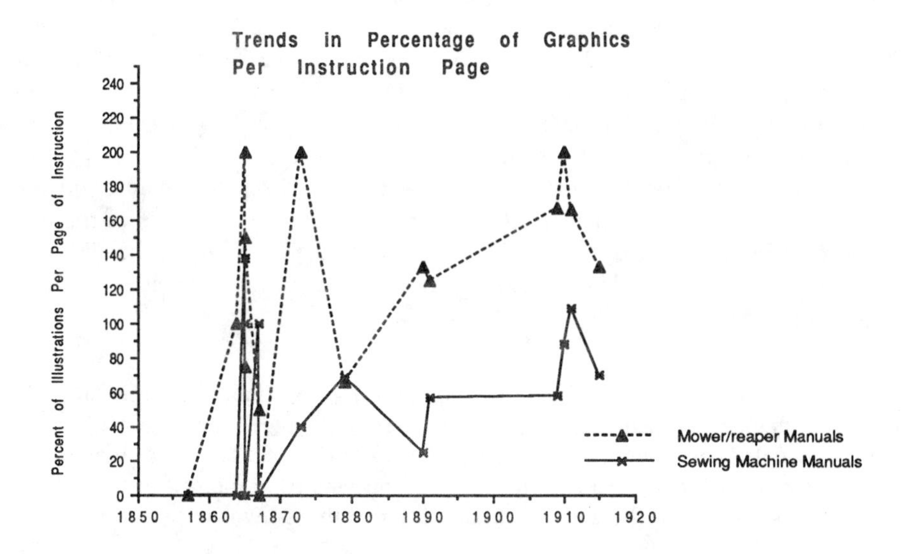

Figure 5.17. Different percentages of graphics devoted to instructions

relative graphics to support instructions, there did seem to be change in both sets of manuals, unlike the consistent stylistic elements displayed in the patent genre.

2. The trends in both sets of manuals were parallel; they changed in similar ways.
3. There were significant differences between the two sets of manuals when examined on the three genre elements. Sewing machine manuals quickly focused on devoting all their pages to instructions whereas mower-reaper manuals continued to devote large portions of their text and illustrations to efforts other than instruction. Mower-reaper manuals were consistently more profusely illustrated as well as contained more ancillary illustrations.

MAKING SENSE OF THE FINDINGS IN THE EXPERIMENTS

Some Genres Change: Stipulated Genres Versus Fluidly Negotiated Genres

The most obvious findings from the four instruction manual experiments was that both the sewing machine and the mower-reaper manuals changed in their relative adherence to the published genre standards as established in the 1880 journals. The change, especially in the sewing machine manuals, represented development in the genre of instruction manuals because they began to more closely adhere to the published standards after the first 20 years of publication. The lines in Figures 5.12, 5.13, 5.14, 5.15, and 5.16 all reveal an increasing normative expression of a genre element, be it brevity, plainness, or illustration. Yet, such development and consolidation was not revealed in the investigations of the patent description genre. Why?

Perhaps the reason for the consistency in the patent genre and the change in the instruction manual genre was because the elements of the two genres were maintained in quite different manners. For example, the elements of the patent were *stipulated* by the government, not negotiated with audiences by authors:[40]

> A Detail Description of the invention and drawings follow the general statement of the invention and a brief description of the drawings. This detailed description, required by Rule 71, section 608.01, must be written in such particularity as to enable anyone skilled in the pertinent art or science to make and use the invention without involving extensive experimentation.

Thus, if they are stipulated and outside the give-and-take of the changing needs of audiences and the changing capabilities of authors, some elements of the patent become almost archaic.[41] For instance, patent drawings basically have not changed even though drafting standards and technology have:[42]

> US patent drawings, which supplement a patent's text, follow Patent Office rules, which are unlike any industrial standards for drafting. To those familiar with conventional engineering drawings, patent drawings have a quaint air. Various views, both orthographic and modified perspective, may be scattered on a single sheet.

All the elements of the stipulated patent genre did not remain fixed; some changed. Working models like those used by Evans, McKay, Griffiths, and Crane, for example, were required of all patent applications up to 1880, at which time the requirement was dropped. However, the change in such a requirement came from actions of the Patent Commissioner who made his own rule to continue requiring working models even after Congress changed the statutory requirement in 1870.[43] Thus, some elements of this stipulated patent genre fossilized, and some were maintained almost as if by the whim of a single individual. By stipulating the patent genre, the genre became set and closed to the feedback of audiences and authors.

The genre of instruction manuals, on the other hand, was not stipulated but *fluidly negotiated*.[44] The authors of instruction manuals were open to the feedback of audiences because they were employed by commercial enterprises that lived and died by the company's relative effectiveness in reaching the marketplace. There were many ways in which the genre of instruction manuals was negotiated among authors, audiences, and the capabilities of the genre itself. For example, negotiations between authors and the genre could be seen as authors pushed and explored the genre's limits and supporting technology. This could supply one explanation for the increasing profuseness of graphics per page of instructions in the mower-reaper manuals as shown in Figure 5.17. As it became less expensive for illustrations to be produced by printing presses over the 60 years studied, more and more illustrations were used by authors, and thus the mower-reaper line in Figure 5.17 shows a very distinct upward movement.[45] Negotiations between authors and the genre and its supporting technology led to very evident change.[46]

Of course, the major direction in which a genre could be said to be fluidly negotiated is between authors and audiences as authors seek to mold the genre to the needs of audiences. A very good example of this is the way that sewing machine instruction manuals changed in response to the shift from an introduction of the technology to users to a more straight forward emphasis on instruction once the technology had been accepted by the public.[47]

Changing adherence to brevity and inclusion of anthropomorphic elements in sewing machine instruction manuals

In August 1885, the *Sewing Machine News* published an editorial describing the change in audience from the time of the introduction of sewing machines in the 1850s to its commonplace acceptance as a consumer item in the 1880s. The editorial was published under the title "The People's Knowledge of Sewing Machinery:"[48]

> Once upon a time, as the children's stories begin, one of the chief obstacles those engaged in the sewing machine trade had to contend with was the ignorance, widespread among all classes, of sewing machinery and its management. Hardly one person in a thousand knew how to run a sewing machine upon even the simplest work, arranging for themselves the required tensions and length of stitch—and not one in ten thousand knew how to adjust a machine rightly when the slightest disarrangement of the mechanism in any of its most simple parts prevent even stitching. In those days the instruction and adjustment departments of sewing machine offices were of the utmost importance, strong in force and kept pretty busy. To maintain and spread the reputation of the sewing machine as an accurate and practical working instrument, destined to supersede hand sewing in a great measure, it was necessary to give full and constant instructions to every customer and to follow up and keep in adjustment for a long time every machine sold. The sewing

> machine was new, and the people did not understand it. It was on trial, as it were, and the world had not quite made up its mind, being as usual slow in coming to such conclusions, as to whether the sewing machine was really a good labor-saving instrument, worthy of endorsement as such, or not. It was necessary for the trade under the circumstances to use unflagging efforts to convince the people that this great innovation in one of the world's universal industries was an article of such simplicity that anyone without special training could run it to perfection and take care of it easily. . . . The fruit of this thorough instruction and adjustment is visible today in the general and accurate knowledge of the management of sewing machines displayed by both sexes. . . . Ladies are by no means scarce who can talk sewing machine with knowledge and intelligence. . . . So general is knowledge of sewing machines that very many women understand how to operate them though they have never owned them. That the sewing machine is no longer a mechanical mystery, but a household instrument which nearly everybody understands, is the great bulwark upon which the trade of the future must rely, and the fact that a vast deal of trouble, expense and advertisement is spared to the trade of the present day is wholly due to the wise foresight of the trade's leaders during the past twenty years who saw that the fate of the business depended upon the people's intimate knowledge of sewing machinery and took care to supply them with it.

What had changed in the sewing machine audience was the acceptance of the technology itself.[49] Thus, perhaps one of the reasons why the percentage of instructions in the sewing machine manuals (see Figure 5.8) took nearly a decade before consistently utilizing 80% for this purpose—a percentage that was never undercut again for 44 years—was that the early manuals had to convince readers as well as instruct them. Technical communication not only had to enable users to operate and use the novel technologies but also had to help to make them willing to adopt the technology in the first place.[50]

Perhaps another reason why it took a decade before sewing machine manuals began emphasizing instructions and ceased devoting large numbers of its pages to making the technology palatable was the time it took the authors to negotiate a clear understanding of their female audiences. Isaac Singer had always considered women to be the primary audience of his company. For example, the front cover of his booklet, "The Story of the Sewing Machine," was draped with the words: "Singer the Universal Sewing Machine, Sold only by the Maker Directly to the Women of the Family."[51]

If women were the target audience of the sewing machine manuals, how did authors initially think women were going to deal with this new technology? One of the most direct comments about how women were supposed to deal with this new technology came in April 1858, the year before the appearance of the Grover and Baker manual. In *I.M. Singer and Co.'s Gazette*, a speech was reprinted that was given before the Shirt Sewers' and Seamstresses' Union in New York City, and the speaker spoke out clearly the prejudices of the time in the following words:[52]

> They [the sewers and seamstresses] might well imagine the disastrous consequences of such a result [the introduction of the sewing machine] to a large portion of the community, and they might also be aware that the sewing machines out of which those houses which manufactured them were reaping such large fortunes, were now diffused to such an extent that ere long it was not improbable that that peculiar branch of industry which exclusively belonged to women—that industry which developed itself in the facile and pliant use of the fingers—would be totally extinguished.

Brandon, in her study of Singer, comments on this speech in the following way:[53]

> The one clear train of thought in this rather muddled tirade is easy to discern; women do not use machines, and will only get work for which no machinery is available. A large part of Singer's subsequent marketing campaigns was specifically directed towards combating this attitude, which was deeply ingrained in the social life and aspirations of the time.

As a consequence, the early writers of sewing machine instruction manuals thought that women, as a sex, were fundamentally unsuited to the working of complex machinery such as the sewing machine.[54] Thus, the words from the "Mr. A." in the short story that began the first instruction manual, Grover and Baker's *A Home Scene; Mr. Aston's First Evening with Grover and Baker's Celebrated Family Sewing Machine Containing Directions for Using* (1859), said perhaps more than was intended about this prejudice: "...I must confess that I was incredulous that you had so good a knowledge of the machine ...."[55]

From 1859 to the early 1880s the authors of the sewing machine manuals felt they had to spend a considerable amount of effort counteracting these female inadequacies with technology and convincing women that they could easily handle such machinery:[56]

> To maintain and spread the reputation of the sewing machine as an accurate and practical working instrument, destined to supersede hand sewing in a great measure, it was necessary to give full and constant instructions to every customer and to followup and keep in adjustment for a long time every machine sold.

However, by August 1885, the Sewing Machine News editorial could proclaim that this incorrect sense of audience was dead:[57]

> Ladies are by no means scarce who can talk sewing machine with knowledge and intelligence. . . . So general is knowledge of sewing machines that very many women understand how to operate them though they have never owned them.

With sewing machines culturally accepted—and with authors successfully coming to understand their audiences—the instruction manual genre in the sewing machine industry was renegotiated and changed.

One of the elements of the early sewing machine manuals that changed in this negotiation process was the presence of testimonials. Testimonials were short product endorsements from individuals with whom the audience could identify. Because the prime market segment for sewing machines was middle-class families, testimonials from lower class seamstresses or upper class famous actresses of the day would not do. Instead, the leading contributors to these testimonials were ministers' wives who were middle class, trustworthy, and domestic.[58] For example, in the first manual studied, Grover and Baker's *A Home Scene; Mr. Aston's First Evening with Grover and Baker's Celebrated Family Sewing Machine Containing Directions for Using* (1859), there were as many pages of testimonials as there were of instructions. Nevertheless, such testimonials in instruction manuals totally disappeared as the instruction manuals moved beyond technology introduction.[59]

Another one of the elements that disappeared as the instruction manuals moved from convincing to instructing is an element not looked at thus far, anthropomorphism—making the nonhuman human or the inanimate animate. Mumford, in *Technics and Civilization*,[60] pointed out that new technologies throughout history have usually had to pass through a phase of anthropomorphism. One historian explained this need for anthropomorphism in technology introductions by noting:[61]

> The familiar dimensions of bodily experience have always provided a reference point for exploring the significance and utility of new and unfamiliar technologies. The social uncertainties created by the introduction of novel, various, and mysterious technologies are reduced and appropriated for a variety of purposes by recasting them in this familiar idiom.

For example, when the sewing machine instruction manual genre standards were being published in the 1880s, electric light technology was also being introduced in America. A humorous example of the use of anthropomorphism used in the electric light's introduction was "electric girls:"[62]

> In 1884 the Electric Girl Lighting Company offered to supply 'illuminated girls' for indoor occasions. Young women hired to perform the duties of hostesses and serving girls while decked out in filament lamps were advertised to prospective customers as 'girls of fifty-candle power each in quantities to suit householders.'

The girls' functioning as servants and hostesses—something familiar to audiences of the 1880s —introduced electric lights—the novel and mysterious technology. The introduction of sewing machine technology was no different and, in fact, used the exact same familiar symbol—the female domestic servant. Thus, in a Wilcox and Gibbs instruction manual written in 1869, the sewing machine was transformed into a "Miss Wilcox-Gibbs:"[63]

> But I have somewhat more to say of Miss Wilcox-Gibbs. She minds her own business, has no company, and is always ready for duty. . . . Miss Wilcox-Gibbs has also a peculiar faculty for making friends. If, on being introduced at a new pace, she happened to meet with a cold reception—as she sometimes does, on account of the prejudices which those in the interest of the Misses Double-thread are active in disseminating—it never disconcerts her, and she never fails to dispel that prejudice on a very slight acquaintance. Her quiet, lady-like, and winning way of receiving and treating strangers, has the effect to put everyone at ease in her company. She is also a great favorite with children. When not otherwise employed, she often amuses them by making frocks and aprons for their dolls; and no matter how roughly they use her, she never gets out of temper, or becomes otherwise unfit for instant service when her mistress calls. She is also on intimate terms with the feeble and the aged, who find in her an agreeable companion and a sympathizing friend.

The 1870 W. G. Wilson Sewing Machine Company's *Directions for Using the New Buckeye Under-feed Sewing Machine*, continued this anthropomorphism in "Suggestions to New Beginners," but focused in on one element of the 1869 "Miss Wilcox-Gibbs," her "peculiar faculty for making friends": "We aim to make every purchaser feel that it [the Buckeye Sewing Machine] is their friend, and that the money invested in it has been wisely expended."[64] By the 1880s, however, this use of anthropomorphism as a technique to introduce sewing machine technology had largely disappeared.

One thing can be observed quite quickly from these few excerpts of early sewing machine manuals: The microcomputer marketing concept of "user-friendly" from the early 1980s and Frude's shocking suggestions implied in her word concerning computers, "The Affectionate Machine" (1983)[65] and *The Intimate Machine*, (1983)[66] both had their analogs over a 100 years earlier in the sewing machine instruction manuals. It is interesting that at a similar early stage of America's acceptance of computers, Charles Collingwood said the following on the national broadcast coverage of UNIVAC 1's prediction of the 1952 election of Stevenson and Eisenhower: [67]

> [UNIVAC] sent me back a very caustic answer. He said that if we continue to be so late in sending him results, it's going to take him a few minutes to find out just what the prediction is going to be. So he's not ready yet with the predictions, but we're going to go to him in just a little while.

The anthropomorphism in the instruction manuals that occurred in the first decade or so of the sewing machine's introduction was accompanied by an anthropomorphic design of the machine itself. Thus, just as the early steam engines had claws, and the test of the steam engine's strength was in "horsepower," so, too, early sewing machines were designed to look like a cherub in 1857, a horse in 1858, and, most amazingly, a squirrel in 1857.[68] Such design antropomorphicism has certainly been echoed in our own computer culture when it "named" and affixed "faces" on Automatic Teller Machines (ATMs) in the 1980s, for example "Harvey Wall-Banker," or "Tillie-the-All-Time-Teller," or when Microsoft felt that it should call its 1995 operating system "Bob" and offer various animated "guides" as operating system aids.

The early 1870s and 1880s did mark a kind of transition in the type of writing used in the sewing machine industry; 10 years elapsed from the publication of the first "convincing and instructing" type manual of Grover and Baker, *Mr. Aston's Evening*, to the advent of the "just instruction" manuals as represented by Finkle and Lyon M'fg Co.'s *Directions for Using Finkle and Lyon M'fg Co's New Sewing Machine, Victor*, the first instruction manual to contain only instructions (see Table 5.4). Furthermore, in the move from "convincing and instruction" to "just instruction," the inclusion of testimonials and anthropomorphism disappeared from the sewing machine manuals because the authors of the sewing machine manuals sensed their audiences no longer needed them. Thus, the genre changed. With various genre elements being initially present and later disappearing, it becomes quite clear that some genres, such as the instruction manual in the sewing machine industry, were fluidly negotiated among audience, authors, and genre capabilities, and change—real change—could occur.[69]

Sewing Machine Instruction Manuals Differed from Mower-reaper Instruction Manuals

In addition to the element of general change in both the sewing machine and mower-reaper manuals, there was also the fact that the manuals in the two industries differed quite a bit. For example, sewing machine manuals began consistently devoting 100% of their pages to instruction beginning with the publication of Finkle and Lyon M'fg Co.'s *Directions for Using Finkle and Lyon M'fg Co's New Sewing Machine, Victor*, 1867–72. Mower-reaper manuals, however, included many more items than instructions right up to the last one in 1915. The earliest mower-reaper manual, Boas and Spangler's *Ketchum's Celebrated Combined Mower and Reaper* (1859), included (see Figure 5.18):

- a half page of names of those farmers who recommended the product (2% of the manual),
- three and a half pages reprinting reports from trials at agricultural fairs (14% of the manual),
- three and a half pages listing and describing Mr. Ketchum's patents and inventions (14% of the manual),
- six and a half pages from the management of Boas and Spangler explaining their product's superiority (26% of the manual),
- ten pages describing other products Boas and Spangler had for sale in Philadelphia (40% of the manual), and
- one page of instructions (4% of the manual).

Moreover, in the final mower-reaper instruction manual studied, International Harvester's *Directions for Setting Up and Operating the McCormick No. 6 and Big 6 Plain Lift Mowers* (1915), a large portion of the manual was still devoted to communicating items other than instruction (see Figure 5.18). There were seven pages of instructions (63%), and four pages giving the price list of repair parts and pictures with part numbers (36%).

Now although there may be a number of ways to explain why the instruction manuals in the two industries deviated from each other, three seem preeminent: (a) perhaps each industry had a different idea of what the genre was, (b) perhaps the learning style afforded by the two technologies caused the variations of the genre, or (c) perhaps the variations in the genres reflected different social relations of men and women in the 19th century.

These possibilities will be considered in turn.

1. The authors of sewing machine instruction manuals worked within a genre, whereas the authors of mower-reaper manuals did not.

One of the reasons, perhaps, for the early divergence of the instruction manuals for the two industries and their later increasing similarity was that each industry identified its written work differently. Only one sewing machine manual was missing the words "Directions" or "Instructions" on its front cover or in its page heading (Perry Mason and Co., 1899, see Table 5.6). On the other hand, none of the mower-reaper manuals, other than Walter

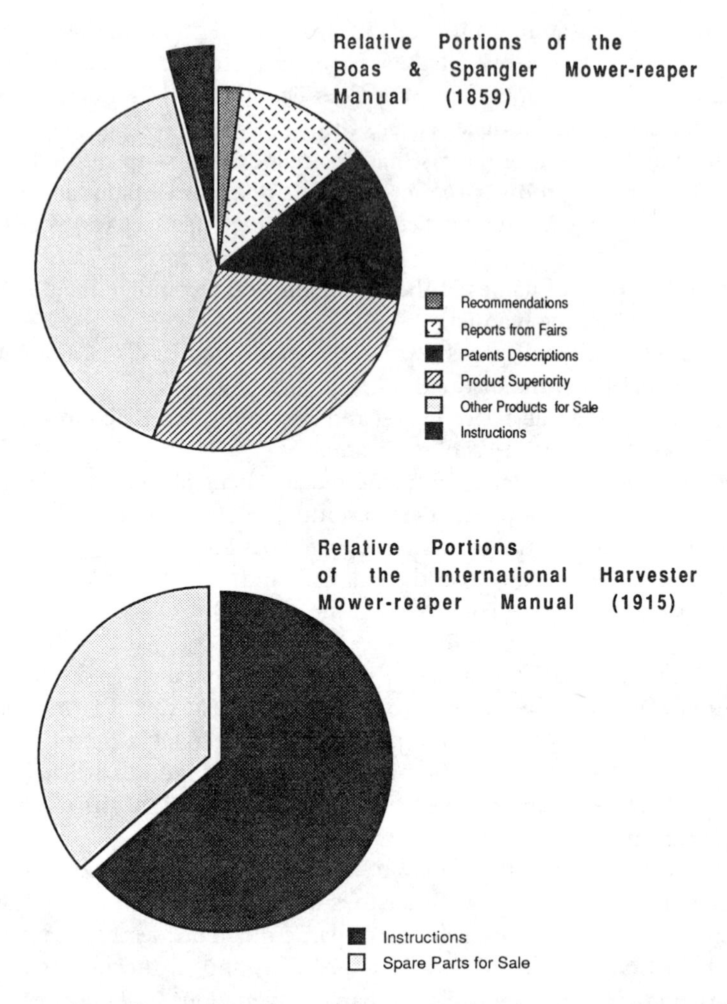

Figure 5.18. Consistently large portions of mower-reaper manuals devoted to content other than instructions

Wood's manual (1873), carried the label "Directions" or "Instructions" until the 1890s (see Table 5.6). Most of the time nothing was on the mower-reaper covers to identify its genre. Twice, however, in 1864 and 1867, a manual was identified as a "circular" by the Alzirus Brown Co., yet a "circular" did not define much that was specific. Then as now, a circular was a business advertisement printed in large numbers for distribution, and the sewing machine industry also had circulars and catalogs which they self-consciously differentiated by design and purpose from their instruction manuals. Evidently the Alzirus Brown mower-reaper company did not so carefully differentiate between types of documents. Hence, if so many of the mower-reaper manuals had no labels on their covers identifying their genre, what did the the mower-reaper companies consider them to be?

Perhaps one way to explain the variations in Table 5.6 and the varied parts of mower-reaper manuals in Figure 5.18 was that the mower-reaper manuals were "pre-paradigmatic" instruction manuals and the sewing machine manuals were "paradigmatic" self-proclaimed instruction manuals. Paradigms are sets of implicit understandings shared by groups of people that order their behavior. Evidently the sewing machine authors must have shared some kind of implicit understandings of instruction manuals after the 1880s in order to produce such roughly similar manuals. However, the variation of content in the mower-reaper manuals between companies continued for quite some time. Thomas Kuhn in his *Structure of Scientific Revolutions* described pre-paradigmatic periods as periods of random fact-gathering:[70]

> In the absence of a paradigm or some candidate for paradigm, all of the facts that could possibly pertain to the development of a given science are likely to seem equally relevant. As a result, early fact-gathering is a far more nearly random activity than the one that subsequent scientific development makes familiar.

Pre-paradigmatic museums contain, for instance, a collection of bubble gum wrappers adjacent to dinosaur bones and china dolls. Pre-paradigmatic scientific journals such as the one Oliver Evans read in the 18th century, Philadelphia's *Transactions of the American Philosophical Society*, contained articles speculating on the nature of electricity juxtaposed to those detailing apparitions seen on Caribbean islands. Without a paradigm to govern what was collected and what was not, everything was collected.

Perhaps, then, the reason for the wide variety of materials in the mower-reaper manuals was that they were examples of a pre-paradigmatic genre. Without a common genre label affixed to their covers or inside pages, could the authors or the audience know what to include or expect? Perhaps the only guidance mower-reaper manual authors had in choosing what to include was to include everything—like the pre-paradigmatic scientists, journals, and museums. The mower-reaper manuals did, indeed, include everything: testimonials, spare parts ordering methods, patent defenses, reports of how well the apparatus performed at county fairs, as well as instructions.

On the other hand, sewing machine manuals seem to have very quickly attained a sense of a genre paradigm for instruction manuals and to have maintained it through the course of the 60 years studied. From the beginning, sewing machine manual covers declared that they were "instructions" or "directions" and that was exactly what they delivered with progressively better and better focus. Additionally, with such a set paradigm, they were able eventually to criticize and compare manuals in their professional journals.

Table 5.6. Genre Labels on the Covers of Sewing Machine and Mower-reaper Manuals (X denotes presence of the header; — denotes absence of the header).

Sewing Machine Company Name	Date	Self-Identified on Cover or Running Header as "Directions" or Instructions	Self-Identified on Cover or Running Header as "Circular"	Date	Mower-reaper Company Name
Grover & Baker	1859	X	—		
		—	—	1857	Boas & Spangler
Singer Sewing Machines	1862	X	—		
		—	X	1864	Alzirus Brown
Howe Machine Co.	1867	X	—		
		—	—	1865	Adriance, Platt & Co.
Singer Sewing Machines	1868	X	—		
		—	—	1865	Williams, Wallace & Co.
Finkle & Lyon M'fg Co	1869	X	—		
		—	—	1865	Nixon & Co.
W. G. Wilson Sewing Machine Co.	1870	X	—		
		—	—	1867	Adriance, Platt & Co.
Davis Sewing Machines	1876	X	—		
		—	X	1867	Alzirus Brown
New American Sewing Machine	1878	X	—		
		X	—	1873	Walter A. Wood Mowing & Reaping Machine Co.

Table 5.6. Genre Labels on the Covers of Sewing Machine and Mower-reaper Manuals (X denotes presence of the header; — denotes absence of the header) (con't).

Sewing Machine Company Name	Date	Self-identified on Cover or Running Header as "Directions" or Instructions	Self-Identified on Cover or Running Header as "Circular"	Date	Mower- reaper Company Name
Florence Sewing Machine Co.	1878	X	—		
		—	—	1879	Eureka Mower Co.
A. G. Mason Mfg. Co.	1880	X	—		
		X	—	1890	Walter A. Wood Mowing & Reaping Machine Co.
Domestic Sewing Machine Company	1880	X	—		
		X	—	1891	Wm. Deering & Co.
Perry Mason & Co.	1899	—	—		
		X	—	1909	International Harvester Co. of America-Champion
Wilcox & Gibbs Sewing Machine Co.	1907	X	—		
		X	—	1910	International Harvester Co. of America-Deering
Singer Sewing Machines	1914	X	—		
		X	—	1911	International Harvester Co. of America-Deering
		X	—	1915	International Harvester Co. of America-McCormick

When agricultural manuals such as those by Wood, Deering, and the International Harvester Company in the 1890s began to print "directions" or "instructions" on their covers and inside pages, these companies were not just affixing ink to paper, but consciously taking on the paradigm of the instruction manual genre pioneered in the sewing machine industry. Consequently, because the two industries both used this paradigm in the 1890s, the characteristics of the two type of manuals converged:

- The right of the lines in Figure 5.14 charts the two sets of manuals indicating that the-percent-of-graphics-for-instruction are much more tightly together later in time;
- The right of the lines in Figure 5.15, shows the sets of manuals again much more tightly together when they achieved similar Reading Ease scores later in time;
- The right of the lines in Figure 5.16 shows the size of paragraphs in the two sets of manuals to be much tighter later in time; also
- Tables 5.4 and 5.5 shows the more universal use of correlative illustrations in both sets of manuals later in time.

When the two sets of manuals converged in a number of ways toward the end of the 60-year time span studied, the mower-reaper manuals became more like the sewing machine manuals rather than vice versa. This would make sense considering the role of paradigms. Given the fact that mower-reaper manuals lacked a coherent understanding in their industry as to what to include and not to include, it would make sense that a group that does have some guiding principles for a genre and an established genre paradigm should provide the framework and the paradigm for the one that lacks it. Thus, the instruction manuals of the two industries may have initially diverged not so much because of the varied technologies or the varied audiences, but because of varied ideas of what the genre of instruction manuals was.[71]

2. Sewing machine manuals differed from mower-reaper manuals because of the variation in learning styles afforded by their technologies.

Perhaps another reason why the types of instruction manuals diverged between industries was because the variations of the manuals responded to the differences in learning styles afforded by their respective technologies. Recall the repeated presence of ancillary illustrations, and the consistently higher proportion of illustrations in all the mower-reaper manuals. Could it be that the farmers learned to assemble and use the mower-reaper with fewer directions because the mower-reapers were easier to use? Or, could they have learned to use the mower-reapers more easily because they were more like previous farm implements than sewing machines were like previous sewing instruments?[72]

Now there may be some truth in this observation because the sewing machine had many of its gears and connections hidden from the operator inside of the machine cover. The mower-reaper, on the other hand, had all of its gears in plain view (compare Figures 5.1 and 5.2). Thus, the mower-reaper operators could more easily get visual feedback on their actions and mistakes than could the sewing machine operators. This directness or hiddenness of the technologies probably called for two different types of cognition, experiential and reflective.

Donald Norman in *Things That Makes Us Smart*[73] described the difference between these two types of cognition:

> The essence of expertise is knowing what to do, rapidly and efficiently. The pilot pushes the throttles forward, controlling the nosewheel and rudder to keep the plane on the runway. . . . All this is done with practiced ease and skill, continually integrating numerous sources of information. . . .These are examples of experiential cognition. The patterns of information are perceived and assimilated and the appropriate responses generated without apparent effort or delay. . . . It appears to flow naturally, but years of experience or training may be required to make it possible. . . .Something happens in the world, and the scene is transmitted through our sense organs to the appropriate centers of mental processing. But in the experiential mode, the processing has to be reactive, somewhat analogous to the knee-jerk reflex.
>
> Reflective thought is very different from experiential thought, even when both are applied by the same people in similar situations. To see this, consider once again our pilots in their cockpit. . . .Suppose some decision making is required. Suppose the pilots have to plan. Now the situation calls for reflection. Reflective thought requires the ability to store temporary results, to make inferences from stored knowledge, and to follow chains of reasoning backward and forward, sometimes backtracking when a promising line of thought proves to be unfruitful. This process takes time. Deep, substantive reflection therefore requires periods of quiet, of minimal distraction.

If sewing machines had much of their mechanisms hidden, direct demonstration or feedback would not have been available. Consequently, sewing machine operators would indeed have had to plan, backtrack, and follow chains of inference. Reflective cognition was perhaps the best way to learn how to use this technology and best served by textual information presented in books. On the other hand, if the gearing of the mower-reapers was more easily seen, and if the reactions to operator activity were much more obvious, then perhaps the knee-jerk, reflex learning style—the experiential learning style—worked best for the farmers.

Historians point out that farmers had many means for finding out about new technology:[74]

> In making his decisions, the midwestern farmer could look for aid in a variety of directions. A number of agencies with varying degrees of impersonality sought to cater to his interests. These were the local press, the agricultural press, the agricultural exhibitions and—of little importance prior to 1880—the agricultural colleges.

Yet, historians also pointed out that farmers were suspicious of book farming:[75,76]

> Especially for those who are not students by training or habit, word of mouth and visual demonstration may well be the most effective method of instruction. Again, the attitude is not wholly extinct that "deep plowin' an' shaller drinkin" is all the book larnin' a farmer needs to know.

Thus, it could have been that the experiential learning for mower-reaper audiences took place, in many cases almost by subterfuge, at trials and demonstrations at the County Fairs and not in the instruction manuals:[77]

> During those decades [1850-70] few other agencies existed to instruct the farmer, and certainly the societies and their fairs were the only means by which large numbers of ordinary farms might be contacted directly. During those years also, the fairs were more educational in nature, with less of the horse racing and carnival atmosphere that later would be the dismay of farmers.

It should come as no surprise, therefore, to find reports of such agricultural fair field trials featured prominently in the mower-reaper instruction manuals. For example, the reports of the agricultural fair field trials were 300% longer than the instructions in *Ketchum's Celebrated Combined Mower and Reaper* (1859) and 200% longer than the instructions in the Adriance, Platt and Co.'s *Buckeye Mower and Reaper* (1865).

However, demonstrations were not given only to the mower-reaper audiences. The sewing machine companies also used demonstrations:[78]

> Many [sewing machine manufacturers] Salons hired pretty girls to demonstrate machines. When a prospect came into the shop, a salesman would escort her to the machine, where a girl would put on a dazzling show. The salesman stayed nearby and at the proper moment, invited her to sit at the machine and see how easy it was to operate.

In fact, in the September 15, 1879 issue of *The Sewing Machine Advance*, a letter to the editor recalled the early astonishment of the American public to such demonstrations:[79]

> The writer can remember in the year 1853, having seen a sewing machine on exhibition in New York City in one of the windows on Broadway, which was operated by a young lady, to the great astonishment of the wondering crowd, who blocked up the sidewalk in order to get a view... [80]

Often the demonstrations in the sewing machine salons were followed up by one-on-one instruction: "Full instructions for operating the machine are given to purchasers gratuitously at the sales-rooms.. . . . Many ladies have the machine sent to their residences after having received a half hour's instruction."[81]

Moreover, the short story initiating the first sewing machine instruction manual by Grover and Baker described how Mrs. Aston received such demonstration instruction:[82]

> It was purchased, and Mrs. Aston sat down to receive instructions for the use thereof. An hour sufficed to accomplish this; and with a gratified smile, she returned home in time to welcome her little assistant, which was to do so much for her relief.

Evidently, however, the demonstration method of transmitting information about sewing machines was never deemed wholly sufficient for the sewing machine audience like it was for the mower-reaper audience. If the demonstration method were wholly effective on its own, why were the sewing machine manuals' sets of instructions so extensive? Why spend so much money on the illustrations for the sewing machine instructions? Why not have only minimal written instructions as presented in the mower-reaper manuals? Could the answer to these questions be that the reflexive learning required to use a sewing machine—a machine that hid its operation from the eye—was not effectively learned by visual demonstration alone? Could it be that for sewing machine audiences using reflexive knowledge, the demonstration method of instruction had to be supplemented by extensive written instructions?

Two audiences with two different learning styles arising from two different technologies produced dramatically different types of manuals with different contents and different styles. Nevertheless, there still may be a third reason why these two sets of manuals diverge from one another.

3. *The sewing machine manuals differed from the mower-reaper manuals because instruction manuals embody the social relations of women.*

One agricultural implement manual that I did not include in the original statistical experiments was written in 1891, *Directions for Setting up and Operating Buckeye Banner Binder* by Aultman Miller and Co. I did not include it earlier because the manual was written for a binder attachment to a mower-reaper rather than for mower-reaper like the others.[83] However, like the mower-reaper manuals of the 1890s (see Table 5.6), this binder manual had the genre identification of "directions" on its cover. More importantly, there was a picture on the cover depicting the cross-fertilization of the instruction manual genre between the sewing machine and mower-reaper industries (see Figure 5.19). An illustration in the center of the cover showed a scene probably taking place between the house and the barn on a typical farm, and three individuals were identified in the captioned dialogue below the picture:

> Ma.—Why it's not as complicated as our Sewing Machine.
> Pa.— It's so simple that a child could set it up.
> Ma.—Bring the horses, Tommy. Pa's got it all set up before the expert got here.

In the foreground of Figure 5.19 stands the male farmer, "Pa," with the completely assembled binder to his right. His left hand holds a wrench, his back is to the audience, and his head is turned to regard his wife at the left. In the midground to the left is a woman of the same age as the farmer, "Ma," wearing glasses, sitting on a stump, and looking up from a book. In the background, Tommy is shown approaching from the barn holding two horses by their bridles. The barn and distant fields form the backdrop for the entire scene.

Much about the genre of instruction manuals is communicated iconically by the elements of this barnyard drama and the captioned dialogue. Nothing like this appeared in any other manual of either industry. Interestingly, this picture just happened to be printed on a cover of a farmer implement instruction manual at the time that the manuals of the two industries were converging in a single understanding of the instruction manual genre, as well as just at the time mower-reaper manuals had begun identifying themselves as "instructions" or "directions." This manual, like only five other mower-reaper manuals, has the word "directions" printed on its front cover. Thus this 1891 binder manual was published just as the genre was going through a transition of understanding in the mower-reaper industry, and the picture probably captures the audience reaction to this transition—that of rejection.

Figure 5.19. Front cover cartoon from *Directions for Setting up and Operating Buckeye Banner Binder*, by Aultman Miller and Co (1891)

On the front cover the binder is shown already fully assembled, implying that before a single page is turned, Pa can claim victory over the novel binder technology: "It's so simple that a child could set it up." He, figuratively, did not need to turn the page. In fact, the picture suggests Pa has completed his task without the need to even read a manual: He has no glasses on and only holds a wrench. Thus, this new "directions" manual suggested on its own front cover that an instruction manual was superfluous: Pa did not need it. If the binder owner was a male farmer, there was probably a good chance he would identify with Pa and, thus, take on his disdain for "direction." Even if the audience were much younger, perhaps a farmer's son like Tommy, they too would not need to read a manual. With his two horses in hand, Tommy is in no position to read a manual nor is one offered to him. Who reads the manual?

Ma is the one shown very prominently in deep concentration reading an instruction book just as she probably would have read a schoolbook. She is not in the middle of a physical action like her husband: She is wearing glasses and sitting. She seems almost to embody Norman's reflexive cognitive mode ("Deep, substantive reflection therefore requires periods of quiet, of minimal distraction").[84] Ma's whole attitude could be appropriate if the manual were for her sewing machine inside the house because, as discussed earlier, the hiddenness of the gearing in a sewing machine made feedback on operating decisions more complicated. Thus, she could exclaim: "Why it's not as complicated as our Sewing Machine." Other alternative interpretations of Ma and her book could be:

- she has been reading the binder manual just out of curiosity and thus noted that the binder was "not as complicated as our Sewing Machine," or
- she has been reading the binder assembly directions to Pa as he puts it together.

In either case, the important fact to note is that it is Ma, not Pa, who is reading.

Nevertheless, something else is being conveyed in the picture in addition to the fact that Pa, possibly embodying experiential cognition, does not need manuals, and Ma, embodying reflexive cognition, does. The artist who produced this picture has positioned Pa with his back to the viewer and his eyes turned into the plane of the picture, into his binder, barnyard, and house. It is almost as if the artist portrayed Pa in a stance rejecting anything to do with outside experts. In contrast, Ma's side is shown in the picture, suggesting that her eyes are taking in the barnyard, her husband, and house, as well as things beyond the plane of the picture—the "beyond" where outside experts exist. What is the identity of the outside experts whom Pa is refusing to acknowledge and for whom Ma is half open? Might they not be the authors of this very binder manual whom Pa has ignored and made superfluous by assembling the binder before the "experts" arrived? Might not the rejected "expert" be the very author of a new transitional binder manual that has the cover label of "directions" (one should remember that *Oxford English Dictionary* defines the word "directions" as "authoritative guidance")?[85]

Thus, in addition to representing iconically the two modes of cognition and the relative appropriateness of reading manuals for each, the picture also represents the acceptance and lack of acceptance of authority rhetorically vested in "directions." All the imperative mood verbs and all the other ways in which "directions" and "instructions" address audiences embody an asymmetric relationship of "experts" and "novices," or those with power and those without.[86] Thus, it was not on the first page of the *Buckeye Mower and Reaper* manual (1865) but on the W. G. Wilson Sewing Machine Co. sewing machine manual, *Directions for Using the New Buckeye Under-feed Sewing Machine* (1870), that the following accusatory pointing finger and large bold letters appeared:

☞ You Cannot use the Machine until you thoroughly. understand the Directions.

The readers of mower-reaper manuals, like "Pa," accepted agricultural implement manuals with their preparadigmatic random collection of material, and the relatively large reliance on demonstrations at fairs or on ancillary graphics open to multiple interpretations and representing reality uninterpreted by another's eyes, because those manuals did not force Pa into a subservient relationship so characteristic of the role assigned to 19th-century women readers of the sewing machine instruction manuals. Women readers of the sewing machine instruction manuals had to take on a rhetorical role that would accept disembodied pointing hands (☞) demanding they refrain from using a machine.

Although Ma and Pa's further discussion is long over, their social relations can be examined by this relative rejection or acceptance of "experts." Thus it is possible today to reexamine Ma's and Pa's 1890 barnyard conversation by looking at the interactional research on the contemporary conversations of men and women reported by Deborah Tannen (1990). Tannen explained the asymmetry of male/female conversation roles in the following way:[87]

> In all these examples, the men had information to impart and they were imparting it. On the surface, there is nothing surprising or strange about that. What is strange is that there are so many situations in which men have factual information requiring lengthy explanations to impart to women, and so few in which women have comparable information to impart to men. The changing times have altered many aspects of relations between men and women. Now it is unlikely, at least in many circles, for a man to say, "I am better than you because I am a man and you are a woman." But women who do not find men making such statements are nonetheless often frustrated in their dealings with them. One situation that frustrates many women is a conversation that has mysteriously turned into a lecture, with the man delivering the lecture to the woman, who has become an appreciative audience. Once again the alignment in which women and men find themselves arrayed is

> asymmetrical. The lecturer is framed as superior in status and expertise, cast in the role of teacher, and the listener is cast in the role of student. If women and men took turns giving and receiving lectures, there would be nothing disturbing about it. What is disturbing is the imbalance. Women and men fall into this unequal pattern so often because of the differences in their interactional habits.

In a follow-up study, Tannen examined how nonexpert men reacted to men who acted as if they were experts:[88]

> Furthermore, when an expert man talked to an uninformed woman, he took a controlling role in structuring the conversation in the beginning and the end. But when an expert man talked to an uninformed man, he dominated in the beginning but not always in the end. In other words, having expertise was enough to keep a man in the controlling position if he was talking to a woman, but not it he was talking to a man. Apparently, when a woman surmised that the man she was talking to had more information on the subject than she did, she simply accepted the reactive role. But another man, despite a lack of information, might still give the expert a run for his money and possibly gain the upper hand by the end.

Following the authority of outsiders, of "experts," of disembodied pointing hands was acceptable for women, the graphic on the front cover of the binder manual seems to say. However, it was not equally acceptable for men or boys. After all, as noted in *Knowledge is Power: The Diffusion of Information in Early America, 1700-1865*:[89]

> Though women's sphere was firmly domestic, women were not to rule the home. Political power, whether by the hearth, at the town-meeting, the county court, or the capital—and even in the parish—was intended exclusively for men.

Men and boys were suppose to figure out novel technology for themselves. They were suppose to be able to just look at the ancillary pictures in the books freed up from the text, or see how to operate the machine at the county fair trial or demonstration.[90] Moreover, if they did learn at the county fair trial or demonstration, they would surely be able to undercut the asymmetric authority role of the mower-reaper demonstrator in myriad ways by using the mower-reaper demonstration trials to place bets and thus changing the focus to gambling and off the demonstrator's authority, stumping the out-of-town mower-reaper demonstrator with odd and unusual bits of local lore, or simply walking away feigning disinterest.

There are yet others way to explain the male farmer's rejection of the instruction manual's outside experts with their asymmetric address. Perhaps the "gender schema" embodied by the instruction manual genre did not fit the internalized "gender schema" of the male readers, thus the male farmers rejected it.[91] However, a more intriguing interpretation using the concept of gender schema and reading may be that the asymmetry of the instruction manual as devised by the sewing machine industry was equally troubling to both male and female readers; however, the female readers had some ways of dealing with such troubling material whereas the male readers at that time did not. For example, Elizabeth Flynn (1986) made the following observation about how men read differently than women:[92]

> The study does suggest, however, that male students sometimes react to disturbing stories by rejecting them or by dominating them, a strategy, it seems that women do not often employ. The study also suggests that women more often arrive at meaningful interpretations of stories because they more frequently break free of the submissive entanglement in a text and evaluate characters and events with critical detachment.

Thus, perhaps women readers of the 19th century, when presented with the disturbing asymmetry of the sewing machine instruction manuals, had the reading techniques to distance themselves from the submissive entanglements of the manuals' rhetoric while at the same time apprehending the content of the manuals.[93] In the same way, perhaps men were unable to apprehend the content because they lacked these reading techniques, and that interfered with their apprehension of the content.

Lubar had earlier noted that technology is not neutral but rather "is the physical embodiment of social order, reflecting cultural traditions. It is part culture, and as such it mediates social relationships."[94] Moreover, Bazerman and Paradis more pointedly suggested that: "Writing is more than socially embedded: it is socially constructive. Writing structures our relations with others and organizes our perceptions of the world."[95] In the same way, the rhetoric of author and audience relations in the documents surrounding a technology, such as these instruction manuals, also mediate and embody social relationships. Thus, perhaps the deepest and most underlying reason why the experiments examining the two sets of manuals showed a difference over and over again between the mower-reaper manuals intended for men and the sewing machine manuals written for women was that the writers understood men's need to reject the subservient interactional role assigned them in "instructions" or "directions."[96] When the mower-reaper manuals of the 1890s, therefore, began to label themselves as "directions," an event in technical communication history happened that was much more than the affixing of ink on a page. It was the beginning of the attempted conversion of men to a rhetorical role of subservience formerly only assigned to women. Hence, the picture in Figure 5.19 records the men's reaction to being assigned such a new and disconcerting role: The reaction was rejection, turning one's back on the expert.[97] (Male audiences continued a rear-guard rhetorical effort even after the 1890s to combat feminine 19th-century roles assigned in this genre. As described in Chapter 8, the instruction books for IBM punched-card technology of the 1920s—1940s were termed "Principles of Operations," not "instruction" or "direction" manuals. Even today, Macintosh and Microsoft instruction books are still called "user manuals," not "instructions" or "directions.")

Thus, this central genre of technical communication—the instruction manual—was created to support the first factory-made technologies. The first audience to be addressed by instruction manuals were women. Moreover, the interactional roles assigned to women by men in the 19th century created the rhetoric of this central genre of technical communication.[98]

ENDNOTES

1. An earlier version of this chapter was published as a conference paper: "Quality is Relative: Quality as a Function of Historical Paradigm." In *Proceedings of University of Waterloo's First Symposium on Quality and Software Documentation, October 1991*. Waterloo, Ontario, Canada.
2. Brandon, Ruth: *A Capitalist Romance: Singer and the Sewing Machine*. Philadelphia: J. B. Lippincott Company, 1977: 116; Danhof, Clarence: *Change in Agriculture*. Cambridge: Harvard University Press, 1969: Chapter 3, "Sharing and Expanding the Fund of Knowledge," as well as Chapter 8 and 9; Rikoon, J. Sanford: *Threshing in the Midwest, 1820–1940: A Study of Traditional Culture and Technology Change*. Bloomington, IN: Indiana University Press, 1988; and David. Paul A.: "The Mechanization of Reaping in Antebellum Midwest." In *Technical Choice Innovation and Economic Growth*. Cambridge: Cambridge University Press 1975: 195–233. However, the break from dependency on word-of-mouth instruction seems to have also occurred at an earlier time: "The content of these technical books suggests that they existed, in many multiple editions, because women read them to learn more about skills they needed in everyday living. After the middle years of the sixteenth century, the oral tradition as a means of transmitting information and skills had become insufficient as knowledge increased and could spread more rapidly by the written text." (Tebeaux, Elizabeth and Lay, Mary: "Images of Women in Technical Books from the English Renaissance." *IEEE Transactions on Professional Communication* 35(4) (December 1992): 197.)

3. Danhof, ibid., 1969: 228.
4. Olmstead, Alan L., Paul Rhode, Carolyn Lawes Brown, Philip Crowe, and Michael Hayes: *A Survey of the Development of California Agricultural Machinery*. Davis, CA: Agricultural History Center, 1981:79–80
5. Finkle and Lyon M'fg Co.: *Directions for Using The Finkle and Lyon M'fg Co's New Sewing Machine "Victor."* location unk., 1867–72.
6. Now most frequently renamed the "user manual."
7. Smith as cited in Gresham, Stephen L.: "Harvesting the Past: The Legacy of Scientific and Technical Writing in America." In *Proceedings of the 24th International Technical Communication Conference, 1977* Washington, DC: Society for Technical Communication, 1977: 306.
8. (vols. 1–35). Chicago, IL, 1879–1913.
9. (vols. 1–16). New York, 1877–1893.
10. (vols. 1–38). New York, 1882–1924
11. For example, at the preeminent national gathering of technical communicators in 1956 (Association of Technical Writers and Editors [TWE] and the Society of Technical Writers [STW]), only one of 20 papers was authored or co-authored by women. At the 1987 Society for Technical Communication convention, 200 of the 312 presentations were authored or co-authored by women.
12. This observation has been cross-checked at the collections of the Textile Division, National Museum of American History (Smithsonian), the Hagley Archives and Library (Wilmington, DE), and the Henry Francis Du Pont Winterthur Museum Library (Wilmington, DE).
13. The first mower-reaper manual in the Smithsonian and in the Agricultural History Center, University of California, Davis, is Boas and Spangler's *Ketchum's Celebrated Combined Mower and Reaper*, published two years earlier in 1857. This observation has been cross-checked at the collections of the Agricultural Division, National Museum of American History (Smithsonian), the Hagley Archives and Library (Wilmington, DE), the Henry Francis Du Pont Winterthur Museum Library (Wilmington, DE), and the Agricultural History Center, University of California, Davis.
14. Grover and Baker: *A Home Scene; Mr. Aston's First Evening with Grover and Baker's Celebrated Family Sewing Machine Containing Directions for Using*. Boston, MA, 1859: 3.
15. Grover and Baker, ibid., 1859: 6.
16. New York: Oxford University Press, 1964.
17. Wilder, Laura Ingalls: *These Happy Golden Years*. New York: Harper and Row, Publishers, 1953: 242–3.
18. *New York Tribune* article quoted in *Wheeler and Wilson Manufacturing Company's Sewing Machines*. Bridgeport and Watertown, CT, 1859 sales brochure: 16.

19. Letter to the editor of the *Independent* quoted in *Wheeler and Wilson*, ibid.: 15. One might also note that Karl Marx in *Das Kapital* discussed the death of Mary Anne Walkley, a milliner, who died of overwork by handsewing before the advent of sewing machines.
20. A very interesting comparison could be made with these 1880s descriptions of instructional manual qualities and the 1990s descriptions of instructional manual qualities. See Teklinski, William J.: "A Style Analysis of Winning Entries in the International Technical Publications Competition." In *1992 Proceedings of the 39th Annual Conference, May 10–13, 1992*. Arlington, VA: Society for Technical Communication, 1992: 337–340.
21. "A Handsome Instruction Book." *Sewing Machine Advance* 4(10) October 15, 1882: 184
22. "The New 'New Home Instruction Book'." *Sewing Machine News* 7(9) September 1885: 7.
23. "The 'Standard' Instruction Book." *Sewing Machine Advance* 8(6) June 1886: 9.
24. "Instruction Dealers and Agents Needed." *Sewing Machine News* 8(8) August 1886: 8.
25. Because these aspects are now quite often the province of the printers rather than the technical communicators themselves, this element of genre expectations will not be further investigated.
26. Deller, Anthony William: *Deller's Walker on Patents* (Vol. 5). Rochester, NY: Lawyer's Co-operative Publishing Co., 1972: 116–7.
27. For a description of the varied type of text and graphics relationships, see Pegg, Barry: "Two-dimensional Features in the History of Text Format: How Print Technology Has Preserved Linearity." *The Technical Writing Teacher* 17(3) Fall 1990: 223–42.
28. Lubar, Steven: "Machine Politics: The Political Construction of Technological Artifacts." In (Steven Lubar and W. David Kingery, eds.) *History from Things: Essays on Material Culture*. Washington, DC: Smithsonian Institution Press, 1993: 207.
29. "Introduction" In (Charles Bazerman and James Paradis, eds.) *Textual Dynamics of the Professions*. Madison, WI: University of Wisconsin Press, 1991: 3. See also Paradis, James: "Text and Action: The Operator's Manual in Context and in Court." In *Textual Dynamics of the Professions*—"The operator's manual [or in this chapter, the instruction manual] is a conceptual framework that infuses human purpose into mechanical devices or their equivalents, thus aligning the neutral products of technology with the value-laden ends of society." (258)
30. The use of the Flesch Reading Ease scale in no way suggests the author's conceptual support of this formula as a guide to effective writing; see Selzer, Jack: "What Constitutes a 'Readable' Technical Style?"; and

Huckin, Thomas N.: "A Cognitive Approach to Readability." In (Paul V. Anderson, R. John Brockmann, and Carolyn R. Miller, eds.) *New Essays in Scientific and Technical Communication: Research, Theory, Practice*. Farmingdale, NY: Baywood Publishing Co., 1983. However, the scale does offer a quantitative and widely known method of measuring "close and terse language." (For a detailed discussion of the formula and its use, see Rudolf Flesch: *The Art of Readable Writing*. New York: Harper and Brothers Publishers, 1949.) The Flesch Reading Ease scale cross references average sentence length with the number of syllables in a 100-word passage. The average sentence length for each of the manuals was determined by averaging the length of three sentences randomly chosen from the beginning, middle, and end of the instruction portion of the manuals. The 100-word passage from which the number of syllables was computed was again chosen from the instruction portion of the manuals. The scale itself runs from 1 to 100 with 1 denoting writing that is "Very Difficult" and 100 denoting writing that is "Very Easy."

31. Pegg, op. cit., 1990: 223–42. See also Brockmann, R. John: "Teaching Nonverbal Technical Communication in the 21st Century Through Emulation." In *Proceedings of the 43rd International Technical Communication Conference, May 5–8, 1996* Arlington, VA: Society for Technical Communication, 1996: 8-13.
32. See Cooper, Grace Rogers: *The Sewing Machine: Its Invention and Development*. Washington, DC: Smithsonian Institution Press, 1976, "Appendix VI. 19th-Century Sewing-Machine Leaflets in the Smithsonian Collections."
33. The elements of sewing machine design had been established a full decade earlier, but initial mass production and mass marketing were delayed by numerous patent infringement suits. The sewing machine that was delivered to the public in the late 1850s needed elements from a variety of patent designs, and it was only with the creation of the "Sewing-Machine Combination," which allowed for the licensing and pooling of multiple patent designs, that the legal groundwork for the mass marketing and mass production of sewing machines was achieved in 1856. See Cooper, op. cit., 1976. However, there were also social reasons for the delay of the sewing machine's introduction. Fenster (Fenster, J. M.: "Seam Stresses." *American Heritage of Invention and Technology*. 9(3) (Winter 1994): 44) also suggested that "social conditions were still [this was 1845] not right for a sewing machine, but they were becoming more favorable. Thimonnier had been ruined by the rage of French tailors and seamstresses [1830]; Hunt and his small circle had been manipulated (for the noblest reasons) by advocates [in 1834 by Mathew Carey who had earlier championed Oliver Evans in 1800]. But in the modern year of 1845, both of these considerations were fading. Machinery in general was

established in so many industries that there was less fear of it than there had been ten years before (though the sentiment persists to this day)."

34. Lemmer, George F.: "Farm Machinery in Ante-bellum Missouri." *Missouri Historical Review* 40 (1946): 473. As with the sewing machines, it was legal delays in settling the patent ownership problem that delayed introduction of mower-reapers until the late 1850s. With the opening of better markets and prices bolstered by the demands of the Crimean War, interest in agricultural technology picked up briskly after 1850. See Paul A. David: "The Mechanization of Reaping in the Ante-bellum Midwest" In (Paul A. David, ed.) *Technical Choice, Innovation, and Economic Growth.* London, 1975: 195–231) also observed that the tradeoff between the cost of purchasing the reapers and the cost of human laborers using scythes occurred when farmers began farming more than 46.5 acres. Furthermore, this tradeoff increasingly occurred with the movement of American cultivation to the the cheap, fertile lands of the west in the 1850s. Finally, the opening of better markets, the achievement of higher prices, and the movement of farming west also all coincided with the expansion of railroads in the 1850s. (Basalla, George: *The Evolution of Technology*. Cambridge: Cambridge University Press, 1988: 153.)

35. The average sentence length and number of syllables for each of these pamphlets that determined the Flesch Formula Score were:

Date	Company	Average Length of Sentences in Words	Syllables Per 100 Words
1878	New American Sewing Machine	31	123
1878	Florence Sewing Machine Co.	54	125
1880	A. G. Mason Mfg. Co.	32	124
1880	Domestic Sewing Machine Co.	41	126

36. See previous footnote concerning the mechanics and use of the Flesch Reading Ease scale in this chapter. The average sentence length and number of syllables for each of these pamphlets that determined the Flesch Formula Score were:

Date	Company	Average Length of Sentences in Words	Syllables Per 100 Words
1873	Walter A. Wood Mowing & Reaping Machine Company	16	116
1879	Eureka Mower Co.	19	130
1890	Walter A. Wood Mowing &. Reaping Machine Company	28	122

37. See previous footnote concerning the mechanics and use of the Flesch Reading Ease scale in this chapter. The average sentence length and number of syllables for each of these pamphlets that determined the Flesch Formula Score were:

Date	Company	Average Length of Sentences in Words	Syllables Per 100 Words
1859	Grover & Baker.	44	127
1862	Singer Sewing Machines	27	128
1867	Howe Machine Co.	15	138
1868	Singer Sewing Machines	44	130
1869	Finkle & Lyon M'fg Co	61	117
1870	W. G. Wilson Sewing Machine Co.	63	123
1876	Davis Sewing Machines	27	126
1878	New American Sewing Machine	31	123
1878	Florence Sewing Machine Co.	54	125
1880	A. G. Mason Mfg. Co.	32	124
1880	Domestic Sewing Machine Company	41	126
1899	Perry Mason & Co.	42	139
1907	Wilcox & Gibbs Sewing Machine Co.	27	136
1914	Singer Sewing Machines	24	128

38. The curves for Figures 5.8 to 5.17 were plotted by the Cricket Graph program (version 1.3.2) running on a Macintosh IIsi. The curve is a third order polynomial curve fit.
39. See previous footnote concerning the mechanics and use of the Flesch Reading Ease scale. The average sentence length and number of syllables for each of these pamphlets that determined the Flesch Formula Score were:

Date	Company	Average Length of Sentences in Words	Syllables Per 100 Words
1859	Grover & Baker.	44	127
1857	Boas & Spangler	17	123
1864	Alzirus Brown	14	125
1865	Adriance, Platt & Co.	15	123
1865	Williams, Wallace & Co.	34	119
1865	Nixon & Co.	57	148
1867	Adriance, Platt & Co.	27	122
1867	Alzirus Brown	16	125
1873	Walter A. Wood Mowing & . Reaping Machine Co	16	116
1879	Eureka Mower Co.	19	130
1890	Walter A. Wood Mowing & . Reaping Machine Co	28	122
1891	Wm. Deering & Co.	20	126

1909	International Harvester Company of America-Champion	21	123
1910	International Harvester Company of America-Deering	21	121
1911	International Harvester Company of America-Deering	41	121
1915	International Harvester Company of America-McCormick	23	138

40. Deller, op. cit.: 116–7.
41. Killingsworth and Gilbertson, op. cit., 1992: 76—"To become effective, the interchange of documents and actions [within a genre] must feed back into itself."
42. Ferguson, Eugene S.: *Engineering and the Mind's Eye*. Cambridge, MA: The MIT Press, 1992: 83–7
43. Ferguson, Eugene S., and Baer, Christopher: *Little Machines: Patent Models in the Nineteenth Century*. Greenville, DE: Hagley Museum, 1979: 11.
44. The development of the instruction manual genre in stages in response to rhetorical negotiations between readers, authors, and the genre itself is much like the development of the experimental article in science in response to changing rhetorical circumstances as sketched by Bazerman (Charles Bazerman: *Shaping Written Knowledge: The Genre and Activity of the Experimental Article in Science*. Madison, WI: University of Wisconsin Press, 1988) or of the memo as sketched by Yates and Orlinkowski (JoAnne Yates and Wanda J. Orlinkowski: "Genres of Organizational Communication: A Structurational Approach to Studying Communication and Media." *Academy of Management Review* 17(2) (1992): 299–326).
45. Beniger, James R.: *The Control Revolution: Technological and Economic Origins of the Information Society*. Cambridge, MA: Harvard University Press, 1986: 356—"Power-driven printing . . . developed rapidly . . . with a steady progression of innovations: the electric press (1839) and rotary printing (1846), wood pulp and rag paper and the curved stereotype plate (1854), paper-folding machines (1856), the mechanical typesetter (1857), high-speed printing and folding press (1875), half-tone engraving (1880), and linotype (1886)." Yates and Orlinkowski (op. cit., 1992) also described the effect that typewriters had in moving the genre of general business letters to the genre of memos (314). Recall also how Evans at the end of the 18th century only very gradually began employing illustrations in his broadsides, and how he only did so after he worked a swap with a newspaper for copperplates which he then proceeded to use later in his book and in all subsequent licenses for his mill.
46. Yates and Orlinkowski (op. cit., 1992: 300) have also described this process of a genre's fluid negotiation as "structuration:"

> Genre is a literary and rhetorical concept that describes widely recognized types of discourse (e.g., the novel, the sermon). In the context of organizational communication, it may be applied to recognized types of communication (e.g., letters, memoranda, or meetings) characterized by structural, linguistic, and substantive conventions. These genres can be viewed as social institutions that both shape and are shaped by individuals' communicative actions. By situating genres within processes of organizational structuration, the proposed framework captures the continuing interaction between human communicative action and the institutionalized communicative practices of groups, organizations, and societies.

See also Berkenkotter, Carol, and Huckin, Thomas N.: *Genre Knowledge in Disciplinary Communication: Cognition/Culture/Power*. Hillsdale, NJ: Lawrence Erlbaum Associates, Publishers, 1995: 20–1.

47. Another example of historical research demonstrating the change in a genre over time due to the change in the audience's background knowledge can be seen in Giltrow's comparison of violent crime sentencing in 1950 and in 1990. (Giltrow, J.: "Genre and the Pragmatic Concept of Background Knowledge." Paper presented at the International "Rethinking Genre" Conference, Carleton University, Ottawa, Canada, April 1992, qtd. in Berkenkotter, Carol, and Huckin, Thomas N.: *Genre Knowledge in Disciplinary Communication: Cognition/Culture/Power*. Hillsdale, NJ: Lawrence Erlbaum Associates, Publishers, 1995: 14–5.)
48. "The People's Knowledge of Sewing Machinery." *Sewing Machine News* 7(8) August 1885: 10. See also Brockmann, R. John: "A Homoculus in the Computer." *Journal of Technical Writing and Communication* 27(2) (1997): 119-145.
49. This acceptance of technology by American society was not just true of sewing machines but also agricultural technology. Such a progressive acceptance and understanding of technology cannot be specifically found for the mowers and reapers but is more generally indicated by the fact that the space allocated to technology ("new machinery, tools, and processes") in agricultural newspapers decreased by 300% between 1860 and 1910. Farrell, Richard T.: "Advice to Farmers: The Content of Agricultural Newspapers, 1860–1910." *Agricultural History* 51(1) (January 1977): 209–17.
50. Kenwood, A. G. and Lougheed, A. L.: *Technology Diffusion and Industrialization Before 1914*. New York: St. Martins Press, 1982.
51. Fenster, op. cit., 1994: 49. "Previous sewing machine promoters were always thinking of men, the ones who owned factories. Singer's mind was more carefully trained. Women would be the ones using the sewing machines in most cases, so Singer marketed the idea directly to dress-

makers, seamstresses, and housewives. When he set up a temporary shop, he decorated it to look like a home, not a store, and certainly not a factory."

52. *I.M. Singer and Co.'s Gazette*, November 1, 1859 (as quoted in Brandon, op. cit., 1977: 70). See also Fenster, op. cit., 1994: 40—"Hand sewers were a national problem in the 1830s and 1840s. Theirs was not just an occupation but a last stop. In good times, stitching for sixteen hours a day, they just barely earned a living; in bad times they were destitute. At the dawn of the Industrial Revolution the sewing machine did not look like a great innovation; it looked like forty thousand sewers with nothing left to do."
53. Brandon, ibid., 1977: 70.
54. Brandon, op. cit., 1977: 122. History has also suggested that this idea that women were unsuited to sewing machine technology by their gender is a particularly curious example of doublethink. For example, many historians are now suggesting that it was not Elias Howe who invented the machine, but Mrs. Elias Howe: "In his most famous lecture, 'Acres of Diamonds,' Russell Conwell tells us that Mrs. Elias Howe completed in two hours the sewing machine her husband had struggled with for years. Conwell's source was impressive—Elias Howe himself, in campfire conversations as they served together in the Civil War: 'He was in the Civil War with me, and often in my tent, and I often heard him say that he worked fourteen years to get up that sewing machine. But his wife made up her mind one day that they would starve to death if there was not something or other invented pretty soon, and so in two hours she invented the sewing machine. Of course he took out the patent in his name. Men always do that.'" And perhaps doubly interesting is the suggestion that it was Ann Manning who invented the mower-reaper: "A third great American invention, the mower and reaper, owes its early perfection to Mrs. Ann Harned Manning of Plainfield, New Jersey, who in 1817–18 perfected a system for the combined action of teeth and cutters, patented by her husband, William Henry Manning, as a 'device for the combined action of teeth and cutters, whether in a transverse or revolving direction.'" Stanley, Autumn: *Mothers and Daughters of Inventions: Notes for a Revised History of Technology*. Metuchen, NJ: Scarecrow Press, Inc., 1993: 433, 114.
55. Grover and Baker, op. cit., 1859: 6.
56. *Sewing Machine News*, op. cit., 1885: 10.
57. *Sewing Machine News*, ibid., 1885: 10.
58. Brandon, op. cit., 1977: 125–6.
59. Another reason why the testimonials disappeared from the instruction manuals was because the sewing machine industry developed different types of documents for different purposes, separating the testimonials

into catalogs or sales brochures. Though not included in this study, the sewing machine industry developed a wide array of catalogues and other sales and advertising documents which, within 10 years, became wholly separate entities from the instruction manuals.

60. Mumford, Lewis: *Technics and Civilization*. New York: Harcourt Brace Jovanovich, 1934: 32–3.
61. Marvin, Carolyn: "Dazzling the Multitude: Imagining the Electric Light as a Communications Medium." In (Joseph J. Corn, ed.) *Imagining Tomorrow: History, Technology, and the American Future*. Cambridge, MA: The MIT Press, 1986: 206.
62. Marvin, ibid., 1986: 208.
63. Wilcox and Gibbs: *Silent Family Sewing Machine*. 1869.
64. W. G. Wilson Sewing Machine Company: *Directions for Using the New Buckeye Under-feed Sewing Machine*. Cleveland, OH, 1870: 2.
65. (December 1983): 24.
66. New American Library, 1983.
67. Shurkin, Joel: *Engines of the Mind*. New York: W.W. Norton and Co., 1984: 252.
68. Cooper, op. cit.: 48–50.
69. Another reason why the genre of instruction manuals may have changed over the last half of the 19th century is because of the principle of the "hermeneutic circle": "A genre, then, reflects in its very character the paradox of the part and the whole, the paradox of the well-known hermeneutic circle. We recognize a genre by the conventions native to it, but to recognize the conventions we must first know the genre. Our recognition that a specific text is a member of a particular genre creates, in turn, certain generic expectations. . . . Our generic perception—our ability to identify texts as members of specific genres—seems, therefore, to operate much the same as our reading experience. To understand a sentence, we must understand the words that constitute it, but to understand the full meaning of the words, we must also understand the sentence. Like our perception of individual sentences, our generic perception requires the interaction of both a text and a context" (Kent, Thomas: *Interpretation and Genre: The Role of Generic Perception in the Study of Narrative Texts*. Lewisburg, PA: Bucknell University Press, 1986: 15–6). Thus, in the first 20 years, perhaps the authors of the instruction manuals needed time within the sewing machine industry to begin to perceive the normative aspects of the instruction manual genre, that is, they were indeed writing in a genre different from the narrative novel genre. Once they perceived such a genre after a number of texts of the 1850s–1870s replicated the same normative aspects, they could more easily reproduce such genre texts.

By the same token, perhaps the mower-reaper industry needed more time to begin to perceive that the genre used in the sewing machine industry had real applicability to their industry: Without the perception, they could not duplicate the normative texts. Moreover, because this negotiation of a genre involves more than just authors, 19th century readers, too, needed to develop a reading competence that could recognize instruction manual genre conventions, and this too took time because of the principle of the "hermeneutic circle." (See Kent, ibid., 1986: 19.)

70. Kuhn, Thomas S.: *The Structure of Scientific Revolutions* (2nd ed.). Chicago: University of Chicago Press, 1970: 15.

71. Yates and Orlinkowski (op. cit., 1992) also had some suggestive comments on this aspect of mower-reaper manuals slowly becoming more like sewing machine manuals: "[G]enres with a more limited normative scope, sometimes subgenera of the more broadly recognized genres, emerge within particular contexts—for example, opinion letters to clients in the accounting profession or customer support calls in service organizations. When genre rules are not mandated, they are likely to emerge and be institutionalized in specific contexts and communities first; they will achieve broader acceptance later only if the emerging genre is perceived by a larger community to respond to a common recurrent situation." (320)

72. Paradis, op. cit., 1991: 264. Paradis suggests that users sometimes learn to operate technologies based on a stock of generic images. "User strategies can be built upon the stock of generic images that, as Boulding has argued, is shared by society. Everyone 'operates' a screwdriver or flashlight without having to be instructed. In a thriving consumer society, this kind of intuitive operational knowhow based upon socially shared imagery must be widely available. Many lawnmower purchasers can by mere inspection decode the fraction of the technology necessary to operate the instrument satisfactorily. Hence, the sport of dispensing with the manual: When all else fails, the saying goes, 'consult the manual.' The highly accessible technologies common in a consumer society are thus based on a social substrate of shared generic imagery, a kind of Platonic world of idealized forms and processes that is presumably the product of elementary and second school education, supplemented by television culture." However, in 1850 rural America, before the consumer society and its "shared generic imagery," was there any "shared generic imagery" appropriate to these new factory-produced technologies? The question is difficult to answer, and, in this chapter, such shared generic imagery is assumed not to have been present to either the farmers or the sewers.

73. Norman, Donald A.: *Things That Make Us Smart: Defending Human Attributes in the Age of the Machine*. Reading, MA: Addison-Wesley Publishing Co., 1993: 22–4.

74. Bogue, Allan G.: *From Prairie to Corn Belt: Farming on the Illinois and Iowa Prairies in the Nineteenth Century*. Chicago: University of Chicago Press, 1963: 195.
75. Bogue, ibid., 203—"But suspicion of book-farming or of the motivation of the editors, the scarcity of money, and the slow, even invisible, return on the investment—all kept the subscription lists of the farm journals within modest dimensions." See also Hayter, Earl W.: *The Troubled Farmer: 1850–1900: Rural Adjustment to Industrialism*. DeKalb, IL: Northern Illinois University Press, 1968: 6.
76. Neely, Wayne Caldwell: *The Agricultural Fair*. New York: Columbia University Press, 1935: 182–3. See also Brown, Richard: *Knowledge Is Power: The Diffusion of Information in Early America, 1700–1865*. Oxford: Oxford University Press, 1989: 158–9—"[F]or most information they relied on direct observation, their own or that of others', and word-of-mouth reports. . . . Indeed it was these ingrained habits that agricultural reformers sought to alter by turning farmers toward print in the first half of the 19th century. Almanacs of the day, a popular medium that reached into farm households on a truly massive scale, preached the doctrine that farmers especially should live by Bacon's dictum that 'Knowledge is power' and explained that accumulating knowledge was an essential use of literacy."
77. Scott, Roy V.: *The Reluctant Farmer: The Rise of Agricultural Extension to 1914*. Urbana, IL: University of Illinois Press, 1970: 16–7. See also Ross, Earle D.: "The Evolution of the Agricultural Fair in the Northwest." *Iowa Journal of History and Politics* 24 (July 1926): 454 —"The whole burden of agricultural experimentation, instruction, extension, and recreation fell upon the agricultural societies whose whole efforts were carried on mainly through state and local fairs."
78. Ewers, William and Baylor, H. W.: *Sincere's History of the Sewing Machine*. Phoenix, AZ: Sincere Press, 1970: 91.
79. Scotchmer, Henry E., letter to the editor. *The Sewing Machine Advance* 1 (5) September 15, 1879: 69.
80. One can almost identify the young lady in that window in 1853 because Ruth Brandon in her book on the sewing machine industry, *A Capitalist Romance: Singer and the Sewing Machine*, discovered that one of the earliest employees hired by Singer was a Miss Brown, a sewing machine demonstrator: "One of I. M. Singer and Co.'s earliest employees was Miss Augusta Eliza Brown, who joined the firm in 1852, learned to operate the machine in two weeks, and from then on spent her time teaching others and demonstrating the novelty." Brandon, ibid., 1977: 124–5.
81. *Wheeler and Wilson*, op. cit., 1859: 4. See also instructions to salespeople on how to demonstrate sewing machines documented in the W. G. Wilson Sewing Machine Co.: *Special Instructions*. Cleveland, OH, circa 1859.
82. Grover and Baker, op. cit., 1859: 3.

83. A binder was an attachment to a reaper, and it was used to bind up the sheaves of grain as they were cut.
84. Norman, op. cit., 1993: 22–4.
85. This rejection of authoritative guidance iconically portrayed in the 1891 cover picture has recently been described as a situational variable causing reader apprehension termed *subordinate status*. See Ferrell, Terri L.: "Applying Communication Apprehension Situational Variables to Receiver Apprehension in Instructional Manuals." In *1992 Proceedings of the 39th Annual Conference, May 10–13, 1992*. Arlington, VA: Society for Technical Communication, 1992: 379–"Another variable that may increase RA [Reader Apprehension] is subordinate status. Although closely related to formality (rigorous adherence to rules), subordinate status is characterized by the perception that the rules are imposed by a person belonging to a superior class or rank."
86. A counter finding to this variance in the instruction manuals based on gender of the addressed audience appeared in Tebeaux and Lay (Tebeaux and Lay, op. cit., 1992). These two authors suggested that "the styles that Renaissance writers selected in preparing books for men and books for women do not differ significantly." (196) Yet, in an earlier article by Tebeaux (Tebeaux, Elizabeth and Killingsworth, M. Jimmie: "Expanding and Redirecting Historical Research in Technical Writing: In Search of Our Past." *Technical Communication Quarterly* 1(2) (Spring 1992): 5–32), she and her coauthor discussed a renaissance book written by Gervase Markham that had two parts: "one for men and one for women." These two parts exhibited the same differences between asymmetric and symmetric rhetoric that appeared in the 19th century American instruction manuals. Consider Tebeaux and Killingsworth's example from Markham's Part One: "For the choice of a good *Stallion*, and which is best for our kingdome, opinion swayeth so farre that a man can hardly giue well received directions; yet surely if men will be ruled by the truth of experience, the best *Stallion* to beget *horses* for the warres is the *Courser*, the *Iennet* or the *Turke*, the best for hunting is the Bastard courser begot of the English. the best for the Coach is the *Flemming*, the best for Trauell or burthen is the *English* and the best for ease is the *Irish-hobbie*." Notice the lack of imperative mood verbs and the self-denigration of the speaker's authority in the phrase "opinion swayeth so farre that a man can hardly giue well received directions." Compare this rhetorical relationship of author and audience to the following passage from Part Two, which is addressed to women: "For them that cannot hold their water in the night time: take a Kiddes hoffe and drie it and beate it into powder, and giue it to the patient to drink either in beare or ale fowre or fiue times. To make a water to heale all manner of wounds: you shall take luphwort flowers, leaues, and rootes, and in March or Aprill, when

the flowers are at the best distill it, then with that water bather the wound, and it will healt it." Even Tebeaux and Lay's article contained such contrasting address if one considers their example from Iohn Mvrell (1617) addressing women (p. 199) with their example from Alexander Read (1632) addressing male surgeons (p. 200). Thus, although this book specifically confines itself to American technical communication, it is of some interest to find the same rhetorical contrasts existing 200 years earlier in England.

87. Tannen, Deborah: *You Just Don't Understand: Women and Men in Conversation*. New York: Ballantine Books, 1990: 124–5. See also Margaret Lowse Benston: "Women's Voices/Men's Voices: Technology as Language" In *Technology and Women's Voices: Keeping in Touch*. New York: Routledge and Kegan Paul, 1988: 26—"Men and women do not, however, communicate as equals about technology. The information flow is almost entirely one-sided: men may *explain* a technological matter to women but they do not discuss it with them; that they do with other men. The education process in technological fields, for example, is heavily dependent on learning from fellow students and this asymmetry becomes a major problem for women. It is very difficult to discuss technical problems, particularly experimental ones, with male peers—they either condescend or they want to simply do whatever it is for you." See also Belenky, Mary Field, Clinchy; Blythe, McVicker; Goldberger, Nancy Rule; and Tarule, Jill Mattuck: *Women's Ways of Knowing: The Development of Self, Voice, and Mind*. NY: Basic Books, Inc., 1986: 45.
88. Tannen, ibid., 1990: 128.
89. Brown, ibid., 1989: 163.
90. The suggestion that the reader or viewer has total freedom in processing the information contained in ancillary illustrations must be tempered with research from individuals like Stone and Crandell, who suggest a rather structured reading process is used. See Stone, D. E. and Crandell, T. L.: "Relationships of Illustrations and Text in Reading Technical Material." *Advances in Reading/Language Research* 1 (1982): 283–307. See also Pedell, Brian J.: "Examining Relationships of Text and Illustrations in Instructional Procedures." *In 1992 Proceedings of the 39th Annual Conference, May 10–13, 1992*. Arlington, VA: Society for Technical Communication, 1992: 437–440.
91. For further discussion of "gender schema" and reading see (Elizabeth A. Flynn and Patrocinio P. Schweickart. eds.) *Gender and Reading: Essays on Readers, Texts, and Contexts*. Baltimore, MD: The Johns Hopkins University Press, 1986: xii–xiii, 15.
92. Ibid., 1986: 285
93. In an analogous way, consider how women in the 1830s and early 1840s in the Lowell textile mills evaded employer edicts (Kasson, John F.:

Civilizing the Machine: Technology and Republican Values in America, 1776–1900. NY: Penguin Books: 78):

> As they [the women] mastered their machines intricacies, they learned to defy the noise and tedium by distancing themselves from their work through private thoughts and daydreams. Furthermore, before operatives were given more looms to attend and the machines speeded up in the mid-1840s, they often had long periods of idleness between catching broken threads. Regulations prohibited books in the mill, but women frequently cut out pages or clippings from the newspapers and evaded the edict. Others worked on compositions in their spare moments or spent the time lost in contemplation. Thus they attempted to give meaning to the time that their work denied and to cultivate a mental separation from their activities and surroundings.

94. Lubar, op. cit., 1993: 207.
95. Bazerman and Paradis, op. cit., 1991: 3.
96. Killingsworth and Gilbertson, op. cit.:, 1992: 214—"The directives [in contemporary instructional texts] flow from a point of greater knowledge to a point of lesser knowledge or authority, from a master of a technique or body of knowledge (author) to an apprentice (audience), from an expert to a non-expert." Ibid.: 218—"[I]nstrumental discourse supports and depends upon faith in the rationality of an authority, a technical expert . . . "
97. Ben and Marthalee Barton have many additional suggestive explanations for why Pa turned his gaze from the viewer and from the experts. See their article "Modes of Power in Technical and Professional Visuals." In (Brenda Sims, ed.) *Studies in Technical Communication: Selected Papers from the 1992 CCCC and NCTE Meetings*. Denton, TX: University of North Texas, 1992: 183–215. Perhaps most interesting of all is their final quote from De Certeau who suggested the need "to bring to light the clandestine forms taken by the dispersed, tactical and makeshift creativity of groups or individuals already caught in the nets of 'discipline.' Pushed to their ideal limits, these procedures and ruses of consumers compose the network of an antidiscipline." (208)
98. Sauer, Beverly A.: "Sense and Sensibility in Technical Documentation: How Feminist Interpretation Strategies Can Save Lives in the Nation's Mines." In Sims, op. cit., 1992: 101—"A feminist analysis demands that technical writers acknowledge the silent power structures that govern public discourse, not because we are interested in theoretical constructs about language, but because those power structures affect the fabric of technology upon which we all depend."

Chapter 6

The Emergence and Development of a Technical Communication Genre, The Instruction Manual

Part 2. Ford and Chevrolet: 1912-1988

A good manual is not a narrative. . . . Nobody ever reads a manual cover to cover—only mutants do that.

—John Greenwald, *Time Magazine* (June 18, 1986)[1]

FROM MA TO PA AS AUDIENCE

As reported in Chapter 5, the following appeared on the first page of the W. G. Wilson Sewing Machine Co., *Directions for Using the New Buckeye Under-feed Sewing Machine*: (1870)

☞ You Cannot use the Machine until you thoroughly understand the Directions.

Further on toward the bottom of this manual's front page, the following sentence also appeared: "The directions furnished with the Buckeye machine are intended for the use of the operators, and it is desirable that they study them carefully before using the machine at all."[2] The A. G. Mason Mfg. Co. manual, *The Improved Common Sense Family Sewing Machine* (1880), had a similar message in its prefatory material on page 2: "The following directions are designed to aid beginners and must receive strict attention."

These words and sentences embodied the social role of Ma, on the stump reading the manual as described in Chapter 5. Ma and other readers modeled on the 19th-century social position of women were not allowed to just read; they were required to study with strict attention. The pointing finger, the boldfaced "you cannot use," and the words "must receive," all marked an attitude of authority demanding a reader's attention. Moreover, this authority would quite naturally assume that readers would read thoroughly line by line, paragraph by paragraph, from the beginning of the manual to its end. Thus in a Howe Sewing Machine Co. manual appearing in the last decades of the 19th century, *The Howe Sewing Machine Instructor*, just above its table of contents—listing the manual's topics, of course, in numerical order from beginning to end—is printed: "To learn to use the machine and attachments, proceed in the following order . . . "[3] In other words, the reader was told in *The Howe Sewing Machine Instructor* to read in sequential order through the book. Selective reading or any kind of choice in reading were simply not offered because readers were required to proceed in numerical page order.

The authority embodied in this rhetoric and the demand to read from beginning to end started to change when the genre of instruction manuals was employed by the new industry automobile in the early 20th century. Why the rhetoric of authority changed probably can never be totally explained, but certainly one explanation lies in the fact that the new automobile instruction manuals did not primarily address women as did the sewing machine manuals. Instead, the automobile manuals were designed for a primarily male audience: "the auto was born in a masculine manger, and, when women sought to claim its power, they invaded a male domain."[4] This primary focus on men as the audience for the automobile manuals occurred despite the fact that women used and had been using automobiles since the turn of the century:[5]

> Despite the traditional association of the automobile as a mechanical object with men and masculinity in American culture, automobility probably has had a greater impact on women's roles than on men's. . . . Because driving an automobile requires skill rather than physical strength, women could control one far easier than they could a spirited team. They were at first primarily users of electric cars, which were silent, odorless, and free of the problems of hand-cranking to start the engine and shifting gears. Introduction of the self-starter in 1912, called

> the 'ladies' aid,' and of the closed car after 1919, which obviated wearing special clothes while motoring, put middle-class women drivers in conventional gasoline automobiles in droves.

However, just as Singer and Howe passed to the authors of their sewing machine manual an image of their audience that arose from their own personal prejudices but which was out of touch with the actual audience (women who were unable to handle sewing machine technology), so, too, did Ford and Chevrolet management pass on their own personal prejudices, which were equally out of touch with the actual audience:[6]

> But Everyman's car [the Model T] went to market accompanied by the customs of the country, including the notion that Everywoman would not use the automobile in the same way that her husband, father, or son did . . . automakers in general and Henry Ford in particular continued to see women as homebodies, neither quite equal to the challenges and inconveniences of the gas-powered car nor particularly interested in the power and range it offered. Ford's own family proved to him the point. The same year that he introduced the Model T, Henry Ford bought an electric car for his wife Clara to use for social calls and short trips around town. "'For visits to Dearborn or more distant points," explained Ford's biographer, Allan Nevins, "'she would go with Henry or [their son] Edsel."

Farmer Pa was the primary audience to be addressed in the automobile instruction manuals—"The Model T was intended as a farmer's car for a nation of farmers."[7] Remember, it was Pa, not Ma, who turned his back on the experts as depicted on the cover of the 1890 binder manual. It was Pa, not Ma, who refused to acquiesce to the social relations being taken on by the mower-reaper instruction manuals when the manuals began labeling themselves as "directions." How then was Farmer Pa to be addressed by the automobile instruction manuals so that he would not turn his back and refuse to listen? The answer lay in diminishing the authority in the rhetoric.[8]

In a large number of automobile instruction manuals, diminishing the authority in the rhetoric meant announcing a new author-reader relationship immediately in the preface. For example, in the 1919-25 version of the Model T manual, *Ford Manual: For Owners and Operators of Ford Cars and Trucks*, neither the words "instruction" nor "direction" appeared on the title page. However, the following did appear in the "Foreword":[9]

> But while it is not imperative, it is however, altogether desirable that every Ford owner should thoroughly understand his car. With such knowledge at his command he is always master of the situation—he

> will maintain his car more economically—prolong its usefulness—and he will also derive more pleasure from it, for it is a truism that the more one knows about a thing the more one enjoys it.

The Ford author no longer demanded a reader's attention. In fact, the author quickly stated that, whatever else, reading the manual and thoroughly understanding the information were "not imperative," but only "desirable." Because the author of the manual could no longer demand a reader's strict attention and study, the author began to bribe the reader to read by offering incentives:

- "he is always master of the situation"
- "he will maintain his car more economically"
- "he will also derive more pleasure"

Although the prevalent use of the male third person singular pronoun may not have meant that all the readers were men, the pronoun strongly suggested that the author of the manual had the recalcitrant Pa's of the world in mind as the intended audience. Thus, the author of the Model T instruction manual explicitly moved away from the "imperative" set of social relations embodied in the rhetoric of the sewing machine instruction manuals.

The Ford authors were not alone in proclaiming their new relationship with audiences. In the competition's contemporary manual, *Instructions for the Operation and Care of Chevrolet Cars (1919–FB Model)*, the following two sentences appeared in its preface:[10]

> But unless the operator of a Chevrolet knows how to properly handle and care for his or her car, the maximum of satisfaction and enjoyment, which our painstaking efforts have provided, will not be realized. It is to give this desirable, if not entirely necessary, knowledge, that this Instruction Book has been prepared.

Again, the author of the 1919 Chevrolet manual diminished the strict authority of the sewing machine authors by suggesting that the information in the manual was "desirable, if not entirely necessary." Like the Ford author, the Chevrolet author cajoled the audience into reading the manual by offering such incentives as "maximum of satisfaction and enjoyment." Absent were the demands of thorough study and strict attention. No pointing fingers of authority graced the prefaces of the automobile manuals in the 1910s.

In fact, the social relations embodied in the instruction manual genre change so much from those of the Buckeye's 1870 sewing machine finger-pointing that by the time of the *1988 Chevrolet Beretta Owner's Manual*, authors felt they could only politely "acquaint" readers with the car and "urge" and encourage readers by pointing out incentives:[11]

> This manual has been prepared to acquaint you with the operation and maintenance of your 1988 Chevrolet, and to provide important safety information. . . . We urge you to read all three publications [Maintenance Schedule booklet and a Warranty and Owner Assistance Information booklet] carefully. Following the recommendations will help assure the most enjoyable, safe and troublefree operation of your car.

"Acquainting" has been taken from the real of personal relationships. In this instance, the author is acquainting the reader with the information. Notice that the locus of action and responsibility has shifted from a reader commanded to study and pay strict attention to an author who now must politely and gently "acquaint" the reader with the information. (This is at least the third time "acquaint" appeared in the automobile manuals; both the earlier Ford *Passenger Car Owner's Manual* (1951) and the *1973 Camaro Owner's Manual* spoke of "acquainting" the reader with information.)

From demanding to acquainting, the power of instruction manual authors diminished to such an extent that eventually they no longer assumed that their readers read all that is written nor that they read the manual from beginning to end. In fact, readers usurped so much authority from authors that Chris Espinosa, a veteran of writing instruction manuals for Apple computers, was quoted in *Time* magazine: "A good manual is not a narrative. . . . . Nobody ever reads a manual cover to cover—only mutants do that."[12]

What were the specific changes in the instruction manual genre caused by this renegotiation of authority from authors to readers? How did the shift to something "not a narrative" manifest itself in the pages and chapters of the instruction manuals? Moreover, how exactly did the instruction manual change when it was used by a new 20th-century industry, the automobile industry? These are the questions I consider in this study of the instruction manuals of Ford and Chevrolet from 1912 to 1988.

WHY FORD AND CHEVROLET?

Ford did not invent the automobile. For that accolade, some point to Seldon's 1879 patent application, others to Benz's 1886 vehicle, to Duryea's 1893 car, or to Olds who claimed that he was "driving regularly through the streets of Lansing" in 1896.[13] Neither was Ford the first to commercially manufacture horseless carriages. For that achievement, some point to the precedent of Duryea Motor Wagon Company which built 13 machines in 1896,[14] others to Olds who turned his steam-powered vehicle into a gasoline-powered model in 1896,[15] or to the Winston Motor Carriage Company which sold a 1400-pound phaeton on March 24, 1898.[16] Nevertheless, Ford has been inscribed in history on the strength of 15,458,781 Model Ts manufactured.[17] Ford's Model T was light, simple, powerful, and inexpensive.[18] Basically what Ford achieved with the Model T was to take automobiles out of the hands of wealthy hobbyists and put them into the hands of the common American.

Ford's long-standing corporate rival, Chevrolet, was not even incorporated until 1911, three years after the introduction of the Model T.[19] However, within 17 years, Chevrolet, just one division of General Motors, was outselling all of Ford Motor Company by 200,000 cars a year.[20] Chevrolet and Ford both aimed for the low-price, high-volume automobile market.[21] Both hoped to sell to "owners who in the great majority of cases have little or no practical experience with things mechanical,"[22] and both engaged in a market-share seesaw battle for decades. Yet, by 1941, Chevrolet trumpeted on the back cover of its instruction manual: "The public names its own car leader . . . Chevrolet . . . holder of first place in motor car sales for 9 out of the last 10 years." Only in 1959 and 1970 did Ford ever again outsell Chevrolet.[23]

Thus, I analyzed the manuals of these two car makers because they addressed similar audiences and markets rather than the higher priced and upscale market focus of Oldsmobile, Pontiac-Oakland, or Cadillac.[24]

WHAT WAS ANALYZED AND WHY

I analyzed 18 manuals from Ford and Chevrolet for each decade from the 1910s to the 1980s.[25] The titles of the manuals are displayed chronologically in Table 6.1. Manuals from two manufacturers rather than just one were used to provide a more general rather than idiosyncratic understanding of the instruction manual genre as it was used by the automobile industry.

Like the guided sampling of sewing machine and mower-reaper manuals, these automobile instruction manuals do not represent a scientific sampling from a larger group of manuals but only those representative of each decade and also readily available from any of the following three sources: Archives and Library of the Henry Ford Museum, Dearborn, MI; Lincoln Publishing. in Lockport, NY (a mail order source of Ford manual facsimile editions); and Classic Motorbooks in Osceola, WI (a mail order company that produces automobile manual facsimiles for a wide variety of car markers).

To discover the specific changes in the instruction manual genre that the automobile industry made, I examined three characteristics in the manuals listed in Table 6.1: brevity, profuseness of illustration, and signs of explicit organization. As frequently as possible, measurements of these three characteristics in the automobile instruction manuals were compared with sewing machine and mower-reaper manuals. Thus, some tables and graphs from Chapter 5 are redisplayed in Tables 6.9 and 6.10 and replotted on graphs with the automobile manuals.

Table 6.1. Timeline of Eighteen Automobile Manuals Analyzed.

Ford	Chevrolet
1910s	
Model T Instruction Book (1919)	Instructions for the Operation and Care of Chevrolet Motor Cars (1919-FB Model)
1920s	
Model T Manual (1919-25)	Instructions for the Operation and Care of Chevrolet Motor Cars (1926-BA, BB Models)
1930s	
Model "A" Instruction Book (1931)	Instructions for the Operation and Care of Chevrolet Motor Cars (1932-Confederate, Series BA: Passenger Models, Series BB; and Commercial Models)
Instruction Book Ford V-8 (1934)	Instructions for the Care of Chevrolet 1937-Master Deluxe, 1937 Master Passenger Models
1940s	
Deluxe and Super Deluxe Reference Book (1941)	Owner's Manual 1941 Passenger Cars
1950s	
Passenger Car Owner's Manual (1951)	Guide to Your New Chevrolet (1955)
1960s	
Mustang Registered Owner's Manual (1965)	Chevrolet Owner's Guide (1964)
1970s	
1971 Passenger Cars: Car Care and Operation	1973 Camaro Owner's Manual
1980s	
1986 Escort Owner's Guide	1988 Chevrolet Beretta Owner's Manual

The first two characteristics of the automobile manuals that I examined were similar to those in the last chapter:

- Brevity. Did the size of the automobile instruction manuals change? Were the automobile manuals similar in design to the sewing machine manuals devoting 100% of their pages to instructions, or was there a mix of different elements as in the mower-reaper manuals? Did the mix of content change?

Brevity was measured quantitatively by counting the number of pages in manuals and computing the percentage of the total number of pages devoted to instructions. Qualitatively, brevity was measured by looking at the types of information included or excluded. For example, in the 1988 Beretta manual, the reader's attention was drawn to three publications, a Maintenance Schedule booklet, a Warranty, and an Owner Assistance Information booklet. Were these items always separate, or were they all originally included in one manual? Additionally, if they were separated out, what could this mean about the instruction manual genre?

- Illustrations. Were the automobile manuals as profusely illustrated as the sewing machine or mower-reaper manuals? Was the nonverbal communication trend that began with antebellum mechanics such as Evans, McKay, and Griffiths continued here? The profuseness of illustrations was measured, as in Chapter 5, by the percentage of instruction pages with graphics.

The third element of instruction manuals I considered was explicit signs of organization affording selective reading. If an instruction manual was not a narrative, and if readers were allowed to read selectively, instruction manual authors must have provided some orientation other than unmarked lines of pages, chapters, and manuals. The question was, then, what exactly were these explicit signs of organization, and did they change over the 76 years studied? The explicit signs of organization affording selective reading were examined in four ways:

1. An author could have explicitly indicated *page-level organization* by breaking up continuous text into paragraphs. The more paragraphs per instruction page, the more selectively a reader could have read. Thus, the number of paragraphs per page was measured by counting the paragraphs on a page at the beginning of the instructions, in the middle, and at the end. These counts were then averaged and plotted.
2. An author could have also explicitly indicated *page-level organization* by labeling paragraphs with words such as "Danger," "Caution," or "Note," or with some other kind of typographic tag (pointing hands, bullets, lists, italic print, tables, boxes, uppercase, underlining, and, in the case of Chevrolet's 1964 manual, blue background screens). The more paragraphs were either labeled or tagged, the more readers were afforded alternative ways of selectively reading at a page level. Thus, the number of labels and typographic tags was counted and then plotted for all the instruction pages.

3. An author could have explicitly indicated *chapter-level organization* by headings and the hierarchy of information signaled by typographically designating them as first, second, third, and fourth level. The more headings present and the more a hierarchy of headings was typographically signaled, the more readers could have read selectively at a chapter level. Thus, the headings per page and the ratios between first-, second-, third-, and fourth-level headings were measured and plotted.
4. An author could have explicitly indicated *whole manual-level organization* by including a table of contents or index. Thus, the presence or absence of a table of contents or index was examined.

Each of these three characteristics of the automobile instruction manuals—brevity, profuseness of illustration, and signs of explicit organization—are presented in turn.

EXAMINING THE GENRE CHARACTERISTICS OF BREVITY IN THE AUTOMOBILE INSTRUCTION MANUALS

The number of pages of instruction for each of the 18 automobile instruction manuals is given in Table 6.2, and the number of pages for each of the 29 sewing machine or mower-reaper manuals is given in Tables 6.9 and 6.10. The percentage of the whole automobile manual that these instruction pages represent is also given in Table 6.2, and the percentage of the whole manual that these instruction pages represent for each of the 29 sewing machine or mower-reaper manuals is given in Tables 5.4 and 5.5 and are graphed in Figure 6.2. (The change in the page count in the later automobile manuals was relatively unaffected by the changing page sizes of the manuals.[26])

Figure 6.1 comparing the raw total numbers of 19th-century manuals to the 20th-century automobile manuals, shows a particular disjunction between the left and right columns: The page size of the automobile manuals was consistently larger than either the sewing machine or mower-reaper manuals. There were only three sewing machine or mower-reaper manuals with more pages than the shortest automobile instruction manual, that of Ford in 1970. Not only were they usually larger, but in the 1970s and 1980s, the automobile manuals skyrocketed in size to 400% and 500% beyond anything ever produced in the sewing machine and mower-reaper industries.

Thus, the automobile manuals were physically larger artifacts than either the sewing machine or mower-reaper manuals, hence there was a definite discontinuity in the size of instruction manuals. However, in this much larger-sized, 20th-century version of the instruction manual, what portion was com-

Table 6.2. Numbers and Percentages of Instruction Pages in Automobile Instruction Manuals.

Instruction Pages in Ford Manuals (% of Total Pages)			Instruction Pages in Chevrolet Manuals (% of Total Pages)	
41	(100)	1910s	48	(94)
60	(100)	1920s	63	(95)
50	(100)			
59	(97)	1930s	66	(94)
62	(95)			
56	(88)	1940s	58	(97)
28	(87)	1950s	31	(94)
56	(77)	1960s	48	(98)
22	(70)	1970s	75	(95)
152	(85)	1980s	121	(98)

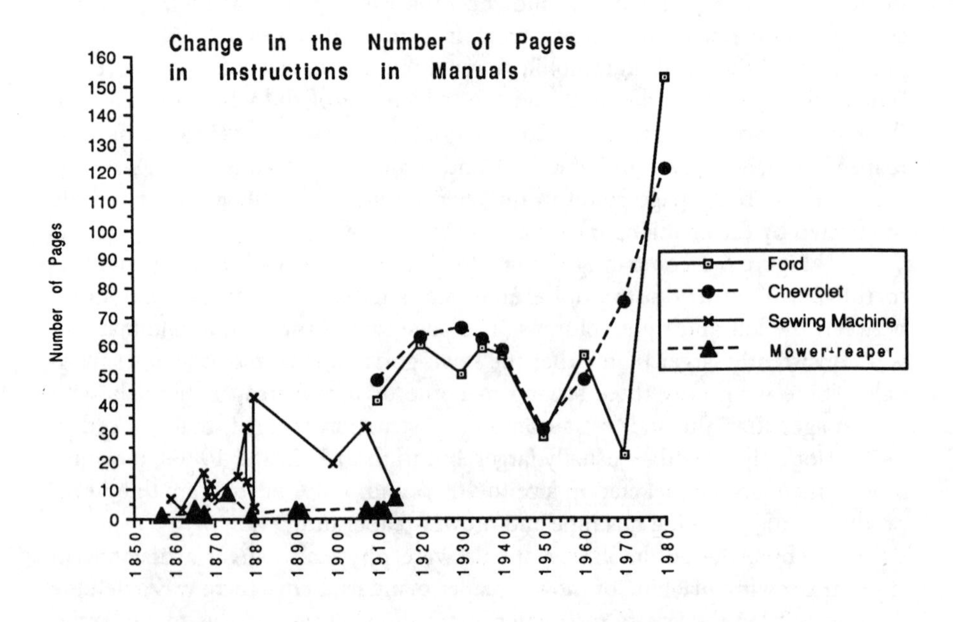

Figure 6.1. The changing size of automobile user manuals

posed of instructions? In considering the portion devoted to instructions, Figure 6.2 depicts an interesting continuity with the sewing machine manuals for at least the first 40 years of the existence of automobile instruction manuals.

From the 1910s to the 1950s, it was almost as if the Ford and Chevrolet manuals simply continued a single line established by the sewing machine manuals from the 1860s to the 1910s. (The Chevrolet manuals continued this continuity through to the 1980s.) The new manuals of the automobile industry maintained an 87% or higher portion devoted to instructions for quite some time, just as did 10 of the 14 sewing machine manuals. It is also interesting to note the discontinuity between the sense of textual brevity in the mower-reaper and the automobile manuals. Figure 6.2 strongly suggests it was not the tradition of textual brevity from the mower-reaper manuals that was continued in the automobile manuals. Only once, in a 1970 Ford manual, did the instructions fail to account for less than 75% of the entire manual. The 1970 Ford manual, like the mower-reaper manuals before it, included such items as coupons and advertisements for do-it-yourself manuals and tools, lists of Ford District Offices, and maintenance record blank pages.

However, the automobile manuals achieved and maintained the genre element of textual brevity with much effort. Why, for example, was there a need to abruptly decrease the page counts in the 1950s as well as the percent-

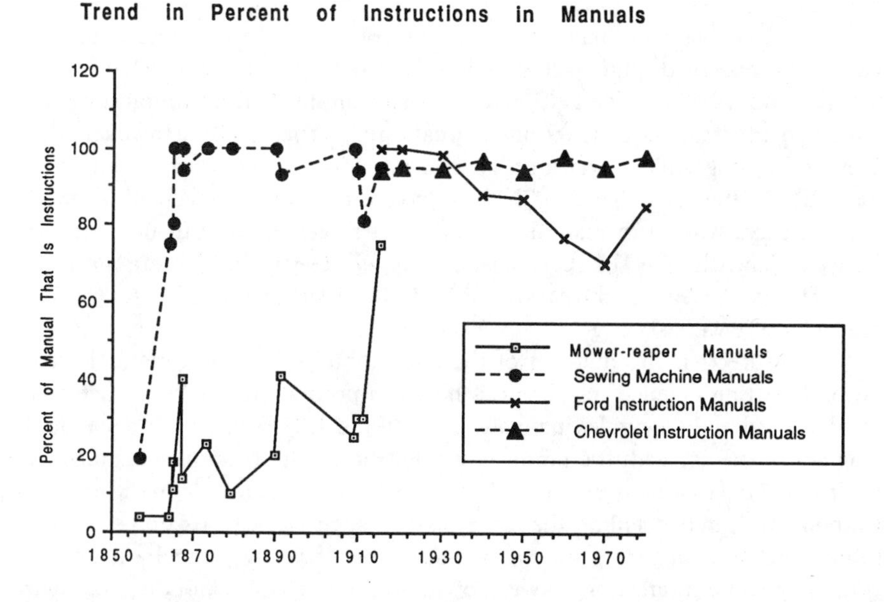

Figure 6.2. Trend in percentage of total instruction manuals pages used for instructions

age of instructions from 1940—1970? If these manuals were indeed so brief and focused, why did Chevrolet decrease its page counts from around 60 pages in 1920—1940 to around 30 in their edition of 1950?

What may have caused the manuals to decrease abruptly in size was, among other reasons, the dispersion of content from a single instruction manual into a number of other manuals and booklets. For example, in the inside cover of the 1941 Chevrolet manual, the author offered to sell another type of manual:[27]

> To anyone desirous of having more mechanical information covering the design and construction of the 1941 passenger car, we will send a copy of the Chevrolet Shop Manual upon receipt of a remittance of 25 cents in coin.

Also, in the Mustang manual, an order form for *The 1965 Comet-Falcon-Fairlane-Mustang Shop Manual* was offered. Thus, one reason the manuals became smaller was because an increasing amount of the mechanical how-to information was separated out of the instruction manuals and put into "shop manuals." Shop manuals were a genre quite different from instruction books whose audience was, as the Model T manual stated: ". . . owners who in the great majority of cases have little or no practical experience with things mechanical . . ."[28]

This division of instruction manual content into various new subgenera can be observed quite visibly in Table 6.3 both by what is deleted from a manual and what is added. Chevrolet, for example, had a Troubleshooting Guide in most of its instruction manuals up to the 1970s. However, this Troubleshooting Guide, even in a general form, was deleted from their 1980 manual. On the other hand, with the decrease of detailed, mechanical information, there was an increase in nonmechanical, general information such as how to clean the car. In the *Model T Manual* (1919-1925) there were no instructions on cleaning the car; however, in the *1988 Chevrolet Beretta Owner's Manual*, there were six pages.

Moreover, the division of the automobile instruction manual went beyond putting detailed, mechanical, how-to information into shop manuals. In both Ford and Chevrolet manuals, up until the 1980s, there was a standard diagram of the car with the lubrication points and a lubrication schedule (see Figure 7.7). However, as early as the 1941 Ford manual, maintenance coupons appeared detailing the same maintenance work to be done and the same schedule, but now the work was to be done by a mechanic. Such coupons invited interaction between owners and mechanics and directed their discussion on lubrication and other maintenance services. By the 1970s, Chevrolet had repackaged such maintenance information into a separate booklet of coupons, and Ford followed suit in the 1980s. Additionally, as early

Table 6.3. Items Included in Automobile Instruction Manuals.

	Includes a Trouble-shooting Guide		Includes a Lubrication Diagram and Chart Maintenance Timing Chart		Includes a Warranty and Explanation of Warranty	
	Ford	Chevrolet	Ford	Chevrolet	Ford	Chevrolet
1910s	No	*	No	*	No	Yes
1920s	Yes	No	Yes	Yes	No	Yes
1930s	Yes	Yes	Yes	Yes	No	Yes
	No	No	Yes	Yes	Yes	Yes
1940s	No	No	Yes**	Yes	Yes	Yes
1950s	No	No	Yes	Yes	Yes	Yes
1960s	No	Yes	No	Yes	Yes	No***
1970s	No	Yes	No	No****	No***	No
1980s	No	No	No****	No	No	No

*The 1919 Chevrolet manual was missing pages.

**Do-it-yourself lubrication diagrams and maintenance charts are joined for the first time by coupons to be given to mechanics detailing services to be performed.

***Warranty has now grown so large it is in a separate booklet.

****Maintenance Schedule chart and coupons to be given to mechanics detailing services to be performed have grown so large they are now in a separate booklet.

as the 1919 Chevrolet manual, there was a warranty that described the legal responsibilities and liabilities of owners and car makers. However, by the 1960s, this warranty material was also split off into a separate manual by Chevrolet, with Ford following suit in the 1970s. Thus, the instruction manual was progressively divided into:

- instructions with general how-to information (called the "Owner's Manual" after the 1950s),
- instructions with detailed, mechanical information (called "Shop Manuals"),
- coupon books organizing and scheduling regular maintenance (called "Maintenance Schedule" booklets), and
- warranty booklets.

Thus, the genre element of textual brevity in the automobile instruction manuals—within a single industry and with quite similar audiences—displayed many more similar characteristics than did the sewing machine and mower-reaper manuals, which were from disparate industries addressing different audiences. The way in which the Ford and Chevrolet data lines seem to follow each other in the right column of Figures 6.1 and 6.2 is in great contrast to the almost mirror image opposites of the sewing machine and mower-reaper manual data lines in the left columns. Interestingly, as much as Ford and Chevrolet are similar to each other, they are not identical: The deviation in the percentage of instructions displayed in Figure 6.2 shows that authors in identical industries, addressing identical audiences, could also vary in the construction of their manuals because of idiosyncratic corporate genre considerations.

One can also observe in these first few figures and tables how the genre element of brevity declared in the 1880s by the sewing machine industry perhaps helped to predispose the 20th-century automobile industry to create instruction manual subgenres such as the shop manual, the warranty booklet, the maintenance schedule booklet, and so on.[29]

Conclusions About the Genre of Instruction Manuals Derived From the Element of Textual Brevity in Automobile Instruction Manuals

Four conclusions can be reached from the information contained in Tables 6.2 and 6.3 and Figures 6.1 and 6.2:

- Automobile instruction manuals were consistently larger than the sewing machine and mower-reaper manuals. Thus, Figure 6.1 shows a large discontinuity between the sizes of instruction manuals in the different industries.
- Automobile instruction manuals revealed much consistency quantitatively for the first 40 years with the genre norm of textual brevity established by the sewing machine manuals.
- Automobile instruction manuals continued to employ the element of textual brevity by progressively splitting off information from the single instruction manual into separate owner's manuals, shop manuals, maintenance schedule booklets, and warranty booklets.
- Instruction manuals written by companies in the same industry varied less than those written by different industries, but they were still differentiated by idiosyncratic corporate concepts of the instruction manual.

EXAMINING THE GENRE CHARACTERISTICS OF PROFUSENESS OF ILLUSTRATIONS IN THE AUTOMOBILE INSTRUCTION MANUALS

The percentage of instruction pages with illustrations in the automobile manuals is given in Table 6.4 and is graphed in Figure 6.3. These findings for the automobile manuals are put into context by comparing them to the percentages in the sewing machine and mower-reaper manuals (see Tables 5.4 and 5.5 in Chapter 5) and are also graphed in Figure 6.4.

Figure 6.3 shows that the manuals of Ford and Chevrolet were quite similar not only in their general percentage of graphics per instruction page, but also in the percentage increases and decreases. (Percentages shown in Figure 6.3 could reach over 100% when there was more than one illustration per instruction page, i.e., during the 1950s and 1960s.)

The contrast, however, between the left and the right segments of the data lines in Figure 6.4 is most dramatic. The profuseness of illustrations in the mower-reaper and sewing machine industries was very inconsistent, whereas those within the single industry (toward the right) reflect each other more closely. The highest and lowest percentages of instruction pages with graphics in the automobile manuals appear to stay within the range established in the last 30 years (from 1880 to 1910) of the sewing machine and mower-reaper manuals. It is almost as if the genre consensus on profuseness of illustration that only slowly developed in these two earlier industries was bequeathed as an unquestioned fact to the new automobile industry and its authors. Rarely did the automobile manuals go below a level of 40% of their instruction pages with illustrations, and, more often than not, they had more—above 100% of their instruction pages had illustrations.

Table 6.4. Percentage of Pages in Automobile Instruction Manuals with Graphics.

	Ford Instruction Manuals % of Pages with Graphics	Chevrolet Instruction Manuals % of Pages with Graphics
1910s	43	96
1920s	43	49
1930s	66	54
	47	70
1940s	69	74
1950s	111	70
1960s	105	145
1970s	40	68
1980s	61	89

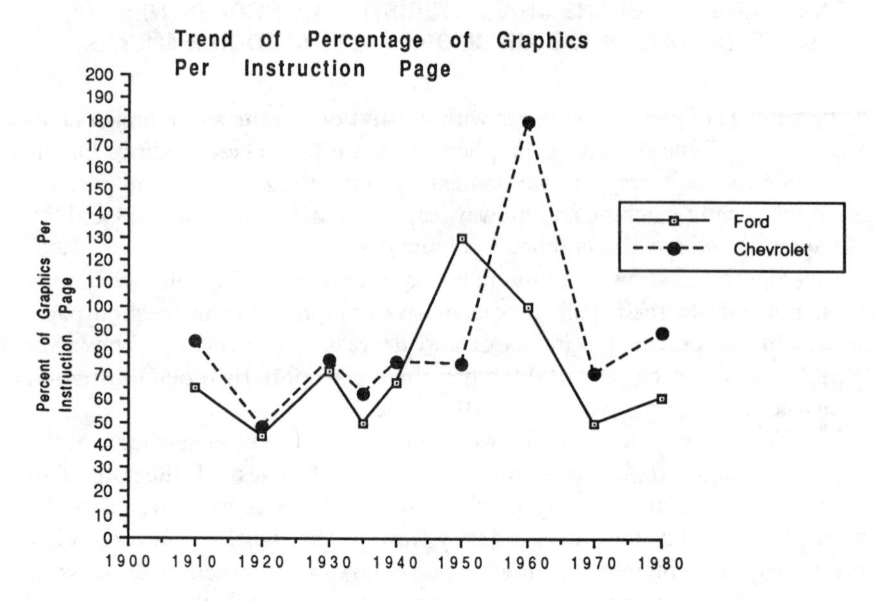

Figure 6.3. Trend in percentage of instruction pages with graphics in automobile manuals

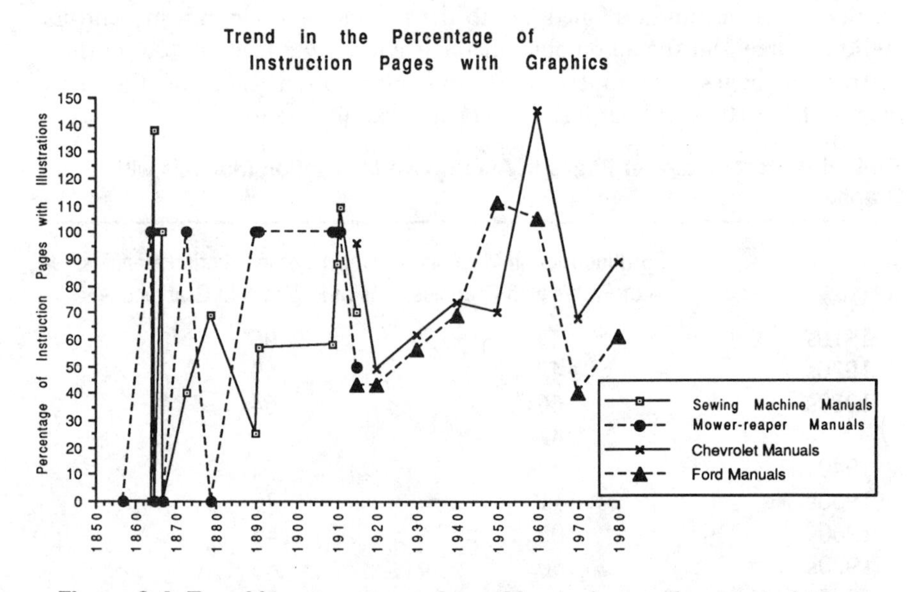

Figure 6.4. Trend in percentage of graphics on instruction pages in four types of instruction manuals

In this element of profuseness of illustration, consequently, there was a continuation of the visual communication tradition of American technical communication. In fact, in some of the automobile manuals of the 1950s and 1960s with more than two graphics per page, much of what is communicated about the automobiles is communicated visually just as in the days of Evans or Griffiths. Moreover, among the graphics that marked the rise seen in Figures 6.3 and 6.4, of much higher percentage were ancillary types of illustrations than had appeared in earlier, 19th century instruction manuals. Many more of the automobile illustrations of the 1950s and 1960s were not specifically referenced in the text and attempted to convey information directly to the viewer without the need to read the surrounding text. From 1910 to 1940, however, most illustrations were correlative and intended to be used with the accompanying text rather than alone. (It is interesting to note that the increased presence of illustrations in the manuals of the 1950s and 1960s was rapidly followed by a return to the established genre norm.)

Conclusions About the Genre of Instruction Manuals Derived From the Element of Profuseness of Illustration in Automobile Instruction Manuals

Three conclusions can be drawn from looking at the genre element of profuseness of illustration listed in Table 6.4 and plotted in Figures 6.3 and 6.4:

- The manuals within an industry were consistent in the genre element of profuseness of illustration.
- The manuals of the automobile industry remained within the range of illustration profuseness established at the very end of the time period studied for the sewing machine and mower-reaper manuals. Thus, unlike the time of experimentation and slow general acceptance of the definitions of the instruction manual genre in the 19th-century, the 20th-century automobile authors began writing within an established genre and understood that the manuals should have a large percentage of instructions and be relatively profusely illustrated.
- The rise above the genre norm in the use of illustrations by both Ford and Chevrolet in the 1950s and 1960s was quickly followed by a return to the general norm.

EXAMINING THE GENRE CHARACTERISTICS OF EXPLICIT SIGNS OF STRUCTURE IN THE AUTOMOBILE INSTRUCTION MANUALS

Authors can construct explicit signs of structure in a number of ways. However, I only examined four here, and I examined them in order of their relatively micro- to macrolevel of revealed structure: paragraphs per instruction page, number of paragraphs labeled or tagged, headings per page, and use of a typographic heading hierarchy, and the inclusion of a table of contents or index in the manual. Each one of these elements in a manual allows a reader to read selectively with greater and greater spans of choice. Accordingly, if the author-reader relationship had indeed changed as the prefaces of the automobile manuals suggested, then each of these four ways of explicitly signifying a text's structure should increase during the history of the automobile instruction manual.

First Explicit Sign of Page-level Structure: Paragraphs per Instruction Page

My first look at structure was a very simple one: How often did the authors of the manuals break continuous text into smaller components, that is, paragraphs? By breaking up a continuous block of information into smaller components, authors visually suggest to readers how the information can be conceptually broken down.

I was able to develop the paragraphs per page findings contained in Table 6.5 by choosing three instruction pages in an automobile manual (one near the beginning, one near the middle, and one near the end), counting the number of paragraphs in each, and deriving the average of the three pages. The findings are graphed in Figure 6.5 and put into the context of the sewing machine manual paragraphing techniques.[30]

Once again, the manuals of two companies within the same industry reflected a common instruction manual genre characteristic by increasing and decreasing in the number of paragraphs per page in similar ways. It is also

Table 6.5. Paragraphs Per Page in Automobile Instruction Manuals.

	Ford Instruction Manuals Paragraphs per Page	Chevrolet Instruction Manuals Paragraphs per Page
1910s	4.6	7
1920s	4.3	6
1930s	5.1	8.1
	8.1	8.6
1940s	6.3	8.1
1950s	5	8.3
1960s	10.1	7.4
1970s	8.3	6
1980s	4	6.9

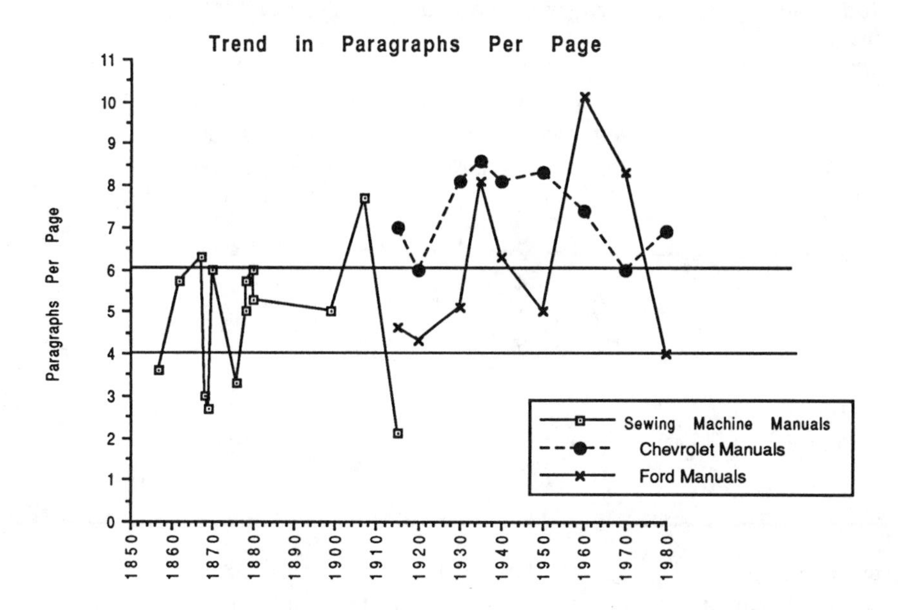

Figure 6.5. Paragraphs per page in instruction manuals

interesting to note that the breaking up of continuous text happened far more often in the automobile manuals than in the sewing machine manuals. None of the automobile manuals had as few paragraphs per page (five of the sewing machine manuals had four or fewer paragraphs per page), and many of the automobile manuals had far more paragraphs per page (seven of the automobile manuals had eight or more paragraphs per page).

Thus, in this first measure of how automobile authors revealed page-level organization in explicit ways, automobile authors did so far more frequently than did the sewing machine manual authors.

Second Explicit Sign of Page-level Structure: Number of Paragraphs Labeled or Tagged

I discovered the findings presented in Table 6.6 by totaling up the number of boldfaced labels "danger," "caution," or "note" in instruction manuals. My findings for the automobile manuals were placed in context by looking for the same labels in the sewing machine manuals. Findings are listed in Table 6.9.[31]

Unexpectedly, until the 1970s, the number of automobile manual paragraphs labeled with "Danger," "Caution," or "Note" did not become dramatically greater than those used in the last 30 years of the sewing machine manuals. In the last four sewing machine manuals examined (Table 6.9), the

Table 6.6. Number of Paragraphs Labeled with "Danger," "Caution," or "Note."

	Ford Instruction Manuals Paragraphs Labeled Danger, Caution, Note	Chevrolet Instruction Manuals Paragraphs Labeled Danger, Caution, Note
1910s	0	3
1920s	2	4
1930s	4	6
	1	2
1940s	2	6
1950s	3	8
1960s	6	6
1970s	6	49
1980s	113	69

average number of paragraphs labeled Danger, Caution, or Note was six. For some 60 years, this number was exceeded in the automobile instruction manuals only once (by Chevrolet in the 1950s). There was no large increase in these labels for some 60 years despite the fact that the automobile instruction manuals told readers to perform dangerous activities never even contemplated in the sewing machine manuals. For example, in the *Instruction Book for Ford Model T Cars* (1913), readers were instructed:[32]

> To ascertain which, if any of the four plugs are fouled with oil, short circuited with carbon or inoperative from some other cause, open the throttle two or three notches to speed up the motor; now hold your two fingers on two outside vibrators so that they cannot buzz. The evenness of the exhaust will show that the other two are working correctly and that the trouble is not there. . . . If this procedure does not locate the trouble, disconnect that particular cylinder wire from the coil and ground the spark plug end to some part of the engine. Hold the other end of the wire near the coil terminal, and if sparks are produced, it is evident that there is nothing wrong with the coil.

In the Instructions for the *Operation and Care of Chevrolet Motor Cars* (1932), readers were told to "be careful":[33]

> Grasp a screwdriver by its wooden handle holding the metal end against the spark plug terminal and at the same time touch the cylinder head, thus short circuiting the plug. Be careful to hold the screwdriver by the wooden handle, or a shock will result . . .

Finally, in Ford's 1941 *Deluxe and Super Deluxe Reference Book*, readers were advised to do the following in case of trouble while "on the road":[34]

> To determine whether or not a spark is being delivered to the spark plug, hold a spark plug wire approximately 1/4 inch away from the cylinder head (see Fig. 25) as the engine is cranked with the starting motor. . . . If a spark is noted from each of the wires, the trouble is not likely to be with the ignition system.

At no point in 1941 did Ford caution the reader against getting shocked or performing this quick-fix while standing in a puddle. Thus, the manuals in this new industry persistently followed the established genre labeling norm set by the sewing machine manuals, despite an actual need for an increased use of "Danger," "Caution," and "Note" labels. In this instance, then, the genre negotiations between readers and authors failed to change and compensate for quite different circumstances.

However, things changed quite dramatically regarding this genre in the 1970s and 1980s. Perhaps the very dramatic upsurge in the use of "Danger," "Caution," and "Note" labels in the 1970s by Chevrolet and a decade later by Ford occurred less because of author-audience-genre negotiations and more because of legislative and judicial stipulations. It may not have been coincidental that Ralph Nader's exposé of the Chevrolet Corvair in *Unsafe At Any Speed* (1965) was succeeded by an 800% increase in the use of paragraph labels by Chevrolet in the 1970s. However, Ford did not similarly increase its use of paragraph labels at that time, but did increase them by some 1880% later in the 1980s. Perhaps Nader's exposé was initially considered by Ford to be Chevrolet's problem. Yet, in the late 1970s when Ford had to pay some $20 million dollars in approximately 125 liability lawsuits—including the famous Pinto gas tank case (concluded March 13, 1980)—liability quickly also became Ford's problem, and, thus, their manuals were redesigned to include many more labels of "Danger," "Caution," and "Note."[35]

These extratextual influences also happened to occur at just the same time that both Ford and Chevrolet began including small print paragraphs in their introductions. The following example is from the *Ford 1971 Passenger Cars: Car Care and Operation*:[36]

> The description and specifications found in this manual were in effect at the time the book was approved for printing. The Ford Companies reserve the right to discontinue models at any time, or to change specifications or design, without notice and without incurring obligation.

Additionally, these extratextual influences also occurred just prior to Ford's 1986 Escort manual which for the first time included a lengthy, eight-page explanation of the Ford Consumer Appeals Board.

In the automobile industry, something akin to the comet collision that ended the ecological equilibrium of the dinosaurs' Cretaceous period happened to the instruction manual genre norms between 1960 and 1980. Moreover, this extra-textual influence did not simply affect the numbers of paragraph labels but, indeed, the entire genre. For example, the size of manuals dramatically jumped beyond the established genre norms at exactly the same dates—1970 and 1980—as the jump in the number of paragraph labels. Moreover, the jump in manual sizes occurred in exactly the same sequence—first Chevrolet, then Ford.

It would be most interesting to look again at this genre element from the vantage point of another 40 or 50 years. Is the dramatic rise in paragraph labels apparent in Figure 6.6 similar to the rapid upturn followed by a rapid downturn in the percentage of graphics per page as seen in Figure 6.3, or does the dramatic rise in Figure 6.6 establish a whole new genre norm? How will automobile instruction manual authors work out the conflict between the new safety labeling element and the genre element of textual brevity? For example, some pages of the *1986 Ford Escort Owner Guide* had more "Dangers," "Cautions," and "Warnings" than they did normal text and illustrations. These

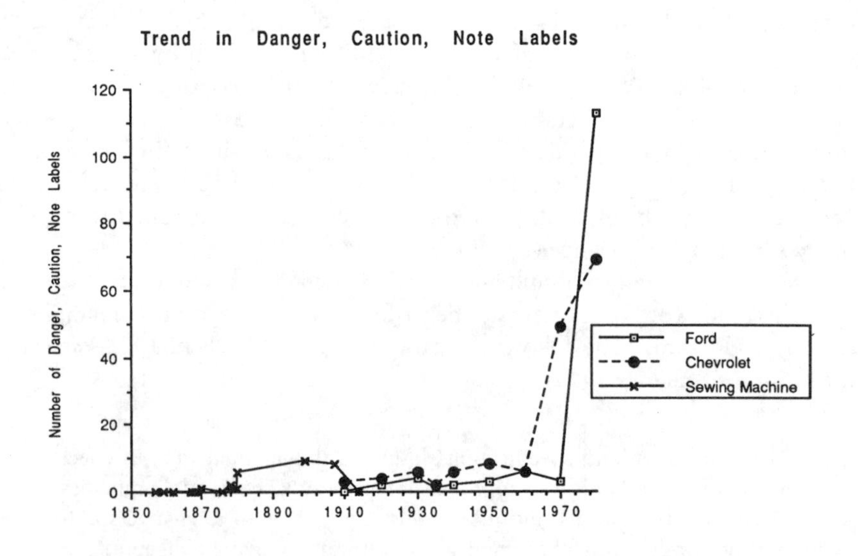

Figure 6.6. Trend in instruction manuals for paragraphs labeled danger, caution, or note

labeled paragraphs fragmented whatever narrative flow there was in the manuals and thus discouraged readers looking for information among all the corporate protective armor of "Dangers," "Cautions," and "Notes." Will all these labeled paragraphs be separated out into another yet-to-be invented subgenre manual like the shop manual, and the warranty booklet?

Another way that authors could present explicit signs of page-level structure is by the use of some kind of typographic tag: pointing hands, bullets, indented lists, italic print, tables, boxes around text, printing text in all uppercase, underlining, or printing text with a background screen (such as a blue tint behind text).[37] In each of these ways, authors can alert readers to qualitative differences between portions of text.

The presence of tagged paragraphs is listed in Table 6.7 and graphed in Figure 6.7.[38] I derived the percentage of tagged paragraphs by counting all the paragraphs in the instruction pages, counting all the tagged paragraphs, and dividing to find the percentage.

Figure 6.7 seems to present the most divergent aspect of the Ford and Chevrolet manuals thus far seen and suggests little industry consensus on this genre element. Beginning in the 1950s, Chevrolet consistently had at least 200% more tagged paragraphs than did Ford. This was also true in the first two decades of the 1900s.

Table 6.7. Automobile Manuals with Tagged Paragraphs.

	Ford Instruction Manuals Percentage of Paragraphs Tagged	Chevrolet Instruction Manuals Percentage of Paragraphs Tagged
1910s	2	35
1920s	3	25
1930s	13	13
	15	10
1940s	5	7
1950s	8	110*
1960s	16	71
1970s	36	72
1980s	31	88

* One can have 110% when there are more typographical tags in a manual than there are paragraphs.

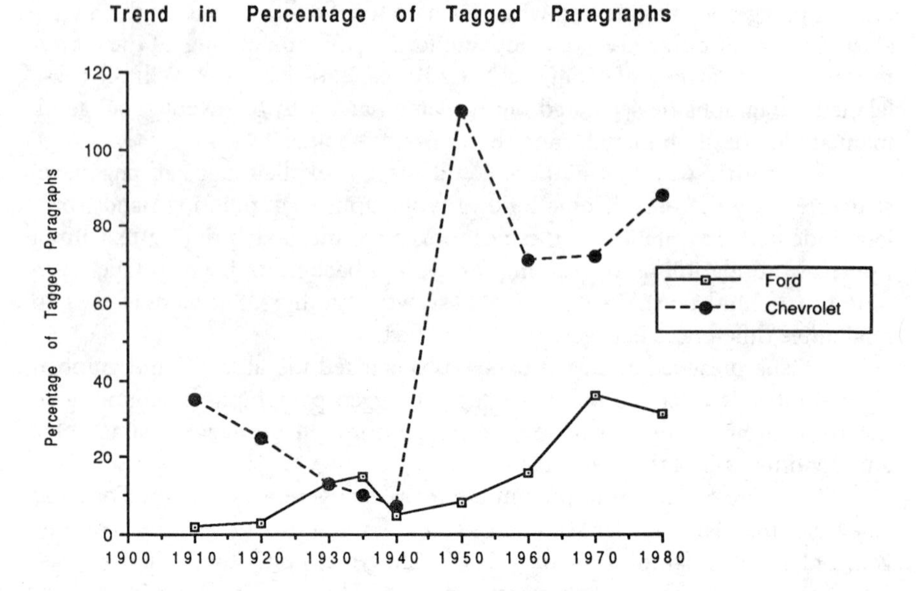

Figure 6.7. Trend in percentage of paragraphs that are tagged in automobile instruction manuals

Conclusions About the Genre of Instruction Manuals Derived From the Explicit Signs of Page-level Structure (Paragraphing, Labeling, and Tagging Paragraphs)

Three conclusions can be drawn from looking at the genre element of explicit signs of page-level structure listed in Tables 6.5, 6.6, and 6.7 and plotted in Figures 6.5, 6.6, and 6.7:

- In the frequency of paragraphing, the automobile manuals did indeed show, as predicted, an increase over the frequency in the sewing machine manuals.
- In the labeling of paragraphs, unlike what was predicted, there was not much change in the genre norm despite the fact that the dangers involved in the actions called for by the automobile manuals were significantly greater than in the sewing machine manuals. However, when the external impetus of exposés and lawsuits occurred in the late 1960s for Chevrolet and in the 1970s for Ford, the genre norm of labeling changed quickly and dramatically.
- In the use of typographic tags to indicate structure, there seemed to be little consensus even within one industry. Chevrolet for 60 of the 80 years made quite dramatically greater use of this element to reveal structure than Ford.

Thus, some signs of explicit structure at a page level were used more often as predicted by the shift in authority from author to reader and in the increase of selective reading. Yet, not all the signs were consistently used, nor can it be ascertained whether the increase in the frequency of signs, such as the labels, came from governmental stipulation as in the case of patent genre or from reader-author-genre negotiations as in the case of the 19th-century instruction manuals.

Third Aspect of Explicit Signs of Chapter-level Structure—Headings Per Page and Use of Heading Typographic Hierarchy

I examined the explicit signs of chapter-level structure by looking at the percentage of headings per instruction page and how structural hierarchy was signaled typographically in these headings. An increasing number of headings per page, and an increasingly deep hierarchy typographically signaled in the headings, should have occurred if the readers were allowed to read selectively within a chapter.

The number of headings per page in the automobile instruction manuals is listed in Table 6.8, for the sewing machine manuals in Table 6.9, and for the mower-reaper manuals in Table 6.10.

Figure 6.8 shows the number of headings per instruction page found in the manuals from both the 19th and the 20th centuries. Also, very much like Figures 6.2 and 6.6, there is a very consistent norm between 1 and 2.5 headings per page, with only the 1980 Ford Escort manual showing a dramatic divergence from the normal range. (Is this again fallout from the legislative and judicial extratextual stipulations?) In fact, for once the automobile manuals seemed to use genre elements from both the sewing machine as well as the

Table 6.8 Headings Per page in the Ford and Chevrolet Instruction Manuals.

	Ford Instruction Manuals Headings Per Page	Chevrolet Instruction Manuals Headings Per Page
1910s	0.97	1.15
1920s	2.6	1.8
1930s	2	1.9
	2.4	1.8
1940s	1	2.5
1950s	2.3	2.2
1960s	2.8	3.4
1970s	5.2	2.9
1980s	10.8	2.8

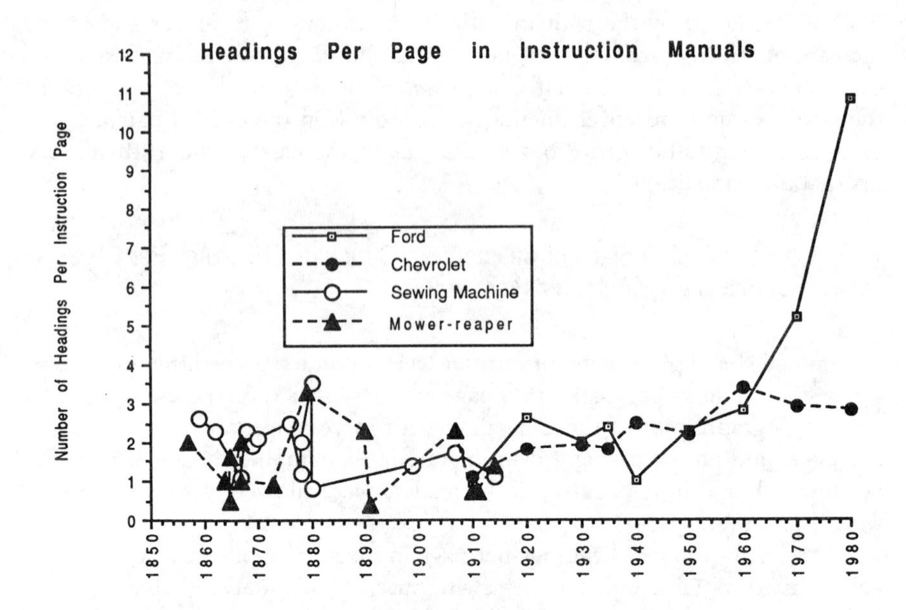

Figure 6.8. Trend in the headings per instruction page found in both 19th- and 20th-century instruction manuals

mower-reaper manuals. Thus, unlike what was expected, on one measurement of the explicit signs of chapter-level structure, the automobile manual authors did not dramatically vary from the norm of headings per page.

Tables 6.9, 6.10, 6.11, and 6.12 list the various percentages of headings typographically signaling whether they are first, second, third, or fourth level. These typographic signals could be typesize, boldness, centering in a column, or distance from the text following the heading. These signals are shown in Figures 6.9 to 6.12.

There was not a very deep hierarchy of structure signaled typographically in the sewing machine manuals (see Figure 6.9). There were no fourth-level headings and very few third-level headings. Most of the headings in the sewing machine manuals projected the same hierarchy level, that is, second-level headings.

However, three manuals stand out from the rest in signaling a deep hierarchy with first-, second-, and third-level headings:

- *Instruction Book for the Howe Sewing Machine Step Feed* (Howe Machine Co., 1867),
- *New American Sewing Machine* (New American Sewing Machine, 1878), and

Table 6.9. Percentages of Various Levels of Headings Found in Sewing Machine Instruction Manuals.

	Number of Instruction Pages	Total Number of Headings Per Instruction Page	% of 1st Level	% of 2nd Level	% of 3rd Level	% of 4th Level	Para-graphs per Page	Total Number of Paragraphs Labeled Danger, Caution, Note
1859	7	18 (2.6)	6	94	0	0	3.6	0
1862	3	8 (2.6)	0	100	0	0	5.7	0
1867	16	18 (1.1)	6	83	0	0	6.3	0
1868	8	19 2.3)	0	100	0	0	3	0
1869	12	23 (1.9)	0	100	0	0	2.7	0
1870	8	17 (2.1)	12	88	0	0	6	1
1876	15	38 (2.5)	5	95	0	0	3.3	0
1878	32	39 (1.2)	5	72	23	0	5	2
1878	13	26 (2)	8	92	0	0	5.7	1
1880	4	14 (3.5)	7	93	0	0	6	1
1880	42	33 (.8)	9	91	0	0	5.3	6
1899	19	28 (1.4)	0	100	0	0	5	9
1907	32	56 (1.7)	8	54	38	0	7.7	8
1914	9	10 (1.1)	0	90	10	0	2.1	0

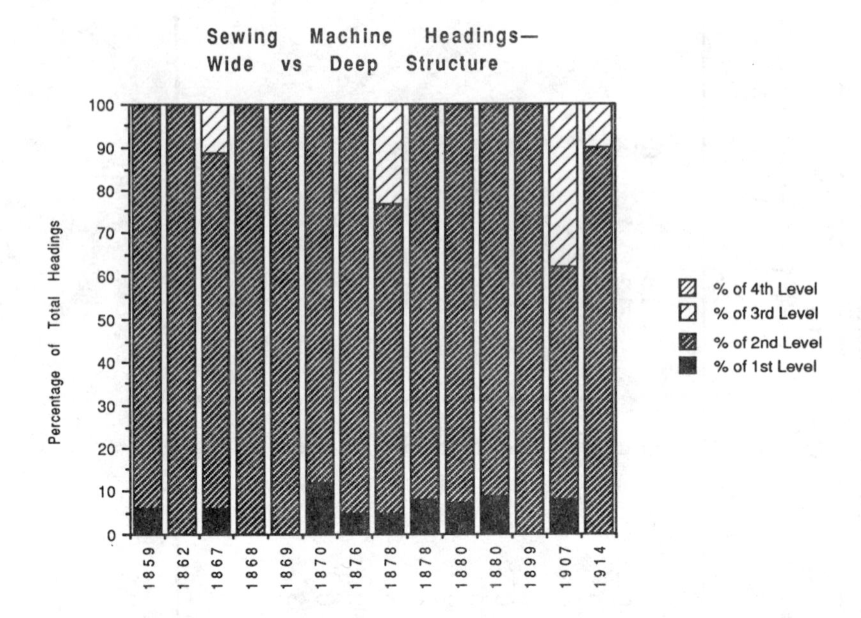

Figure 6.9. Percentages of heading types in sewing machine manuals

- *Directions for Using the Wilcox and Gibbs Automatic Noiseless Sewing Machine* (Wilcox and Gibbs Sewing Machine Co., 1907).

Interestingly, these were three of the five sewing machine manuals with the most pages (note the left-most column of Table 6.9): the 1867 manual had 16 pages, and the 1878 and 1907 manuals each had 32 pages.

Thus, there seems to be a correlation between number of pages in a sewing machine manual and the presence of a deep heading structure. However, such a correlation does not fit the fact that the other two sewing machine manuals with the most pages, the 42-page *Domestic Sewing Machine Without a Teacher* (Domestic Sewing Machine Co., 1880) and the 19-page manual *The New Companion* (Perry Mason and Co., 1899), did not have such a deep hierarchy. In fact, the longest sewing machine manual, *Domestic Sewing Machine Without a Teacher* (1880), had as much explicitly signaled hierarchy as a 4-page manual, *Improved Common Sense Family Sewing Machine* (A. G. Mason Mfg. Co.,1880), published in the same year.

Thus, for the most part, the sewing machine manuals typographically signaled a flat hierarchy that did not very effectively aid readers in selectively choosing what to read at a chapter level. However, there did seem to be an incipient notion by the sewing machine manual authors that as the number of pages of a sewing machine manual increased, so should the depth of the signaled heading hierarchy.

Most mower-reaper manuals, on the other hand, presented their readers with even fewer explicit signals for chapter-level structure. Of course, the need to reveal structure in the mower-reaper manuals was a relatively small one because most had only a few instruction pages (see Table 6.10, left-hand column).

However, two mower-reaper manuals did not fit this flat hierarchy generalization, *Wilber's Eureka Mower, Direct Draft* (Eureka Mower Co., 1879) and *Directions for Setting Up and Operating the McCormick No. 6 and Big 6 Plain Lift Mowers* (International Harvester Company of America, 1915). Furthermore, trying to apply the possible correlation from the sewing machine manuals of size and depth of signaled hierarchy simply did not hold in these mower-reaper manuals. The correlation of the presence of signaled heading hierarchy to the size of the manual did fit the 1915 mower-reaper manual because, with three levels of headings, it also contained the second most numbers of pages (seven). However, the 1879 manual also had a three-level heading hierarchy, yet it was one of the smallest mower-reaper manuals. By the same token, the largest of the mower-reaper manuals, the nine-page *Directions for Putting Together and Using Walter A. Wood's Harvesting Machines* (Walter A.

Table 6.10. Percentages of Various Levels of Headings Found in Mower-reaper Instruction Manuals.*

	No. of Instruction Pages	Total No. of Headings (Per Instruction Page)	% of 1st Level	% of 2nd Level	% of 3rd Level	% of 4th Level
1857	1	2 (2)	50	50	0	0
1864	1	1 (1)	100	0	0	0
1865	2	3 (1.6)	33	66	0	0
1865	4	2 (.5)	100	0	0	0
1865	2	1 (.5)	100	0	0	0
1867	2	4 (2)	50	50	0	0
1867	1	1 (1)	100	0	0	0
1873	9	8 (.9)	88	12	0	0
1879	1.5	5 (3.3)	40	40	20	0
1890	3	7 (2.3)	29	71	0	0
1891	2.5	1 (.4)	100	0	0	0
1909	3	7 (2.3)	29	71	0	0
1910	3	2 (.7)	100	0	0	0
1911	3	2 (.7)	100	0	0	0
1915	7	9 (1.4)	11	22	67	0

*No column on danger, caution, or note is included because only two such items appeared in the entire corpus

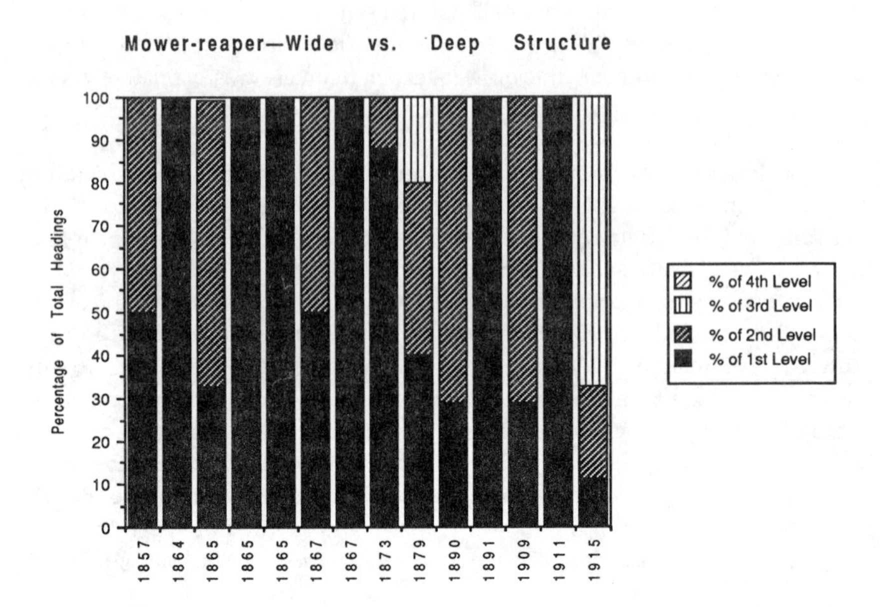

Figure 6.10. Percentages of heading types in mower-reaper manuals

Wood Mowing and Reaping Machine Co., 1873) had a very flat hierarchy with few second-level headings and no third-level headings. Thus, the notion of correlating the size of the manual and the depth of heading hierarchy illustrated in the sewing machine manuals seems not to apply to the mower-reaper manuals.

When considering the depth of heading hierarchy in the automobile manuals, remember the relative size of instructions in the various manuals (Figure 6.1). The mower-reaper and sewing machine manuals were consistently smaller than the automobile manuals; on only four occasions did the 29 sewing machine and mower-reaper manuals grow in size larger than 20 pages, yet none of the automobile manuals ever had less than 20 pages. Perhaps because of this growth in size of the manuals, the correlation of size and signaled hierarchy structure was present in the Ford manuals' hierarchies listed in Table 6.11 and graphed in Figure 6.11.

As displayed in Figure 6.11, the majority of the Ford manuals (five out of nine) do signal a heading structure that includes first-, second-, third-, and even fourth-level headings. Also, in as shown Table 6.12 and Figure 6.12, seven of the nine Chevrolet manuals have a structure with first-, second-, third-, and even fourth-level headings.

One interesting element to note is how the correlations of size and the presence of a deep heading hierarchy that initially seemed to work as an

Table 6.11. Percentages of Various Levels of Headings Found in Ford Instruction Manuals.

	% of 1st Level	% of 2nd Level	% of 3rd Level	% of 4th Level
1910s	77	23	0	0
1920s	10	90	0	0
1930s	18	81	1	0
	15	58	3	24
1940s	44	56	0	0
1950s	25	75	0	0
1960s	29	70	1	0
1970s	4	32	64	0
1980s	4	21	66	9

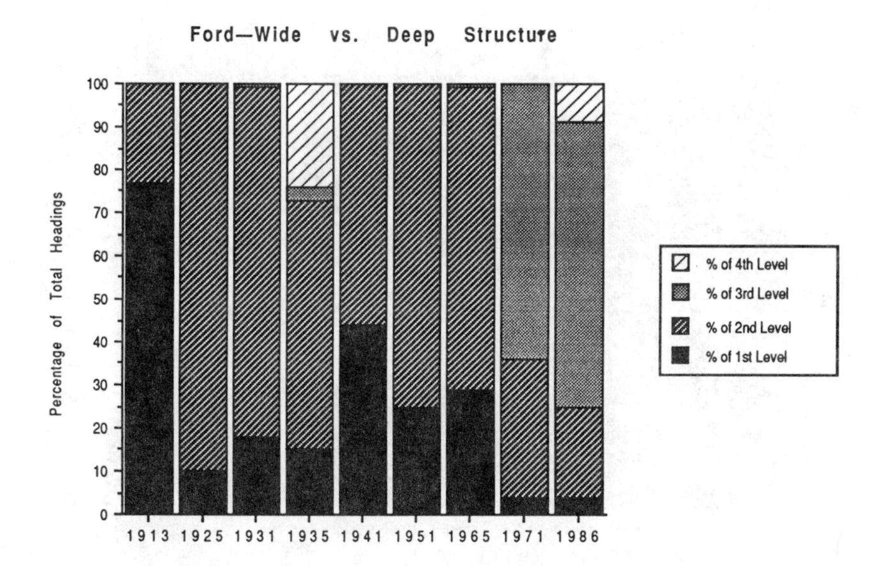

Figure 6.11. Percentages of heading types in Ford manuals

explanation were later lost. For example, the Chevrolet manuals became longer in the 1920s–1940s, and the heading structure in the 1930s–1940s, following the correlation, became more detailed and deeper. However, when the manuals of Chevrolet got much smaller by some 50% in the 1950s, Chevrolet continued to signal a deep hierarchy. In other words, the structure

Table 6.12. Percentages of Various Levels of Headings Found in Chevrolet Instruction Manuals.

	% of 1st Level	% of 2nd Level	% of 3rd Level	% of 4th Level
1910s	85	15	0	0
1920s	93	7	0	0
1930s	4	88	8	0
	3	70	27	0
1940s	9	43	46	2
1950s	16	37	47	0
1960s	7	50	35	8
1970s	2	17	62	19
1980s	2	42	53	3

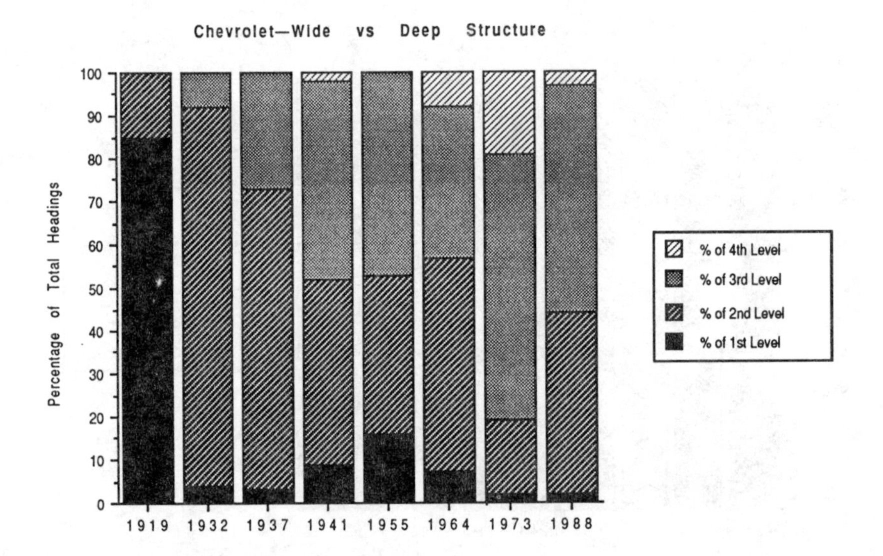

Figure 6.12. Percentages of heading types in Chevrolet manuals

signals continued to be used even though their purpose, to allow readers to handle the large number of pages, was actually diminished or made unnecessary by the fewer pages in the manual.

Despite the fact that the majority of Ford and Chevrolet manuals had at least a three-level structure, automobile manuals for the first two decades of the 20th century did maintain rather flat heading structures not unlike those in the sewing machine and mower-reaper manuals. Only in the 1930s did the automobile manuals display a consistently deep hierarchy of headings. Thus, the precision with which a reader could select alternative texts at a chapter level of structure did not change very much for some 20 years despite the fact that the industry and audience had changed.

Conclusions About the Genre of Instruction Manuals Derived From Chapter-Level Explicit Structure (Headings and Heading Hierarchy)

Four conclusions can be drawn from the genre element of explicit signs of chapter-level structure shown in Tables 6.8 through 6.12 and plotted in Figures 6.8 through 6.12:

- There seems to be a consistent use of between 1 and 2.5 headings per page for all the instructions manuals in all industries throughout the 130 years of instruction manuals studied.
- Few of the sewing machine and mower-reaper manuals typographically signaled a deep hierarchy of headings that included first-, second-, third-, and even fourth-levels of headings. There did seem to be some incipient correlation between the size of the sewing machine manuals and the presence or absence of a deep heading structure (the bigger the manual, the deeper the heading structure). Yet, the correlation was inconsistent, and no correlation between size and heading hierarchy seemed present in the mower-reaper manuals.
- The correlation between size and heading hierarchy did appear to explain the growth of heading hierarchy in automobile manuals of the 1930s and 1940s, but did not seem to apply when the automobile manuals decreased in size in the 1950s and 1970s.
- The use of a deep heading structure did not occur in the first two decades of the automobile manuals.

Fourth Aspect of Explicit Signs of Whole-Manual-Level Structure—The Inclusion of a Table of Contents or Index in the Manual

The final and fourth way of explicitly signaling structure is the incorporation of the most macrolevel, that is, inclusion of a table of contents, index, or bleeding edge tabs. This allows readers to choose alternative paths through a manual not just within a chapter as with headings or within a page as with paragraphing.[39]

Table 6.13 records the presence of tables of contents or indexes in sewing machine and mower-reaper manuals, and Table 6.14 includes the same information for automobile instruction manuals. (None of the mower-reaper manuals had a table of contents or index. Again, this might make sense because the number of pages of instructions never were more than nine.)

The correlation of these signs of structure to the size of manuals was more consistent than the correlation of heading hierarchy to manual size. When the sewing machine manual was small—whether it be in 1914 or 1859—neither a table of contents nor an index was included. However, when the manual grew in number of pages, these items were included. For example, each of the four largest sewing machine manuals had either an index or a table of contents (compare Tables 6.9 and 6.13):

- the 1880 Domestic Sewing Machine manual had 42 pages and a table of contents;
- the 1907 Wilcox and Gibbs manual had 32 pages and an index of caution notices;
- the 1878 New American Sewing Machine manual had 32 pages and a table of contents; and
- the 1899 Perry Mason and Co. manual had 19 pages and an index.

The automobile instruction manuals echoed the sewing machine manual correlation of size and the presence of these macrostructure elements. Averaging nearly twice the size of sewing machine manuals, the automobile instruction manuals were very consistent in their use of indexes and tables of contents. In fact, not only did the automobile manuals have them but, beginning in the 1960s, they had both. (In the 1970s and 1980s, Chevrolet even included another explicit sign of whole-manual-level structure, bleeding-edge tabs.) Like the deep heading hierarchies, these indexes and tables of contents were retained by authors even when the size of the automobile manuals decreased in the 1950s for Chevrolet and in the 1970s for Ford.

Also, like the heading hierarchy, the inclusion of tables of contents and indexes lagged behind a decade from the time the size of the manuals increased to the time of an initial use of tables of contents or indexes. For example, the first Ford manuals did not have an index or table of contents despite the fact that they were as long as the longest sewing machine manual that did have an index or table of contents.

Table 6.13. The Presence or Absence of Tables of Contents or Indexes in Sewing Machine and Mower-reaper Manuals.

Sewing Machine Company Name	Date	Manual With Table of Contents or Index?	Mower-reaper Company Name	Date	Manual With Table of Contents or Index?
Grover and Baker	1859	—	Boas and Spangler	1857	—
Singer Sewing Machines	1862	—	Alzirus Brown	1864	—
Howe Machine Co.	1867	—	Adriance, Platt and Co.	1865	—
Singer Sewing Machines	1868	—	Williams, Wallace and Co.	1865	—
Finkle and Lyon M'fg Co.	1869	—	Nixon and Co.	1865	—
Buckeye	1870	—	Adriance, Platt and Co.	1867	—
Davis Sewing Machines	1876	—	Alzirus Brown	1867	—
New American Sewing Machine	1878	Table of Contents	Walter A. Wood Mowing . and Reaping Machine Co	1873	—
Florence Sewing Machine Co.	1878	—	Eureka Mower Co.	1879	—
A. G. Mason Mfg. Co.	1880	—	Walter A. Wood Mowing and Reaping Machine Co.	1890	—
Domestic Sewing Machine Company	1880	Table of Contents	Wm. Deering and Co.	1891	—
Perry Mason and Co.	1899	Index	International Harvester Co. of America-Champion	1909	—
Wilcox and Gibbs Sewing Machine Co.	1907	Front index of caution notices	International Harvester Co.	1910	—
			of America-Deering		—
Singer Sewing Machines	1914	—	International Harvester Co.	1911	—
			of America-McCormick	1915	—

Conclusions About the Genre of Instruction Manuals Derived From Looking at the Explicit Signs of Whole-Manual-Level Structure (Tables of Contents and Indexes)

Three conclusions can be drawn from looking at explicit signs of whole-manual-level structure as listed in Tables 6.13 and 6.14:

- There were no whole-manual-level explicit signs of structure in the mower-reaper manuals. This is logical because the mower-reaper instructions never exceeded nine pages.
- There were some tables of contents and indexes in the sewing machine manuals, and there seemed to be a consistent correlation between their presence and the size of the manual; the longer the manual the more often these appeared.
- Most of the automobile manuals had indexes and tables of contents, and their presence seemed independent of the size of the manual.
- The use of indexes and tables of contents in manuals appeared to lag; it took a decade after an increase in the size of a manual for indexes and tables of contents to be included.

WHAT DO THE FORD AND CHEVROLET MANUALS TELL US ABOUT THE INSTRUCTION MANUAL GENRE?

Based on a close examination of the instruction manual genre as it was used by the automobile industry, there are five general conclusions:

- the instruction manual genre was more consistent in the 20th than in the 19th century and more consistent within a single industry than in disparate industries,
- the elements of the instruction manual genre were persistent and resisted change,
- the instruction manual genre did change, but most rapidly and dramatically from extratextual influences,
- the industry that bequeathed the most to the instruction manual genre as it is used by the automobile industry was the sewing machine industry, and
- the elements of the genre that were most altered were also those from the sewing machine industry.

Each of these five conclusions is considered in turn.

Table 6.14. The Presence or Absence of Tables of Contents or Indexes in Automobile Manuals.

Ford Model	Date	Manual With Table of Contents or Index?	Chevrolet Model	Date	Manual With Table of Contents or Index?
Model T	1913	—	FB	1919	?*
Model T	1919-1925	Index	BA, BB	1926	Index
Model A	1931	Index	Confederate Series	1932	Index
V-8	1934	Index	Master Deluxe, Master Passenger	1937	Index
Deluxe and Super Deluxe	1941	Index	Passenger Car	1941	Index
Passenger Car	1951	Table of Contents	New Chevrolet	1955	Table of Contents
Mustang	1965	Both	Chevrolet	1964	Both
Passenger Car	1971	Index	Camaro	1973	Both
Escort	1986	Both	Beretta	1988	Both

*The 1919 Chevrolet manual was missing pages.

The Instruction Manual Genre Was More Consistent in the 20th than in the 19th Century and More Consistent Within a Single Industry Than in Disparate Industries

The first document in this examination of the instruction manual genre, Grover and Bakers manual, *A Home Scene; Mr. Aston's First Evening with Grover and Baker's Celebrated Family Sewing Machine Containing Directions for Using* (1859), was an odd manual with both a six-page narrative and five pages of instructions. Nevertheless, the document was invaluable for how it physically represented both the past and the future of the instruction manual genre—narration was "instrumentalized" in this 1859 manual.[40] However, this change did not happen overnight. For example, in both Figure 6.2, which plotted the percent of instructions in the manual as a measure of textual brevity, and Figure 6.4, which plotted the percentage of instruction pages with graphics as a measure of profuseness of illustration, there are numerous changes in the data lines in the left column where the 19th-century manuals' data points lie, and far fewer changes in the data lines in the right column where the 20th-century automobile manuals' data points lie. Even within a single industry, such as the sewing machine industry, the range of high and low data points decreases quite dramatically in the 20th century.

Thus, the dimensions and the techniques that together created the instruction manual genre took time to hone and for consensus to be reached by the authors and by all the companies employing them.[41] This evolution and consensus all took place from 1850 to 1950 at a time of little or no professional manual writer infrastructure—classes, societies, or handbook literature. It is as if it took time for good examples of the genre to circulate within American society before they were picked up and used. By the time professional writing classes and societies began in the 1950s, the instruction manual genre was mature and well defined. Yet, this genre's evolution had not ended by 1950.

It is also true that there was far more consensus within the automotive industry, which was addressing a single audience, than there was between the two industries addressing two different audiences. Figure 6.2 on brevity and Tables 6.13 and 6.14 on the use of indexes and tables of contents show much more agreement within the single industry. This consensus may have occurred because the manuals are from only one industry or because the genre had matured. It may also be a combination of the two.

The Instruction Manual Genre was Persistent and Resisted Change

Despite the overall change in the instruction manual that occurred because of its use within a single industry or its maturation, there was also a remarkable persistence and resistance to change. For example, despite having manuals nearly twice as large that had to document vastly more dangerous activities in automobile maintenance and operation, the automobile manual writers kept the number of paragraphs labeled "Danger," "Caution," or "Note" at the same level as the 1880 sewing machine manuals. Similarly, the decade-long lag time in the automobile manuals between increasing in size and effectively signaling a deeper heading hierarchy also revealed persistence of the genre norms in the face of changed circumstances. Finally, the diffusion of the unified automobile instruction manual into the new subgenres of shop manuals, warranty booklets, and maintenance schedule booklets suggested the persistence of the genre norm of textual brevity now carried to a very deep level of document design.[42] (Such a diffusion of material from an instruction manual had occurred in sewing machine manuals of the 1860s when the testimonials were separated from instructions and operator information separated from maintenance information.[43])

The instruction manual genre also revealed remarkable persistence in how automobile manual authors pursued experimentation with elements of the genre followed by a quick return to genre norms. For example, Figure 6.3 shows a large rise in the use of graphics for instruction by both Ford and Chevrolet in the 1950s and 1960s. This departure from the genre's norm was followed almost immediately by a return to previous norms in 1970. In like fashion, Ford in the 1960s and 1970s departed from the genre norm of paragraphs per page (Figure 6.5) but returned to it in the 1980s. Finally, the departure from the norm of a four-level hierarchy of headings by Ford in 1935 (Figure 6.11) was followed six years later with a return to the normal heading hierarchy of only two levels.

This persistence in the power of the instruction manual genre to guide the choices authors made concerning elements ranging from paragraphing to illustrating distinguished the whole negotiation process of author, reader, and genre. This persistence even in the face of substantially changed circumstances suggests that technical communication is a conservative rhetoric ready to remain at, or quickly return to, normative writing styles.[44]

The Instruction Manual Genre Changed Most Rapidly and Most Dramatically From Extra-textual Influence

The extratextual comet that hit the genre of instruction manuals in the 1960s and 1970s—the exposés by Nader and the Ford lawsuits—appeared to have had the most impact in changing the direction of the genre.[45] The publication of *Unsafe at Any Speed* and the 120 lawsuits against Ford in the 1970s coincided quite closely with the substantial growth of manuals (Figure 6.1), the rapid increase in the labeling of paragraphs (Figure 6.6), and even the dramatic rise of headings per page in Ford manuals in the 1970s and 1980s (Figure 6.8). The extratextual comet could also have explained the emergence of the "warranty booklet" and maintenance coupon booklet subgenres when the warranty and its explanation as well as the lubrication and maintenance timing diagrams disappeared from the Chevrolet manuals in the 1970s after being consistently a part of them for 60 years.

If a genre normally lives and changes through acts of incremental reader-author-genre negotiations, perhaps after an initial burst of experimentation and creativity in discovering the boundaries and norms of a genre, a genre settles into a state of equilibrium. When readers' expectations for a genre become set; when authors begin to sense consistent success in reaching an audience with a particular configuration of genre elements; and when examples of a genre accumulate in the public domain, each member of the genre negotiation triangle helps to create inertia against change, even when the change is appropriate or even necessary. Yet, the instruction manual genre did change. But it changed most dramatically because of events beyond the reader-author-genre negotiations. It changed most dramatically from extratextual events.

How often do these extratextual comets collide with genres to altar their courses? If such an extratextual event disrupted the rhetorical equilibrium of the unified instruction manual creating the numerous subgenres of automobile manuals and booklets, wasn't it also the extratextual comet called "the factory production of domestic technology" that first split off instructions from narratives? Wasn't it such a comet in the 1850s that disrupted the equilibrium of word-of-mouth communication for women and men alike and gave birth in the first place to the instruction manual genre?

The Sewing Machine Industry Bequeathed the Most to the Instruction Manual Genre

If the instruction manual genre can be said to have only two parents, the sewing machine instruction manuals and the mower-reaper manuals, the parent who bequeathed the most to the automobile manual writers was without question the sewing machine industry. Only the sewing machine industry produced manuals of sufficient size and consistency to establish a norm on textual brevity (Figure 6.2), effectively use paragraph labels such as "Danger," "Caution," and "Note" (Figure 6.6), and employ indexes and tables of contents correlated to the size of a manual (Table 6.9).

The Elements of the Genre That Must be Most Altered Were Also Those From the Sewing Machine Industry

Most of the changes in the elements of the instruction manual genre can be explained, yet there is still a missing explanation for the headings, indexes, and tables of contents. Table 6.9, Figure 6.9, and Table 6.13 suggest an incipient correlation between the increasing size of sewing machine manuals and the presence of a deep heading structure signaled typographically and the inclusion of indexes and tables of contents. These correlations seem to be borne out by the obverse fact that the mower-reaper manuals with few pages also had few deep heading structures and no indexes and tables of contents.

When the automobile industry authors began writing within the instruction manual genre and producing documents larger than even the largest sewing machine manuals (Figure 6.1), it made sense that they would create manuals displaying a deep heading hierarchy as well as including a table of contents and index. However, the correlation between the use of a deep heading hierarchy, table of contents, and index with document size broke down in the 1950s with Chevrolet when their page count dropped in half and in the 1970s with Ford when their page count dropped by two thirds (Table 6.2). It would seem logical that with smaller document sizes the hierarchy would get flatter with just first- and second-level headings, but instead both companies created manuals with three levels of headings. It would have been logical that with smaller manuals an index or table of contents would be deleted, but, in 1955, in its smallest manual, Chevrolet included its first table of contents and, in 1971, in its smallest manual, Ford included an index just as it had since 1919.

This unlinking of the correlation between document elements and size derived from the sewing machine industry suggests that automobile manual authors picked up and followed the correlation but for wholly other reasons.[46] The automobile instruction manual authors used these elements independent of the size of the manuals because they allowed readers to read selectively. They not only simplified audience reading techniques and afforded better use of their reading time but also better embodied the social relations acceptable to the Pa's of America. These elements diminished the "imperative" authority in the rhetoric of instruction manuals that the Pa's so resented. These ways to selectively read gave the audience, symbolically, a way to turn their backs on the experts such as Pa did in 1891. Remember what Deborah Tannen noted about male-female patterns of dialogue:[47]

> Furthermore, when an expert man talked to an uninformed woman, he took a controlling role in structuring the conversation in the beginning and the end. But when an expert man talked to an uninformed man, he dominated in the beginning but not always in the end. In other words,

> having expertise was enough to keep a man in the controlling position if he was talking to a woman, but not if he was talking to a man. Apparently, when a woman surmised that the man she was talking to had more information on the subject than she did, she simply accepted the reactive role. But another man, despite a lack of information, might still give the expert a run for his money and possibly gain the upper hand by the end.

Could these genre elements of selective reading be the consensus of instruction manual authors on how to handle male audiences who "might still give the expert a run for his money and possibly gain the upper hand by the end?" At the county fair when mower-reapers were demonstrated to men, they liked to receive their information in that manner because there were many ways to give the expert a run for his money—including heckling the demonstrator! Signs of selective reading were exactly that same accommodation in print for readers; They could walk away from the expert author or pick and choose what they wanted when they wanted it.

So if one could reword the opening quote of this chapter "A good manual is not a narrative. . . . Nobody ever reads a manual cover to cover—only mutants do that,"[48] it might read that a "good" manual was a manual acceptable to an audience who resented the authority vested in the rhetoric of instruction manuals. The readers of automobile manuals resisted the imperative rhetoric of authors who as teachers demanded strict attention as well as assigned reader's roles as students. Reading a manual cover to cover was not only an inefficient use of time but a pattern of behavior that was rejected in favor of one that granted readers the freedom to walk away from experts. That the instruction manual genre gave readers an increasing ability to read selectively in the 20th century may have much less to do with time efficiency and much more to do with diminishing the authority in the rhetoric. After all, instruction manual authors could only "acquaint" readers with information; they could no longer demand thorough study and strict attention.

ENDNOTES

1. Greenwald, John: "How Does this #%$@! Thing Work?" *Time Magazine.* June 18, 1986: 64.
2. W. G. Wilson Sewing Machine Co., *Directions for Using the New Buckeye Under-feed Sewing Machine*. Cleveland, OH, 1870: 1.
3. Howe Machine Company: *The Howe Sewing Machine Instructor*. New York: nd: table of contents.
4. Scharff, Virginia: *Taking the Wheel: Women and the Coming of the Motor Age*. New York: Free Press, 1991: 13.

5. Flink, James J.: *The Automobile Age*. Cambridge, MA: MIT Press, 1988: 162
6. Scharff, op. cit., 1991: 53: "Innovative as they were in some ways, Ford and his automotive colleagues were generally conservative when it came to gender. To them and to most Americans, 'driver' meant male; the modifier 'woman' meant interloper." (166)
7. Flink, op. cit., 1988: 229, 131—"By the time the Model T was introduced, the early luxury market for cars among the wealthy in large cities was saturated. Rising farm incomes until the post-World War I recession and declining Model T prices combined to make midwestern farmers yearning "to get out of the mud: the mainstay of the developing automobile market. By the mid-1920s the Model T had become a rural necessity. Few farmers by then remained autoless."
8. Killingsworth, M. Jimmie and Gilbertson, Michael K.: *Signs, Genres, and Communities in Technical Communication*. Amityville, NY: Baywood Publishing Company, Inc., 1992: 102: "To anticipate and counteract the surliness of their readership, many manual writers strive for . . . a conciliatory tone . . ."
9. *Ford Manual: For Owners and Operators of Ford Cars and Trucks* (1919-25 version): 2.
10. *Instructions for the Operation and Care of Chevrolet Cars* (1919–FB Model): 5.
11. *1988 Chevrolet Beretta Owner's Manual*: 1.
12. Greenwald, op. cit., 1986: 64.
13. Nevins, Allan: *Ford: The Times, the Man, the Company*. New York: Charles Scribner's Sons, 1954: 127–164.
14. Nevins, ibid., 1954: 162.
15. Nevins, ibid., 1954: 164
16. Nevins, ibid., 1954: 164
17. Nevins, Allan and Hill, Frank Ernest: *Ford: Expansion and Challenge*. New York: Scribner's Sons: 432.
18. The Model T initially cost $850 in 1908 and dropped to $290 by its final year of production in 1927. See Wik, Reynold M.: *Henry Ford & Grass Roots America* Ann Arbor, MI: University of Michigan Press, 1972: 230. See also Flink, op. cit., 1988: 38.
19. Dammann, George: *75 Years of Chevrolet*. Sarasota, FL: Crestline Publishing, 1986. See also Kimes, Beverly R. and Ackerson, Robert E.: *Chevrolet: A History from 1911*. Princeton, NJ: Princeton Publishing Co., 1984.
20. Nevins, op. cit., 1954: 476.
21. Flink, op. cit., 1988: 241.
22. Ford, op. cit., 1915–25: 2.
23. Dammann, op. cit., 1986: 122.

24. The similarities between Ford and Chevrolet did not end with the audience they battled for and their common business strategy, but carried over into their common management styles. For example, when Alfred Sloan of General Motors hired the former Ford executive William S. Knudsen, Knudsen used much of his experience with Ford to manage Chevrolet.
25. Note that there are four manuals from the 1930s (two each from Ford and from Chevrolet—early 1930s and middle 1930s) included in the 18.
26. Later, larger manuals had larger page sizes or at least were fairly consistent in the page size.

	Chevrolet Sizes		Ford Sizes	
1910	9 by 5 1/2	(49.5 sq. in.)	6 by 3 3/4	(22.5 sq. in.)
1920	8 1/4 by 5 1/4	(43.3 sq. in.)	8 by 5 1/4	(42 sq. in.)
1930	8 1/4 by 5 1/4	(43.3 sq. in.)	7 3/4 by 5 1/2	(42.6 sq. in.)
1935	8 1/4 by 5 1/4	(43.3 sq. in.)	7 1/2 by 5 1/2	(41.3 sq. in.)
1940	8 3/8 by 5 1/4	(43.9 sq. in.)	6 1/4 by 4 1/2	(28.1 sq. in.)
1950	8 1/4 by 5 1/4	(43.3 sq. in.)	6 by 4 1/2	(27 sq. in.)
1960	5 1/4 by 8 1/4	(43.3 sq. in.)	4 by 9	(36 sq. in.)
1970	5 1/4 by 8 1/4	(43.3 sq. in.)	4 by 8 1/2	(34 sq. in.)
1980	7 1/4 by 5 1/4	(38 sq. in.)	7 1/2 by 4	(30 sq. in.)

27. It is not until the mid-1930s that there is evidence that Ford and Chevrolet systematically collected what was known as "shop bulletins," which became Mechanic's Repair Manuals and, later, Shop Manuals.
28. Model T manual, *Ford Manual: For Owners and Operators of Ford Cars* and Trucks (1915–25): 2.
29. Other extratextual reasons why these subgenres may have occurred include the growth and development of the automobile corporate infrastructure (e.g., nationwide distribution of repair facilities in car dealerships), and the development of improved automotive designs, systems, and mechanisms that solved many of the problems the text in the manuals formerly helped owners solve.
30. The mower-reaper manuals were not included. One cannot get a wide sampling range for measuring paragraphs per page because most of the mower-reaper manuals were less than three pages.
31. Mower-reaper manuals are not included because only two such labels appear in any of the 14 manuals.
32. *Instruction Book for Ford Model T Cars* (1913): 17.
33. *Instructions for the Operation and Care of Chevrolet Motor Cars* (1932): 46.
34. *1941 Deluxe and Super Deluxe Reference Book*: 43.

35. Cullen, Francis T., Maakestad, William J., and Cavender, Gray: *Corporate Crime Under Attack: The Ford Pinto Case and Beyond.* Cincinnati, OH: Criminal Justice Studies, Anderson Publishing Co., 1987: 297.
36. *Ford 1971 Passenger Cars: Car Care and Operation*: 1.
37. This tagging of text can also be described as the evolution of "visible text" in terms best defined by Berhardt in his article, "Seeing the Text," *College Composition and Communication* 37(1) February 1986: 66–78.
38. This element of tagging is not placed in context with its 19th-century predecessors because there are only eight such examples (eight bold-faced words) in all of the mower-reaper and sewing machine manuals. Five paragraphs are presented in an indented list, and eight paragraphs have pointing hands.
39. One problem with measuring the presence or absence of these items is that the distinctive functions of an index and a table of contents were not clear to earlier manual authors. For example, in the New American manual of 1878, a table of contents is included with the chapter headings listed in the usual numerical order. However, it is titled an "index" on the page. Two items suggest that early manual authors had a difficult time discriminating between tables of contents and indexes:
    1. In some manuals, the index is broken up into a collection of minichapter indexes so that the functions of a table of contents to reveal numerical order by chapters and an index to reveal alphabetical order by page are combined.
    2. Most manuals do not have both a table of contents and an index; thus, most do not recognize their complementary but contrasting purposes. It might be interesting to note that this confusion of tables of contents and indexes continues to this day in Australian and Israeli computer-user manuals.
40. Killingsworth, M. Jimmie and Gilbertson, Michael K.: *Signs, Genres, and Communities in Technical Communication.* Amityville, NY: Baywood Publishing Company, Inc., 1992: 226—"[T]echnical communication has the smell of work and the material world. It represents the instrumentalizing [goal-oriented action] of narrative."
41. JoAnne Yates and Wanda J. Orlinkowski: "Genres of Organizational Communication: A Structurational Approach to Studying Communication and Media." *Academy of Management Review* 17(2) (1992): 305—"[G]enres 'are by-products of a history of negotiation among social actors that results in shared typifications which gradually acquire the moral and ontological status of taken-for-granted facts'."
42. Yates and Orlinkowski, op. cit.: 308—"The result of such ongoing challenges [to genre rules] is 'that the set of genres is an open class, with new members evolving, old ones decaying'."

43. The division of operator instructions from mechanical information was foreshadowed in some of the sewing machine manuals. For example, in *The Howe Sewing Machine Instructor* under the heading "To The Learner" states:

> Never attempt to take the machine apart, or you will be sure to get into trouble; we frequently hear ladies boast that they "have taken their Sewing Machines all to pieces, and put them together again." Nothing worse can be done; it is a mistake to suppose that the machine can be benefited by being tampered with by inexperienced persons.

(Howe Machine Company, New York: nd: 5. Similar warnings appeared in the 1870 W. G. Wilson Sewing Machine Co. instruction manual [page 2], the 1867 Howe manual [page 19], and the 1876 Davis manual [page 9].)

The operator, a woman, was just supposed to operate the machine, not take it apart. And, nowhere were ladies given the possibility of getting shop manuals. In fact, it was almost as if John Patterson's marketing directive at NCR, "Don't talk cash registers, talk the prospect's business," (Johnson, Roy W. and Lynch, Russell W.: *The Sales Strategy of John H. Patterson. [Sales Leaders Series]* Chicago and New York: The Dartnell Corp., 1932: 146–8), were taken to an extreme, and the inner workings of the sewing machine were never revealed to owners. If internal mechanical problems occurred, sewing machine manuals directed operators to simply apply to the knowledgeable company agent:

> Operators are cautioned not to attempt to adjust the machine, unless its sewing qualities are impaired and not then, unless they are perfectly familiar with its principles and mechanism. Any agent of the Company will willingly do this, and in such case parties should apply to the nearest agent or to someone familiar with the mechanical construction of the machine. The attempt by any unskillful person to repair a machine, often occasions more derangement than years of ordinary wear could accomplish.

(Davis Sewing Machine Co.: *Directions for Operating the New Davis Sewing Machine*. Watertown, NY: 1876: 9.)

44. Such persistence could also reflect how authors come to regard genres as having "the moral and ontological status of taken-for-granted facts" (Yates and Orlinkowski, op. cit.: 305–6). For example, in the 1890s the Scovill Manufacturing Company wanted to require all internal correspondence to have subject lines and a single subject in order to aid the

company's new efforts at vertical filing. There was, however, some resistance to this change in the genre of business letters: "One week after the mandate was issued, headquarters wrote to reprimand some lapses in providing the requested subject lines, ending with this statement: 'We are changing our system of filing, and we must INSIST that you pay particular attention to this matter.'" (Yates and Orlinkowski, ibid.: 314)

45. Yates and Orlinkowski, ibid.: 306—"Thus, on occasion, individuals modify (deliberately or inadvertently, whether by mandate or spontaneously) some of the established genre rules of substance and form. These modifications may be triggered by material or perceptual changes in the recurrent situation. That is, changes to the social, economic, or technological context (e.g., changed organizational forms, new or less expensive electronic media, revised reporting requirements), or changes in how social groups recognize and respond to situations (e.g., an ad hoc group's redefinition of itself into a regular task force) may occasion a deviation from habitual use of genre rules."
46. Could these elements be skeuomorphs? See Basalla, George: *The Evolution of Technology*. Cambridge: Cambridge University Press, 1988: 106–7—"The regularity with which new materials are handled and worked in imitation of displaced, older ones had led archaeologists to coin a word to designate the phenomenon: *skeuomorphism*. A skeuomorph is an element of design or structure that serves little or no purpose in the artifact fashioned from the new material but is essential to the object made from the original material."
47. Tannen, ibid., 1990: 128.
48. Greenwald, op. cit., 1986: 64.

THEME 3

The Individual American Technical Communicator and the Possibilities of Innovation

> Men make their own history, but they do not make it just as they please; they do not make it under circumstances chosen by themselves, but under circumstances directly encountered, given, and transmitted from the past.
>
> —Karl Marx (1852)[1]

Two extratextual events left their imprints on the course of technical communication history and, most especially, on the development of the instruction manual genre. In the 1850s, the arrival of factory-produced domestic technologies such as the sewing machine and the mower-reaper seem to have propelled the emergence of step-by-step instructions from narratives. Grover and Baker's *A Home Scene; Mr. Aston's First Evening with Grover and Baker's Celebrated Family Sewing Machine Containing Directions for Using* (1859) recorded the impact of this extratextual event in its pages which contained both a short story and step-by-step instructions. A little over 100 years later, in the 1960s and 1970s, the exposés and lawsuits surrounding the automobile industry also seem to have propelled step-by-step instructions to increase in size dramatically and to take on a whole new aura of legal stipulation. The *1986 Escort Owner's Guide* recorded the impact of this extratextual event in its pages: The paragraphs labeled "Danger," "Caution," and "Note" quite often outnumber the paragraphs of step-by-step instructions.

However, the stories have not been told of how the technical communicators during both of these events offered readers novel solutions to these new technological-political-sociological problems. None of the authors' names of the instruction manuals from the sewing machine industry, the mower-reaper manufacturers, or the automobile industry have been recorded. Thus, stories of how they developed their instruction manuals have been lost among the ranks of history's "anonymous" technical communication authors.

Yet, much of history is made up of the stories of individuals. Simon Schama, a great recent writer of history stories, pointed our the need to unearth the stories of the past:[2]

> History's mission, then, is to illuminate the human condition from the witness of memory. Yet the truths likely to be yielded by such histories will always be closer to those disclosed in great novels or poems than the abstract general laws sought by social scientists.

Donald Norman similarly described the power of stories:[3]

> There is something important and compelling about stories that bears considering in greater detail. Stories are marvelous means of summarizing experiences, of capturing an event and the surrounding context that seems essential. Stories are important cognitive events, for they encapsulate, into one compact package, information, knowledge, context, and emotion.

Surely one of the greatest technical stories since World War II has been the invention of the digital computer and the creation of the computer industry. History has recorded the stories of the engineers and managers such as Mauchly and Eckert with ENIAC, EDVAC, BINAC, and UNIVAC, as well as Watson, Hurd, and Rochester with IBM's Type 701. Yet, where are the stories of the technical communicators who wrote the instruction manuals, created the first computer patent disclosures, and trained the original computer operators? Is it possible to unearth the stories of their individual creative processes from the ranks of unrecorded anonymous authors?

David Dobrin (1983) observed that corporations would be loath to save the documentation produced to support a product after the product was economically passé, and, more to the point, even if the published documents were to be found, there would be little to examine about what went on behind the scenes during the process of document design and development.[4] Henrietta Shirk was even more specific than Dobrin in her sense of the inability to recover the early history of computer documentation because computer documentation is "by definition designed and destined for destruction." She continued: "Neither the widely read nor the unread have survived. . . . One must not look for the relics of the past (which have almost without exception been destroyed)."[5]

Yet, quite to the contrary, not only can computer documentation of the past be recovered as were the patents and instruction manuals, but so can the behind-the-scene steps in the document design and development taken by two of the very first writers of computer documentation in the late 1940s and early 1950s, Joseph Chapline and Sidney Lida.

Joseph Chapline's adventures offer a rare, detailed look behind the scenes at how the writer of the earliest computer manuals discovered what it was to write, and later manage, computer documentation for the Eckert Mauchly Computer Company from 1948 to 1954. This unusual look behind the scenes is possible because of a quirk of fate: The lawsuit over who invented the computer.[6] In the lawyers' desire for support for their cases, letters, memos, rough drafts, and all the documents detailing the early history of computers were recovered and collected. In 1986 these documents were made available to the general public at the Hagley Archives and Library in Wilmington, DE.[7] The second quirk of fate was that in 1990 Mr. Chapline received the RIGO Award for significant lifetime achievement from the Association for Computing Machinery, Special Interest Group in Computer Documentation and, at the time of the award, gave a lengthy oral interview. Hence, Mr. Chapline himself lent a human voice to the pieces of paper collected in the corporate archives.

Sidney L. Lida was the author[8] of *Principles of Operation: Type 701 and Associated Equipment* (1953)—the first external IBM computer instruction manual for the IBM Type 701. Tom Watson, Jr., President of IBM, described the Type 701 as "the machine that carried us into the electronic business."[9] Fewer of the materials documenting the document design and development steps of Mr. Lida are available to the public, because IBM's corporate archives are closed to the public, and Mr. Lida disappeared from the computer industry over 20 years ago. However, many of the key engineers and managers have described their work surrounding the Type 701 computer and preserved these memories in various books and oral histories. Also, many of the preliminary working draft documents for the Type 701 computer have been collected by the Computers, Information and Society Division of the National Museum of American History (Smithsonian Institution).

Because the work of Chapline and Lida occurred at roughly the same moment of history in the late 1940s and early 1950s, there is a wonderful opportunity to discover the story of how two writers in two competing companies forged creative solutions to the new technological-political-sociological problems posed by the electronic digital computer and the birth of the computer industry. Nevertheless, did Chapline and Lida know at the time that they were solving radically new technological-political-sociological problems for a new industry? Was the introduction of the electronic digital computer an event of a magnitude similar to those that altered the development of the instruction manual in the 1850s, 1960s, and 1970s?[10] Were Chapline and Lida really called on to create innovations in technical communication rhetoric? The press at the time labeled Eckert and Mauchly's ENIAC as "a tool with which to rebuild scientific affairs on new foundations,"[11] and the IBM Type 701 project, IBM's first digital computer, was the most expensive project in all of IBM history to that point.

Yet, couldn't the heralding of the birth of a new industry and a new information age simply be the selective vision of hindsight? For example, just prior to beginning work on the Type 701, IBM management was reluctant to enter the electronic digital computer race because the company "saw no large potential market for the computer in 1949" and supposed that "world demand could be satisfied by just 10 or 15 computers."[12] Isn't it possible that the computer revolution was largely "unforeseen" at the time by its participants:[13]

> [T]he inventors and the first commercial manufacturers underestimated the computer's potential market by a wide margin . . . computer pioneers understood the concept of the computer as a general-purpose machine, but only in the narrow sense of its ability to solve a wide range of mathematical problems. Largely because of their institutional backgrounds, they did not anticipate many of the applications computers would find . . .

Even if the birth of the computer was such a historical event as to cause technical communicators to create new innovations, were Chapline and Lida able to develop such innovative responses? The description of the technical writer's "creative flair" was what Joseph Chapline focused on in "What Is a Technical Writer" (1955):[14]

> The final attribute of a technical writer is a flair for creativity. . . . He is not a mere mechanic who performs always according to a set of prescribed rules. Instead, he views his assignment with the eye of a craftsman ready for any momentary flash that may guide him in a slightly better way for solving his problem than some mere rote following of organized rules. . . . Here is the unknown quantity that makes technical writing a challenge.

This passage from his article highlighted the creative, path-breaking role technical communicators like to think they play. However, the emphasis on the "creative flair" might miss some of the effects of genre persistence and audience conservatism already explored in the chapters of Theme 2. Missing from Chapline's passage is the understanding that the rhetoric of technical communication is a social activity involving not only a writer but an audience and genera, each with set expectations and inertia against change. Hence, any kind of rhetorical creativity, or path-breaking role involves more than just change in a single individual writer; rhetorical creativity requires change in an entire discourse community.[15]

Without a community understanding and observation of an author's rhetorical actions, the "creative flairs" are for naught. Did not the author of Grover and Baker's *A Home Scene; Mr. Aston's First Evening with Grover and Baker's Celebrated Family Sewing Machine Containing Directions for Using* (1859) include a familiar artifact—a short story—as well as a novel artifact, the step-by-step directions, in the same document? Individual technical communicators must wrestle over and again with audiences and communities that demand the preservation of familiar forms, tones, voice, methods, and metaphors in the face of technological and corporate changes. For example, in 1980, a technical communicator involved in the initial introduction of an unfamiliar document design in IBM reported:[16]

> One obstacle to this method was that customers and field personnel had adapted so well to the existing forms of documentation that they could not conceive of anything else.

Perhaps the very preservation of these communication aspects in the face of great technological change allows technological change to be interwoven into established American values and societal expectations.

How did Chapline and Lida respond to the birth of the computer? Was it of sufficient technological-political-sociological magnitude to propel these two men to overcome the inertia of genre, audience, and community to create innovative solutions? Or, did rhetorical inertia force these two men, as Marshall McLuhan put it, to "[g]o into the future looking through the rearview mirror,"[17] the rhetorical rearview mirror, that is. What is the role technical communicators play between the opposing forces of technological innovation and rhetorical inertia?

ENDNOTES

1. *The Eighteenth Brumaire of Louis Bonaparte*. NY: International Publishers, 1963: 15.
2. Schama, Simon: "Clio Has A Problem." *The New York Times Magazine* September 9, 1991: 32.

3. Norman, Donald A.: *Things that Make Us Smart: Defending Human Attributes in the Age of the Machine*. Reading, MA: Addison-Wesley Publishing Co., 1993: 129.
4. Dobrin, David N.: "What's Difficult about Technical Writing?" *College English* 47 (1985): 137.
5. "Technical Writing's Roots in Computer Science: The Evolution from Technician to Technical Writer." *Journal of Technical Writing and Communication* 18(4) 1988: 305–23.
6. *Honeywell Inc. v. Sperry Rand Corporation and Illinois Scientific Development, Inc.*, US District Court, District of Minnesota, Fourth Division (19 October 1973).
7. Some documents on Eckert-Mauchly's work can also be found at the Babbage Institute, University of Minnesota; National Museum of American History; Smithsonian Institution, Washington, DC; and the Computer Museum, Boston, MA.
8. Hurd, Cuthbert C.: " Sales and Customer Installation." *Annals of the History of Computing* 5(2) April 1983: 163. A letter from Mr. Hurd (9/21/90) explained that Mr. Lida was last heard of working in Oklahoma on bovine artificial insemination.
9. Hurd, Cuthbert C.: "Prologue." *Annals of the History of Computing* 5(2) April 1983: 110.
10. The whole question of does technology drive history is perhaps apropos in considering the relationship of genera evolution, the extratextual events of the 1850s, 1960s, 1970s, and Chapline and Lida's work at the inception of a new technology. See Smith, Merrit Roe & Marx, Leo. *Does Technology Drive History: The Dilemma of Technological Determinism*. Cambridge, MA: MIT Press, 1994.
11. "Electronic Computer Flashes Answers, May Speed Engineering." *New York Times* (February 1946) qtd. in Shurkin, Joel. *Engines of the Mind*. New York: W. W. Norton and Co., 1984: 75, 196.
12. Rosenberg, Nathan: "Inventions: Their Unfathomable Future." *The New York Times*. August 7, 1994: F-9.
13. Ceruzzi, Paul: "An Unforeseen Revolution: Computers and Expectations, 1935–1985" In (Corn, Joseph J., ed.) *Imagining Tomorrow: History, Technology, and the American Future*. Cambridge, MA: MIT Press, 1986: 188–201.
14. Chapline, Joseph: "What is a Technical Writer." Unpublished article, June 1955.
15. Harrison, Teresa: "Frameworks for the Study of Writing in Organizational Contexts." *Written Communication* 9 (January 1987): 3–23—"Communities of thought render rhetoric comprehensible and meaningful. Conversely, however, rhetorical activity builds communities that subsequently give meaning to rhetorical action". (9) See also Harris,

J.: "The Idea of Community in the Study of Writing." *College Composition and Communication* 40 (1989): 12—"[I]t is only through being part of some ongoing discourse that we can, as individual writers, have . . . points to make and purpose to achieve."

16. Mitchell, Georgina E.: "Solving Problems of Information Gathering with Task-oriented Library Design." *In Proceedings of the 29th International Technical Communication Conference, May 1982*. Washington, DC: Society for Technical Communication, 1982: C-77. See also Harvey, Patrick: "An Experiment in Changing Styles." In *1992 Proceedings of the 39th Annual Conference, May 10–13, 1992*. Arlington, VA: Society for Technical Communication, 1992: 262—"There seemed to be three . . . reasons for the initial mixed reception: [1] Resistance to the new format. Despite the fact that the new format was in response to user comments and despite explanations that went out with the documents, some users objected to the manuals being different." See also Mirel, Barbara: "The Politics of Usability: The Organizational Functions of an In-House Manual." In (Doheny-Farina, Stephen, ed.) *Effective Documentation: What We Have Learned From Research*. Cambridge, MA: MIT Press, 1988: 277–97.
17. McLuhan, Marshall. *Medium is the Message*. New York: Random House, 1967: 75.

Chapter 7

The Story of Joseph D. Chapline, First Computer Documentation Writer and Manager, 1948-1955

A discussion with Mr. Frank Bell of Northrop Aircraft, Inc. today, indicated an understanding on the part of Northrop that the operating and maintenance instructions would correspond to those issued to the purchaser of a new motor vehicle. That is, there would be included a description of the location and function of controls, together with directions for such preventive maintenance as deemed necessary. In the case of a motor vehicle, such preventive maintenance information takes the form of a lubrication chart, while computer maintenance included recommendations for the periodic replacement of vacuum tubes, cautions to be observed in the replacement of crystal diodes, and other such periodic adjustments as may be recommended by the Engineering Dept.... With the relatively simple maintenance instructions which are contemplated by Northrop, it may be possible to materially simplify the discussion of the theory behind the maintenance instructions.

—Letter to Joseph Chapline, June 21, 1949[1]

SETTING THE SCENE: THE RACE TO THE COMPUTER MARKETPLACE

Joseph Chapline's story in the late 1940s and early 1950s began before electronic computers were invented at a time when Chapline himself was a "computer" assigned by the U.S. Army to operate a "differential analyzer."[2] Differential analyzers were mechanical calculating machines comprised of bus shafts, gears, motors, and analog disc integrators invented by Vannevar Bush in 1925 and used primarily for complex engineering and scientific calculations.[3] "Instructions" to the analyzer machines were "entered" using plotting pens connected to bus shafts, and changes to such instructions were difficult and time consuming to make:[4]

> Experience, mathematical skill, and sometimes days were required to translate difficult problems into forms that could be attacked by the calculator and then to connect the machine to perform the calculations. When Lemaitre and Vallarta calculated the trajectories of cosmic rays under the influence of the earth's magnetic field, it took five staff members and eighteen students thirty weeks to obtain the solutions.

Analyzers were also large—some second-generation systems weighed 100 tons. Moreover, when hundreds of calculations were required to support the war effort in World War II—specifically, in Chapline's story, the artillery firing tables developed by the U.S. Ballistics Research Laboratory in Aberdeen, MD—it was apparent that analyzers were too slow. Many companies were interested in developing fast, new, electronic calculators in the 1940s and 1950s: General Electric, Bell Laboratories, Philco, National Cash Register, Westinghouse, and RCA.[5] However, history has usually focused in on the efforts to create fast electronic calculators made by Thomas J. Watson Sr.'s company, International Business Machines (IBM), and Joseph Chapline's employer, the Eckert-Mauchly Computer Corporation (EMCC).

In 1944, IBM donated the Automatic Sequence Controlled Calculator (ASCC), or Mark 1, to the Harvard Mathematics Lab.[6] Unlike the differential analyzers based on Bush's designs of the 1930s, the Mark 1 was an electromechanical paper punched-card calculator based on Herman Hollerith's earlier designs of the 1880s that were perfected and marketed by IBM in the 1920s–1940s. Electromechanical paper punched-card calculators used punched-cards and trays of plug-in cables as their input media rather than the plotting pens of the analyzers. Also, the calculators had been primarily used by business and government for repetitive, relatively simple calculations—such as those for the Census Bureau—rather than by engineers for complex single calculations such as Lemaitre's and Vallarta's description of the trajectories of cosmic rays. Yet, even if powered by electricity, both the differential analyzer and the electromechanical paper punched-card calculators mechanically performed their calculations.

The Mark 1 could have been IBM's entry into the race for developing an electronic digital computer, a computer that could have performed its calculations in nonmechanical, electronic ways. Nevertheless, IBM conceived of itself as in the business of developing and marketing calculators based on Hollerith's 1880s patents. The CEO of IBM, Thomas J. Watson, Sr., thought that computers were simply toys for large government bureaucracies and of no practical day-to-day use to IBM's business customers.[7] Why, the directors of IBM reasoned, should IBM spend millions of dollars on unpredictable research and development at a time when it was already making millions ($10.9 in 1945[8]) on punched-card, electromechanical, calculating technology?[9] Thus, until the early 1950s and the advent of the first IBM electronic computer, the Type 701 (see Chapter 8), IBM did not seek to develop computers.

At the time of IBM's donation of the Mark I to Harvard, John Mauchly and J. Presper Eckert, Jr. were not corporate leaders of a multimillion dollar corporation. John Mauchly was only a faculty member on the staff of the Moore School of Electrical Engineering at the University of Pennsylvania, and "Pres" Eckert only a graduate student. In the university building where they had their offices were also two differential analyzers then being rented by the U.S. Army Ballistics Research Laboratory in Aberdeen, MD,[10] and Joe Chapline, a former student of Mauchly's at Ursinus College, maintained and serviced the analyzers. In 1945, it was the professor and the grad student, not IBM, who announced the birth of the first nonmechanical, electronic digital computer, ENIAC (see Figure 7.1).[11]

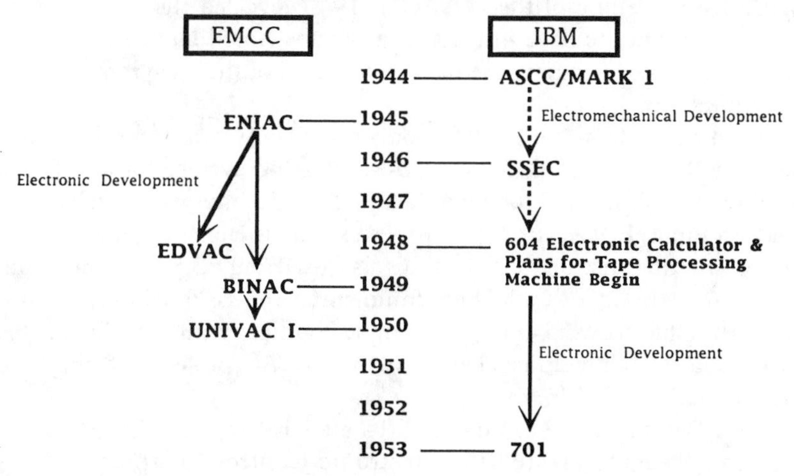

Figure 7.1. EMCC's and IBM's timelines for creating electronic computers

IBM's Mark 1 would take 840,000 microseconds on average for a calculation, but Mauchly and Eckert's ENIAC was much faster, taking only 9,000 microseconds. ENIAC was the first *large* electronic digital computer,[12] and its size was indeed prodigious: 40 panels, 10,000 capacitors, 6,000 switches, and 17,468 vacuum tubes. ENIAC cost approximately $500,000, was about 1,800 square feet, and, like the IBM's Mark I, was a one-of-a-kind machine.[13] The *New York Times* cover story describing ENIAC's public unveiling in February 1946 was headlined, "Electronic Computer Flashes Answers, May Speed Engineering."[14] The story continued:[15]

> One of the war's top secrets, an amazing machine which applies electronic speeds for the first time to mathematical tasks hitherto too difficult and cumbersome for solution, was announced here tonight by the War Department. Leaders who saw the device in action for the first time heralded it as a tool with which to rebuild scientific affairs on new foundations.

However, ENIAC's promise often fell short of its performance. For example, ENIAC could run for perhaps only five minutes at a time before one of the 17,000 vacuum tubes would burn out. Moreover, a single moth shorting out one circuit on one vacuum tube could cause the entire computer to stop dead—the dreaded beginning of "computer bugs." Furthermore, like the instructions to the differential analyzers, the instructions to ENIAC were both difficult to enter and prone to error because they were entered via a large bank of switches and trays of plug-in cables. Eckert and Mauchly began to address these problems in a supplement to the ENIAC contract, a project called EDVAC. The unveiling of the EDVAC in 1949 revealed the first large-scale electronic computer to have instruction programs stored internally to run the calculations thus overcoming the input difficulties of the analyzers, the calculators, and ENIAC.[16]

Like the Mark I and ENIAC, only one EDVAC was built.[17] Moreover, whereas the Mark 1 took 840,000 microseconds on average for a calculation and ENIAC 9,000, EDVAC took only 2,221. Also, to improve the reliability of instruction input, EDVAC used paper tape as well as Hollerith paper punched cards for entering instructions. EDVAC also used only 3,600 vacuum tubes rather than ENIAC's 17,000, thus diminishing the risks of tube burnout. However, unable to work out patent right conflicts with the University of Pennsylvania, Mauchly and Eckert quit the EDVAC project and the Moore School on March 31, 1946.[18]

In December 1946, nine months after leaving the EDVAC project, Eckert and Mauchly created the Electronic Control Company (ECC) in Philadelphia. At ECC, they and their associates began improving on the ENIAC and EDVAC projects and set their sights on building large numbers of computers for commercial applications—*general-purpose computers*. Such general-purpose computers could work out not only the calculations for a single complex numerical problem, *scientific computing* (the focus of the differential analyzers, ENIAC, and EDVAC), but also handle repetitive, simple calculations, such as the computing of monthly payrolls. In a word, Eckert, Mauchly, and their associates had their eyes set on UNIVACs—"Universal Automatic Computers." A year later, in December 1947, ECC—renamed the Eckert Mauchly Computer Company (EMCC)—signed an agreement with the National Bureau of Standards to develop UNIVAC 1 for the Census Bureau's use in the then upcoming 1950 census.[19]

However, money for developing such commercial, general-purpose computers was tight, and thus EMCC was reluctantly persuaded to develop another one-of-a-kind computer whose profits, they hoped, would tide the company over until the money from the sale of UNIVACs arrived. This final prototype computer was called the BINAC—"Binary Automatic Computer."[20] On October 9, 1947, ECC agreed to develop BINAC for Northrop. The BINAC was primarily designed to be a smaller version of the EDVAC. In fact, BINAC was going to be only 20 cubic feet (versus ENIAC's 1,800 square feet).[21] Twenty cubic feet may seem a rather odd dimension to settle on for an early computer, however, it was the size of the bomb bay on a B-52: The BINAC was to be used for the airborne control of guided missiles and thus needed to fit through the bomb bay doors.

BINAC was officially delivered to Northrop on August 31, 1949.[22] Like EDVAC, BINAC had internally stored programs to overcome the earlier instruction input problems, but BINAC used magnetic tape to enter instructions rather than EDVAC's paper tape or paper punched cards. The number of BINAC's vacuum tubes diminished to only 1,400 in comparison to ENIAC's 18,000 and EDVAC's 3,600, and BINAC's calculating speed was on average 524 microseconds per calculation in comparison to ENIAC's 9,000 and EDVAC's 2,221.[23]

However, as good as BINAC was technically, it did not prove to be the hoped-for financial bridge to the time of UNIVACs. Consequently, EMCC was sold to Remington Rand (which later became Sperry Rand and is now UNISYS).[24] The EMCC division (renamed the UNIVAC subsidiary) of Remington Rand remained very much an independent subsidiary because Remington Rand had long focused on card filing systems and typewriters from which they had made most of their money, and thus had little feel for the new electronic digital computers. Nevertheless, it was Remington Rand in 1950 who announced that the sale of the first commercially available electronic computer produced by their UNIVAC subsidiary.

Unlike the earlier electronic computers, UNIVAC was not a one-of-a-kind prototype; 46 of the UNIVAC 1s were sold. UNIVACs were sold not only to the government and military development companies, but also to a number of insurance firms and others such as the A. C. Nielsen company. In the early 1950s UNIVAC was the premiere electronic computer:[25]

> The first UNIVAC was delivered on June 14, 1951. For almost five years after that it was probably the best large-scale computer in use for data processing applications. . . . Remington Rand was launched into the computer field with a product that was five years ahead of any of its competitors.

The reaction of the American public to UNIVAC was immediate and widespread, and from books to animated cartoons,[26] "UNIVAC" came to mean "computers"—much like "Xerox" for copiers and "Kleenex" for tissue paper. Some of this public reaction was due to a famous public relations coup on television in which UNIVAC 1 bested Charles Collingwood and a young Walter Cronkite in successfully predicting the outcome of the 1952 election.[27] "UNIVAC" was so synonymous with "computer" in the public's mind that two years later, when IBM finally did come out with their first computers, IBM salespeople found their computers being referred to by customers as "IBM's UNIVAC."[28]

Nevertheless, for all its success with the general public and the press, Remington Rand failed to market UNIVACs successfully to deep-pocket purchasers, and, by 1957 Remington Rand had only a 10% share of the computer market.[29] When IBM finally decided to enter the electronic digital computer market in 1953 with its Type 701, it quickly caught up with Remington Rand's market share and surpassed them by offering five different models of the 700 series computers in quick succession. Thus, IBM captured over 39% of the market by 1957 and created a relationship in the industry with UNIVAC and six other competitors characterized by many at the time as "IBM and the Seven Dwarfs."[30] For UNIVAC in particular, one industry observer coined an enduring phrase: "UNIVAC snatched defeat from the jaws of victory."[31]

Yet, in July 1947 when Joseph D. Chapline joined the Eckert-Mauchly Computer Corporation for the BINAC project, defeat simply did not appear on the horizon.

JD CHAPLINE: BINAC TECHNICAL WRITER, 1947–1949

Joseph D. Chapline joined EMCC largely because he had had John Mauchly as his teacher and mentor at Ursinus College on the outskirts of Philadelphia.[32] At Ursinus, Chapline had taken two years of advanced composition while majoring in history and political science. Luckily, he had also taken courses in physics, optics, and calculus. On his graduation from Ursinus, Chapline followed Mauchly to the Moore School of Electrical Engineering at the University of Pennsylvania where he worked for him as a research assistant operating and maintaining the differential analyzers.

When Mauchly left the Moore School and started ECC and then EMCC, Chapline again followed him and joined EMCC in July 1947. In the summer of 1947, Joe's first informal task was to edit the EMCC engineer's "theses" into a proposal that eventually landed the work on the Census Bureau's UNIVAC. With this success under his belt, Chapline began that fall to work formally as a writer and editor, and his first official task was to write the BINAC documentation.[33]

Portrait of Joseph Chapline taken by an EMCC photographer on his staff in 1954

The Influence of Genera: The Scylla and Charybdis of Chapline's Voyage

The central problem that Chapline and his Editorial Department faced on the BINAC project was to decide on exactly what genre to use for the computer documentation.[34] Precedents, consultants, audiences, and experience pulled Chapline in many directions as he made his decision because each genre he considered created different relationships between EMCC as a company, the BINAC computer, and readers.

Similar to the technical communicators in Themes 1 and 2, from Evans to the automobile instruction manual writers, one very strong influence on Chapline's genre considerations was precedent. Many at EMCC thought that the BINAC documentation should be like that delivered with ENIAC and written at that time for EDVAC; both were primarily *technical reports* (see Figures 7.2 and 7.3 for a table of contents page and a sample page from the *Operating and Programming Manual for EDVAC*).[35] In the ENIAC and EDVAC technical reports, the tables of contents and organization were dependent on a numbered series of paragraphs rather than on page numbers. There were also no indexes or section tabs because the manuals were designed to be exhaustive technical descriptions for engineers on how the computers worked, not how to operate them. Furthermore, the technical descriptions in these reports built on each other in detail from page to page. Such a cumulative design assumed that the engineer readers were going to read these reports like a textbook, cover to cover and from front to back.

For example, in the EDVAC report, page 3.1-1 introduced the fact that EDVAC used a binary code which had to be entered into the system in binary numbers by an operator. Based on this assertion, page 3.2-1 introduced the binary number system by giving examples, then page 3.32-4 illustrated how binary division was accomplished. Finally, on page 5.19-4, the EDVAC report showed how binary fractions were to be entered as instructions to the computer. Such an example illustrates how the design of the EDVAC as a technical report assumed that a reader would read three consecutive chapters before a single action was taken. The 119-page EDVAC report also had only 17 graphics, and most were technical circuit patterns. Thus, to understand the EDVAC information, lengthy study and reading was required, and even then the reader had to deduce the concrete actions to be taken with EDVAC from the textual-mathematical report descriptions.

Yet, the EDVAC documentation evidently worked for its audience of knowledgeable, military engineers and scientists because it was used successfully for working out calculations in the hydrogen bomb project. Perhaps one of the rhetorical reasons for its success was that it was not written with the rhetoric of authority used in the sewing machine instruction manuals; a rhetoric that created an asymmetric relationship between a knowledgeable

OPERATING AND PROGRAMMING MANUAL FOR THE EDVAC

Contents

1. Introductory
2. Operation
    - 2.1. General Description
    - 2.2. Sequence of Operation
    - 2.3. Handling of Input
        - 2.31. Preparing Preliminary Tape
        - 2.32. Verifying
        - 2.33. Preparing Input Wire
    - 2.4. Operation of EDVAC Proper
        - 2.41. The Wire Drives
        - 2.42. The Manual Controls
        - 2.43. The Indicators
        - 2.44. Shutdown and Restart
    - 2.5. Handling of Output
        - 2.51. Preparation of Final Tape
        - 2.52. Printing
3. Coding
    - 3.1. General
    - 3.2. Binary and Octal Numbers
    - 3.3. Standard Coding Forms
        - 3.31. Numerical Data
            - 3.311. Decimal Numbers
            - 3.312. Octal Numbers
            - 3.313. Addresses

Figure 7.2. Sample page from *Operating & Programming Manual for the EDVAC* (Hagley Museum and Archives, Wilmington, DE: Accession 1825, Box 24, Page 0-1). Notice how the manual's table of contents has no page numbers and instead uses a paragraph numbering scheme

author and an ignorant reader. Rather, in the EDVAC technical report, the role of author and reader was a somewhat symmetrical one between equals because actions were left ambiguous, and the operational implications of the text often had to be deduced by the audience. The EDVAC technical report could assume such a balanced author-reader relationship because there was, after all, only one EDVAC rather than hundreds of Howe sewing machines.

2.32. Verifying

The Verifier is electrically connected to the Perforator-Printer and serves to check the perforations in tape run through it against the key operation of the latter. The Preliminary Tape is inserted in the Tape Reader on the Verifier and a blank paper tape is inserted in the Perforator-Printer. The latter is also provided with a roll of paper, with or without carbons, as desired. This paper is standard Teletype paper, 8½ inches wide. The Operator then types from the original Code manuscript using the keyboard of the Perforator-Printer. This keyboard is similar to that of the Preliminary Tape Perforator except that the blanking character / is omitted and that the end of each Code "word" corresponds to carriage return and line feed as well as a typed character.

When the key pressed by the Operator is in agreement with the perforations in the Preliminary Tape at the point being read, the Perforator-Printer prints the character corresponding on the paper inserted in it and simultaneously perforates the type passing through it to correspond to the type character. The paper carriage, the new tape, and the Preliminary Tape all advance one position in readiness for the next symbol. The new or Verified Tape is not of "Chadless" type, but is perforated with full holes, and does not have superposed type characters; otherwise it is similar to the Preliminary Tape. The printed sheet is typed 10 characters to the inch, each "word" in the Code being printed on a separate line with a spacing of 3 or 6 lines to the inch as desired. The Word-End symbol (*) types between lines during normal operation.

Figure 7.3. Sample page from *Operating & Programming Manual for the EDVAC* (Hagley Museum and Archives, Wilmington, DE: Accession 1825, Box 24, Page 2.32-1). This is one of the pages describing the operation of the verify operation. Notice how the text uses none of the physical layout devices—indentation, two columns, numbered steps—that have become customary to use when giving steps. It uses a description rather than an instruction imperative mood approach in addressing the reader: "When the key pressed by the Operator"

Also, EDVAC was going to be used by atomic physicists and their skilled technicians rather than by women in the 1850s who mostly lived on farms and who had little or no previous technological experience.[36]

In addition to the technical reports, another strong genre precedent for Chapline was *military specifications*—"AN specifications" (now known as MIL Specs)—especially those that had been developed during the war for radar. AN specifications were rigidly organized technical descriptions and operational directives issued by the military to ensure consistent use of equipment by all military personnel. Moreover, because most of the principals of EMCC had worked with the military during World War II, they were familiar with this genre.

EMCC principals were also quite familiar with radar AN specifications. For example, Eckert supplied Chapline with the radar AN specifications and suggested that Chapline use this genre as a model for computer documentation. Eckert perhaps arrived at this idea because he had been working on radar systems for the military during World War II at the M.I.T. RADLAB just prior to joining with Mauchly for the ENIAC project. In fact, not only were early computers linked to radar by shared personnel (e.g., Eckert in the US and Trevor Pearcey in Australia—the man who directed the building of the first Australia computer, CSIRAC—both had prior careers in the development of radar), but computers and radar also shared some of the same hardware; for example, the mercury delay lines vital to EDVAC's and BINAC's memory systems were re-engineered from World War II radar systems. Thus, the pressure on Chapline to use the AN genre model may have been prompted as much by EMCC's familiarity with military methods of documentation as much as by the numerous hardware similarities between radar and computers.

On September 9, 1948 Chapline wrote to his boss, George V. Eltgroth,[37] indicating that he had successfully collected two AN specifications: "AN-H-20a Handbooks: Radar and Radar Equipment Maintenance" and "AN-H-21a Handbooks and Catalogs: Radio and Radar Equipment" (a "typical Page" is shown in Figure 7.4). Chapline concluded this memo noting: "They may also be valuable to us as a guide in the preparation of similar documents both for our commercial accounts and other government agencies."[38]

Chapline was further encouraged to use the genre of military specifications as his model in creating the new computer documentation because of proposals for manual preparation submitted to EMCC by two experienced New York technical writing firms, TECHlit Consultants Inc. and the Jordanoff Corp. For example, TECHlit wrote in their November 1948 proposal: "These manuals will be complete in every respect, in accordance generally with the requirements of current AN specifications such as AN-H-20, AN H-7b, No. 4017 and No. 41018."[39]

In December, the Jordanoff Corp.'s proposal also offered to prepare manuals, "in accordance with specifications AN-H-20a and AN-H-21A"—the

AN-H-21a -6
(May, 1944)

Section VI
Paragraphs 4-8

RESTRICTED
AN 00-00-0

c. Place cowl flaps, auxiliary cowl flaps, and oil cooler flaps in "CLOSED" position before starting engines.

d. Except in an emergency, connect a battery cart of ample capacity (135 amperes) to the socket on the lower right side of the fuselage skin. This cart plug-in is quickly identified by the decalcomania "BATTERY PLUG-IN" above the opening.

e. With the battery and ignition switches "OFF," turn the engines over four or five revolutions by pulling the propeller through by hand in the direction of rotation, if the engines have not been running for 2 hours or more.

f. PRIMING.

(1) The lower the temperature, the greater the amount of priming which will be required. Experience will dictate how much priming is necessary to obtain good starting in various temperatures.

(2) When the starter switch is placed in the "ENGAGE" position, simultaneously place the pri[illegible] switch in the "ON" position and prime while [illegible] is turning over.

(3) Overpriming is indicated by raw g[illegible] the exhaust, collector ring, or on spark plugs. [illegible] a fire hazard and washes oil from the cylind[illegible] which may cause scored pistons and cylinders. If engine is overprimed, open throttle wide and clear out engine by pulling through several times by hand with battery and ignition switches "OFF." Be sure battery cart is not plugged in when clearing out engine.

Note

In cold weather, fuel may be discharged through the blower drain pipe and still leave the engine underprimed.

g. MIXTURE CONTROL.

(1) If starting difficulty is encountered in sub-zero weather, move mixture control out of "IDLE CUT-OFF" to "AUTOMATIC RICH" at the same time the starter is engaged.

(2) If the engine does not start almost immediately (3 seconds), return mixture control to "IDLE CUT-OFF" position.

(3) If the engine does not start within the next 5 seconds with mixture control in the "IDLE CUT-OFF" position, continue to turn the engine with the starter and repeat the procedure. One to three repetitions of this procedure will usually start the engine.

(4) As soon as the engine starts, move the mixture control to "AUTO-RICH" and observe the engine instruments for normal operation. Do not GUN the throttle.

(5) If the engine does not "take," return the mixture control to the "IDLE CUT-OFF" position immediately to avoid flooding the supercharger. Allow the starter switch to return to the "OFF" position. Energizing the booster coil longer than 15 seconds may burn out the coil winding.

5. ENGINE WARM-UP.

a. Keep cowl flaps and auxiliary cowl flaps closed until the engines are actually running, then open them.

b. Close the surface control booster shut-off valve located at the base of the control pedestal.

c. Place the control booster heater switch located on the control pedestal to "ON" position.

6. TAKE-OFF.

a. Open the surface control booster shut-off valve.

b. Do not take off on soft snow. Taxi along the runway a few times to pack it down.

c. Regardless of the degree of cold weather encountered, take-off should be made with the cowl flaps open. The hazard of taking off with partially closed flaps is too great to risk, and the engines will not cool off excessively during the take-off and rated power climb.

CAUTION

[illegible]en taking off on muddy or snow-covered [illegible], there is danger of the landing gear re-[illegible] or mechanism being clogged with mud [illegible]t snow, and then becoming inoperative [illegible]etraction because of freezing. If this is likely to occur, do not retract the landing gear.

TYPICAL PAGE

7. DURING FLIGHT.

If icing conditions exist:

a. Turn on wing and empennage de-icer system. Check de-icer system operation by the pressure gage on the lower right instrument panel.

b. Turn on carburetor, windshield, and propeller anti-icing fluid.

c. Place the carburetor heat control on "HOT."

d. Install the defroster windshield panels.

8. LANDING AND TAXYING.

a. AVOID EXCESSIVE COOLING. — Temperature inversions are common in winter and the ground air may be much colder than at altitude. To avoid excessive cooling when letting down, use considerable power and regulate cowl flaps to keep engine temperatures up.

b. CARBURETOR. — Make approach and landing with carburetor heat on, if there is a tendency toward carburetor icing during the approach. Turn on carburetor anti-icer if necessary.

c. COWL FLAPS.—If cowl flaps are closed to maintain satisfactory cylinder head temperatures, remember to open them if power is to be used for an extended period of time.

d. ENGINES. — In approach for landing in cold weather, do not idle engines at low speed. Check engines frequently for ability to accelerate.

e. LANDING WITH ORDINARY TIRES.—When landing with ordinary tires in temperatures between −18°C (0°F) and −35°C (−32°F), allow the airplane to roll to a stop, if practical, rather than apply the brakes, as application of the brakes builds up friction

6–22 RESTRICTED

Figure 2A. (Sheet 3 of 3)
Mechanical Specifications for Typeset Copy (Reduced to 86% of Correct Size)

Figure 7.4. Typical page from MIL Spec AN-H-21a (Hagley Museum and Archives, Wilmington, DE: Accession 1825, Box 41)

identical specifications Chapline had collected three months earlier.[40] To both Chapline and Eltgroth, proposals like these seemed to indicate a professional consensus that the genre of AN specifications was the correct model for computer documentation.[41]

Despite all these sources of encouragement to use the AN specification genre, Chapline felt uneasy about using a genre derived from old radar systems for this new invention, the computer. Using the old genre for a new technology just seemed, he said in an interview, like trying to pound a square peg into a round hole. Also, he was uneasy about using rigid specifications for a computer that was still a prototype. He may also have been uneasy about the asymmetric relationship of author and reader enacted in the military specification genre in comparison with the relatively symmetric relationship of author and reader in the technical reports. Look again at Figure 7.4 and note the rhetorical authority embodied in the step-by-step directions. A rhetoric of authority, of course, would make perfect sense for the original audience of AN specifications who were uniformed, regimented soldiers. Chapline must have wondered if an audience of engineers and scientists would respond as favorably to such a rhetoric of authority.

Chapline's uneasiness through the Fall 1948 can be observed when he queried his audience, Northrop Corp., about exactly what they wanted in the BINAC manuals. On October 27, Chapline asked George Gore, Secretary of Northrop, through his own boss, George Eltgroth:[42]

> Specifically, is this report intended for consultation by operating personnel so that only operating instructions are required; or is it to be reviewed by engineers so that an engineering report would be preferable; or do you desire a general handbook of operation and maintenance, including theory and description of circuit operations.

Gore responded within a fortnight on November 10:[43]

> We think the contents of the final report are up to you, but we suggest that it include the theory and description of circuit operations, the theory behind the operating and maintenance instructions, a description of the final tests and the performance of the computer, a compilation and extension of the monthly progress reports, and a statement that the article submitted complies with the requirements of the contract. This report should be an engineering report, and it will be reviewed and used by engineering personnel.

Northrop, as audience and customer, had certainly defined the "final report" for Chapline as in the engineering report genre like that of ENIAC and EDVAC, not in the military specification genre. However, in the same November 1948 memo, Gore also asked for a separate "operating and maintenance instructions" document.[44] What were these operating and maintenance instructions; how should the document differ from the familiar engineering report?

In December 1948 Chapline first offered his plan for an "instruction book."[45] It called for five parts:

- a description of terms
- an operating procedure description
- a theory of operation
- adjustment procedures displayed via "test points, oscillographic patterns, and variable resistance and capacitance adjustments"
- programming problems

Eltgroth, Chapline's boss, approved his plan the same day[46] and then passed a summary of it onto Northrop for their review and approval.[47]

Chapline's five-part plan differed from engineering reports in its explicit directives of "operating procedure description" and "adjustment procedures." Yet, by including "description" and "theory" in parts 1, 2, and 3, Chapline still suggested the ambiguous linking of the "operating and maintenance instructions" to actual concrete operator actions like those presented in the EDVAC technical report. In fact, Chapline specifically wrote in this memo: "This second section will not describe the problems of programming, but will assume that a problem is already in terms of computer language. . . . The third section will be the theory of operation considered from the non-morbid angle. This section will not cover any adjustments." Thus Chapline's five-part plan straddled both the AN specification genre and the technical report genre.

Six months later, on June 21, Chapline received Northrop's response to his five-point plan. In the letter, Eltgroth conveyed the fact that Northrop agreed to the earlier design of the engineering report, but he then asked Chapline to design the "operating and maintenance instructions" for BINAC using as his model the genre of instruction manuals developed by the automobile industry:[48]

> A discussion with Mr. Frank Bell of Northrop Aircraft, Inc. today, indicated an understanding on the part of Northrop that the operating and maintenance instructions would correspond to those issued to the purchaser of a new motor vehicle. That is, there would be included a description of the location and function of controls, together with directions for such preventive maintenance as deemed necessary. In the case of a motor vehicle, such preventive maintenance information takes the form of a lubrication chart, while computer maintenance included recommendations for the periodic replacement of vacuum tubes, cautions to be observed in the replacement of crystal diodes, and other such periodic adjustments as may be recommended by the [EMCC] Engineering Dept. . . . With the relatively simple maintenance instructions which are contemplated by Northrop, it may be possible to materially simplify the discussion of the theory behind the maintenance instructions.

What Northrop probably had in mind in this directive can be seen in Figures 7.5, 7.6, and 7.7 taken from a contemporary automobile instruction manual, the *1941 Ford Deluxe and Super Deluxe Reference Book*—one of those analyzed in Chapter 6.

Northrop's suggestion of this radical change in documentation genera must have had an effect on Chapline because he delivered a new and different outline for the Operating and Maintenance Manual (Figure 7.8) the very next day on June 22.

This new outline was quite a change from Chaplain's five-part plan six months earlier. Gone were the "descriptions" and "theory" deriving from the influence of the technical report genre. For example, Chapline's new outline differed from the table of contents for the EDVAC technical report (Figure 7.2), and he anticipated many contemporary approaches to instructions. Whereas the EDVAC's report assumed the computer was already setup and required its audience to deduce appropriate prior actions, Chapline explicitly began with BINAC's installation and used such task-oriented words as "uncrating," "cabling," "fusing," and so on. Nothing in the setup was assumed. Also notice how the outline began to convey more of a rhetoric of authority: a knowledgeable author addressing a less knowledgeable reader, much like the Ford manual authors when they addressed their readers to "Keep the clutch

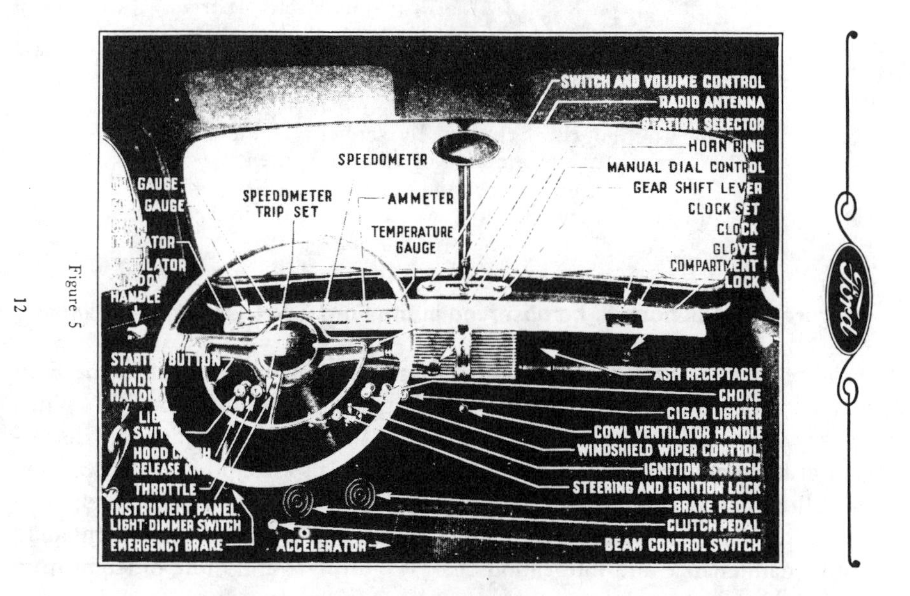

Figure 7.5. Description and location of controls in the *1941 Ford Instruction Manual*

From low to second gear below five miles per hour.

From second gear to high gear at twenty-five miles per hour.

Clutch • Don't ride the clutch pedal.

Driving with the foot resting on the clutch pedal will result in premature wear of the clutch facings and will necessitate frequent adjustment.

Keep the clutch pedal properly adjusted. This is indicated by having from 1″ to 1¼″ free movement of the pedal before the clutch starts to disengage.

Figure 13

The amount of movement can easily be determined by moving the clutch pedal down with a slight hand pressure. This will become less as the mileage on the car increases. If free movement less than 1″ is noted, the clutch pedal adjusting rod should be shortened. (See Figure 13.) Decreasing the length of the rod by screwing the clevis in increases the amount of free movement.

Your dealer can make this adjustment for you very quickly at a nominal charge.

23

Figure 7.6. Cautions to be observed in the *Ford 1941 Instruction Manual*

pedal properly adjusted" (Figure 7.6). Compare such a statement in the Ford manual with EDVAC's more descriptive approach shown in Figure 7.2: "When the key is pressed by the Operator." The EDVAC manual excluded any discussion of maintenance, and Chapline's earlier plan also excluded the discussion of "adjustments." Yet, Chapline's new plan for "Adjustments," "Tests," and "Tube Replacement Recommendations" resulted in a section quite similar to Ford's maintenance and lubrication chart (Figure 7.7) and quite different from any prior manual or plan.

The memo traffic between Northrop and Chapline in 1948 and 1949 is crucial to observing Chapline wrestling with what genre was most applicable to

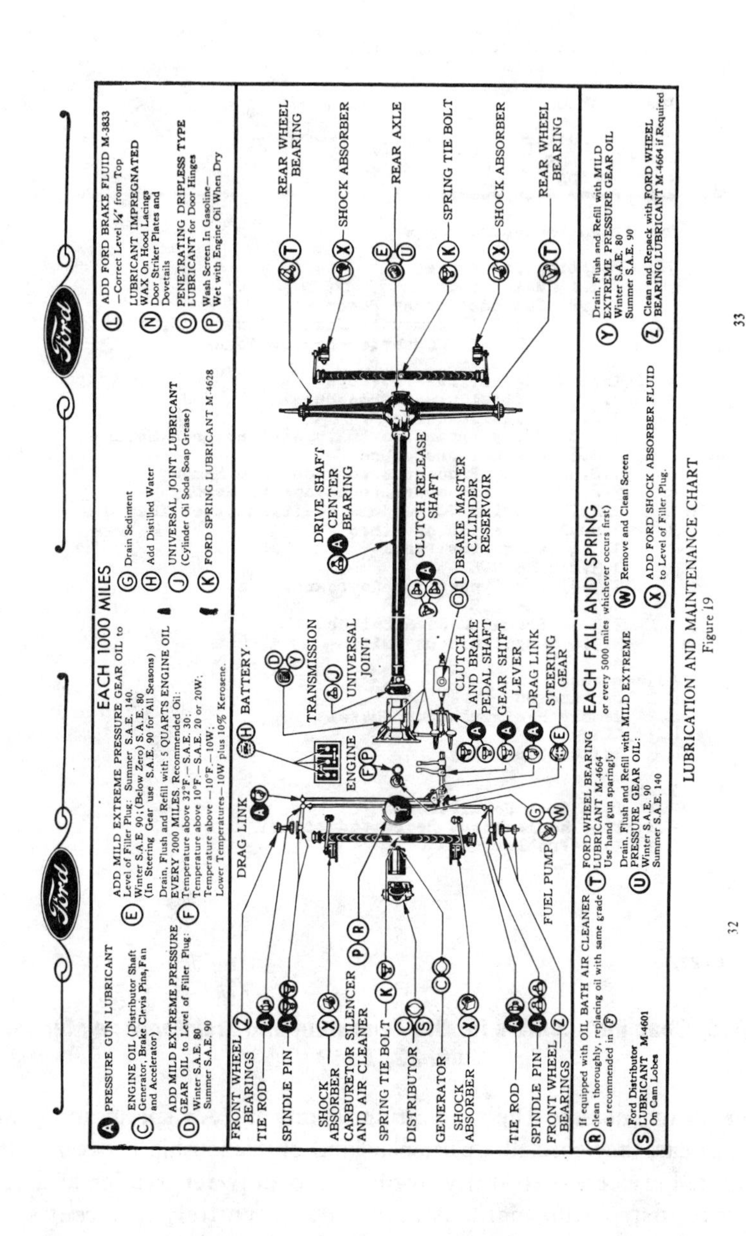

Figure 7.7. Lubrication and maintenance chart in the 1941 *Ford Instruction Manual*

OUTLINE -- INSTRUCTION MANUAL

1. Introduction

   1.1 Specifications
   1.2 Description of Components
   1.3 Spare Parts

2. Installation

   2.1 Uncrating
   2.2 Cabling
   2.3 Fusing
   2.4 Power Requirements

3. Operating Procedure

   3.1 Starting Procedure
      3.1.1 Main Power
      3.1.2 Cathode Heater Power
         3.1.2.1 Computer Heater Power
         3.1.2.2 Converter Heater Power
      3.1.3 DC Power
   3.2 Initial Operating Procedure
      3.2.1 Initial Clear Procedure
      3.2.2 Fill Procedure
      3.2.3 Empty Procedure to Typewriter from Memory
      3.2.4 Compute Procedure
      3.2.5 Empty Procedure to Tape from Memory
      3.2.6 Fill Procedure from Tape to Memory
      3.2.7 Setting CC to Some Position Other Than 000
   3.3 Manual Operation of Binac
      3.3.1 Converter Panel
      3.3.2 The RCB
      3.3.3 The Typewriter Keyboard
      3.3.4 The LCB
      3.3.5 Even-Odd-Even Switch
      3.3.6 Oscilloscope Switch

4. Adjustments
   4.1 PFR and CG
   4.2 Memory Recirculation Chassis
   4.3 Clock
   4.4 Adjustable Delays

5. Tests
   5.1 Tube Test Requirements
   5.2 Tube Replacement Recommendations
   5.3 Test Problem

JDC:fm
6/22/49

Figure 7.8. Chapline's outline for the *Operating and Maintenance Manual* of June 22, 1949

computer documentation. The engineering reports and AN specifications were not just formats; they enacted a particular kind of relationship between author and reader, between text and graphics, and between reader and text. Automobile instruction manuals conveyed an entirely different set of reader–author–text relationships. Both the Ford and Chevrolet instruction manuals of the 1940s had information conveyed visually, for example, the lubrication diagram and maintenance timing charts and 60% of the Ford and Chevrolet pages had illustrations and graphics. Paragraphing was especially frequent in the 1940 automobile manuals; moreover, the 1940 manuals had indexes. In short, the change from the engineering report and AN specification genera to the automobile instruction manual genre was a change from exhaustive technical coverage as the prime function to a simplified presentation of instructions to be used: "It may be possible to materially simplify the discussion of the theory behind the maintenance instructions"[49] Gore wrote in his June 1949 letter to Chapline. It was a shift from the needs of a trained audience who would study the technical specifications to an untrained audience who would perform the given actions. The change to the automobile instruction manual model meant readers could follow the instructions, not just read about them, because the connection of text to action was explicit and clear. The change meant random access of information rather than front-to-back reading. Finally, the change meant less emphasis on theory and more emphasis on clarity of presentation. In other words, the change in genre models was a whole new way of writing computer documentation. But, Joseph Chapline was not ready to accept it.

Chapline Verifies His Course

Chapline in his interview noted that, even with this rather unambiguous statement from Northrop urging him to use the automobile instruction manual as the genre for the *Operating and Maintenance* manual, he was still unsure of what genre to use. Admittedly, he no longer considered using radar AN specifications. Nevertheless, he reportedly reasoned, was using an automobile instruction manual any better? After all, the experienced technical writers at TECHlit and Jordanoff suggested using the AN specification genre. Why should he—an untrained, inexperienced writer on his first big job—change genera because of a single memo from his audience at Northrop? How could Northrop know what would be best? How could he receive verification for a change in genera for computer documentation?

Donald Norman offered an interesting psychological explanation as to perhaps why Chapline had such a hard time shifting his writing to the genre suggested by Northrop. Norman called the explanation "cognitive hysteresis:"[50]

> [Cognitive hysteresis] was derived from the phenomenon in magnetism called hysteresis, which refers to the property that once a material has been magnetized in one direction, it is difficult to make it change to being magnetized in the opposite direction. Simple magnetic fields that earlier would have caused the unmagnetized substance to be magnetized in either direction no longer have any effect. Once the magnetic state has been set, it is stable and resists further change . . . There are other names for this phenomenon in the psychological research community: functional fixedness, cognitive narrowing, tunnel vision. Whatever the name,

> they all describe pretty much the same phenomenon: People tend to focus upon the active hypothesis and, once focused, find it very difficult to change, even in the face of contradictory information.

Caught in the bind of his own "cognitive hysteresis," Chapline soon received verification for using the new genre from an unexpected source—students. On July 11, 1949, two weeks after Chapline received Northrop's automobile memo and after he drafted the new outline to the manual, seven Northrop employees (Floyd Steele, Fred Stevens, George Fenn, Irving Reed, Irving Wieselman, Jerry Mendelson, and Dick Baker) arrived in Philadelphia for a training course under a contract just signed with EMCC.[51] According to the contract, this course was supposed to focus on the "operation, maintenance, circuitry, diagnosis and correction of stoppages, and remedies and prevention of operations troubles of the BINAC."[52] On that July 11, Chapline recalled that he was busy at his desk editing drafts of the *BINAC Engineering Report* when Weiner, the head of EMCC Engineering, called him to a conference room. On arriving in the conference room, Chapline noticed that Weiner was talking to seven strangers.

"What's going on?" Chapline asked.

"Joe, you're going to teach a class on BINAC to some Northrop engineers."

"Fine. How much time do I have to prepare?"

"You don't, Joe. Here are your students."

At that instant, Joe Chapline was launched into a career as a BINAC instructor.[53]

By this time, Chapline had nearly completed the *Engineering Report on BINAC*. The purpose and audience had been clearly defined in Gore's November 1948 letter, and the genre of an engineering report was familiar to Chapline. However, he still had difficulty incorporating the instruction manual genre into the operating and maintenance manual that was meant to accompany the *Engineering Report*. The training contract with Northrop called for a class to focus on the "use of the machine as a computer and not the operation of the machine,"[54] and that was exactly what the operating and maintenance manual was supposed to deliver. Thus, Chapline took advantage of the classroom situation and asked his students, "What do you want to know about BINAC, and why do you want to know it?".[55] As a result Chapline used the class to help him verify the direction of the genre of computer documentation.

Within no time, the Northrop students knew Chapline was fishing for information from them. Moreover, according to a letter written by a Northrop engineer in the course, Chapline apparently got answers that surprised him.[56] He was surprised to learn, for instance, that "there is quite a spread in the ability of the students, [only] two of whom seem to be doing well."[57] However, Chapline had the information about the BINAC audience that he needed; in fact, he described it in his interview as a most "inspiring moment." Later, in the minutes of an EMCC meeting held just after he finished teaching the course, there was the note that the operating and maintenance manual was "in draft only" prior to the course, but after the course, he "started over to do the manual — final 6 to 8 weeks — pad at a time."[58]

Years later Chapline recalled that his month-long writing block in June 1949, that cognitive hysteresis concerning the incorporation of the automobile instruction manual genre, was broken by the question-and-answer repartee with his Northrop students. In the class, Chapline got to know the specific needs and backgrounds of the Northrop readers, and he took on the role of knowledgeable teacher-expert while they the roles of novice-student—the typical roles enacted rhetorically in the 19th-century instruction manual bequeathed to the automobile industry. Thus, Chapline was persuaded that the choice of the instruction manual genre was correct. "When I finished with them [the students], I knew exactly how to write the manual."[59]

The complete *Operating and Maintenance Manual for the BINAC*[60] was both similar and dissimilar to the 1940 instruction manuals for automobiles. Similar to the automobile manuals, the *Operating and Maintenance Manual for the BINAC* had

- 37 pages—a little smaller than the average of 57 pages for the 1940s automobile instruction manuals.
- 10 paragraphs per page—many more than the average of 7 paragraphs per page in the automobile manuals; however the 8 1/2" by 11" size of the BINAC manual pages versus pages half that size in the automobile manuals may account for the difference.
- five paragraphs labeled with the words "Caution" or "Warning"—nearly the same number as Chevrolet's 1940 manual.
- 1.9 headings per page—close to the average of headings per page of the automobile manuals.
- 69 total headings of which 4% are first-level headings, 29% are second-level headings, 57% are third-level headings, and 10% are fourth-level—this distribution and deep hierarchical structure was similar to the 1941 Chevrolet manual which had a heading distribution of 9% for first-level headings, 43% for second-level, 46% for third-level, and 2% for fourth-level.
- action-oriented operating procedures like "Switching Procedure," "Initial Clear Procedure," "Verify Procedure," and "Typing a Word to a Particular Memory Location Without Changing CC."
- actual names of buttons and switches in which the Engineering Report described only the logical operation of the computer (as seen in Figure 7.13).
- a much more skilled use of page layout that revealed subordination and parallelism in the text.

Yet, unlike the automobile manuals and more like the *EDVAC Technical Report*, the *Operating and Maintenance Manual for the BINAC*

- was produced on a mimeograph machine in the same way as the EDVAC report and not typeset—thus not nearly as "finished" as the automobile manuals of the 1940s.
- had only seven illustrations like the sparsely illustrated EDVAC report—significantly fewer than the average in the automobile instruction manuals. (However, whereas EDVAC had complex technical timing diagrams [see Figure 7.11 from the *Engineering Report*], the *Operating and Maintenance Manual for the BINAC* had photograph-like representations of operational controls as shown in Figure 7.12. The "BINAC-Local Control Box," "BINAC Converter Panel," and the "BINAC Keyboard" look just like the automobile manual's graphics and locations of controls.)
- had what is titled a "table of contents"—all the headings are listed with heading numbers indicating the hierarchy of information, but no page numbers (Figure 7.9). In fact, there were no page numbers in the entire BINAC manual similar to the EDVAC technical report (Figure 7.3) and the *BINAC Engineering Report* (Figure 7.10) and very unlike the automobile instruction manuals .[61]

The 37 pages of the *Operating and Maintenance Manual for the BINAC* embodied a genre compromise much like that struck in 1859 by the author of Grover and Baker's *A Home Scene; Mr. Aston's First Evening with Grover & Baker's Celebrated Family Sewing Machine Containing Directions for Using*, who included both a short story and step-by-step instructions. Joe Chapline's *Operating and Maintenance Manual for BINAC* was a genre compromise among the pressure of precedent, the engineering report, and the audience's call for innovation; thus, it contained elements of everything.[62]

This genre compromise in Chapline's thinking can be observed by taking a look at two incidents occurring just prior to his July 11 class. While the *Operating and Maintenance Manual for the BINAC* was still in a "draft only" form,[63] and the day after Northrop's letter concerning the automobile manuals was received, Chapline informed his boss, Eltgroth, that he was having difficulties with the design engineers:[64]

> It has been learned that only very slight use (incidental only) can be made of the engineers concerned. Therefore, the resulting report will be based only on those facts recorded in the notebooks and then only on those which form a complete story or which can be made complete by a question or two put to the appropriate engineer.

TABLE OF CONTENTS

1. Introduction
   - 1.1 Power Requirements
   - 1.2 Components of System
   - 1.3 Memory Chassis

2. Specifications for a Binary Computer
   - 2.1 Representation of Information
   - 2.2 Input and Output System
   - 2.3 Internal Memory
   - 2.4 Units of the Computer
     - 2.4.1 Accumulator
     - 2.4.2 L Register
     - 2.4.3 R Register
     - 2.4.4 Control Counter
     - 2.4.5 Control Register
     - 2.4.6 Program Counter
     - 2.4.7 Cycle Counter
     - 2.4.8 Function Table
     - 2.4.9 Cycling Unit
     - 2.4.10 Clock (Timing Pulse Generator)
   - 2.5 Instruction Code
   - 2.6 Manual Controls

3. Manual Controls
   - 3.1 The Typewriter Keyboard
   - 3.2 The Converter Panel
     - 3.2.1 The Operation Switch
     - 3.2.2 The Control Buttons
     - 3.2.3 The Neon Indicator Lights
   - 3.3 The Local Control Box (LCB)
     - 3.3.1 Static Register
     - 3.3.2 Sign Sensing Indicator Lights
     - 3.3.3 FF_{TO} and FF_{TS}
     - 3.3.4 Program Counter (PC) and FF4
     - 3.3.5 Cycle Counter
     - 3.3.6 Computer Starting Controls
     - 3.3.7 Scope Tank Selector Switch
     - 3.3.8 Scope Synchronizer Selector Switch
   - 3.4 Remote Control Box (RCB)
     - 3.4.1 The Function Switch
     - 3.4.2 The "DC-on" Indicator Light
     - 3.4.3 The Stop Flip-Flop
     - 3.4.4 CC and CY Clear
     - 3.4.5 T Clear
     - 3.4.6 Start Computing
     - 3.4.7 The Break Point Switch
     - 3.4.8 One-Operation Switch
   - 3.5 Operating Procedure
     - 3.5.1 Starting Procedure
     - 3.5.2 Initial Clear Procedure
     - 3.5.3 Tape Manipulation
     - 3.5.4 Fill Operation
     - 3.5.5 Verify Procedure (for each computer)
     - 3.5.6 Compute Procedure
     - 3.5.7 Empty Procedure
     - 3.5.8 Setting the CC to a Value other than Zero
     - 3.5.9 Observing Contents of a Memory Location not Specified by CC
     - 3.5.10 Typing a Word into a Particular Memory Location without Changing CC

4. Adjustments and Other Maintenance Details
   - 4.1 Adjustment of Pulse Formers
   - 4.2 Tuning Procedure for the Memory Recirculation Amplifier
   - 4.3 Adjustment of Adder Delays
   - 4.4 Tube Specifications
   - 4.5 Miscellaneous Precautions
   - 4.6 Maintenance

Figure 7.9. Table of *Contents of the Operating and Maintenance Manual for the BINAC* as it was delivered in December 1949

TABLE OF CONTENTS

1. Introduction

1.1 About the Corporation
1.2 About the BINAC
1.3 Elements and Organization of BINAC System

1.3.1 Keyboard-Printer Unit
1.3.2 The Converter
1.3.3 The Computers
1.3.4 The Memory Unit
1.3.5 Power Supplies
1.3.6 Blower System

1.4 BINAC and ENIAC Contrasted
1.5 The BINAC and UNIVAC
1.6 Power Requirements
1.7 Organization of this Report

2. Introduction to the Description of Logical Operation

2.01 Half Adder
2.02 Complementer
2.03 Unit Adder
2.04 Binary Adder

2.1 Clock and Cycling Unit
2.2 Cycling Unit
2.3 High Speed Bus Amplifier
2.4 Control Circuits

2.4.1 The Time Out Flip-Flop
2.4.2 Stop FF or FF5
2.4.3 The Start Circuit
2.4.4 Ending Pulse

2.5 Cycle Counter and Function Switch
2.6 Control Counter
2.7 Distributor and Static Register
2.8 Memory

2.8.1 Recirculation Principle
2.8.2 Memory Switch
2.8.3 Temperature Control

2.9 Control Register
2.10 Instruction Register
2.11 Function Table

2.11.1 Decoding Function Table
2.11.2 Main Encoding Function Table

2.12 The Accumulator Register
2.13. The "R" Register

Figure 7.10. Table of contents of the *Final Engineering Report* (Hagley Museum and Archives, Wilmington, DE: Accession 1825, Box 28)

Note how Chapline was still using the word "report" to describe the *Operating and Maintenance Manual for the BINAC* even after acknowledging receipt of the directive urging the automobile instruction manual genre. Notice also that Chapline was being forced to rely on notebook information from engineers rather than his own experience using BINAC or his observation of

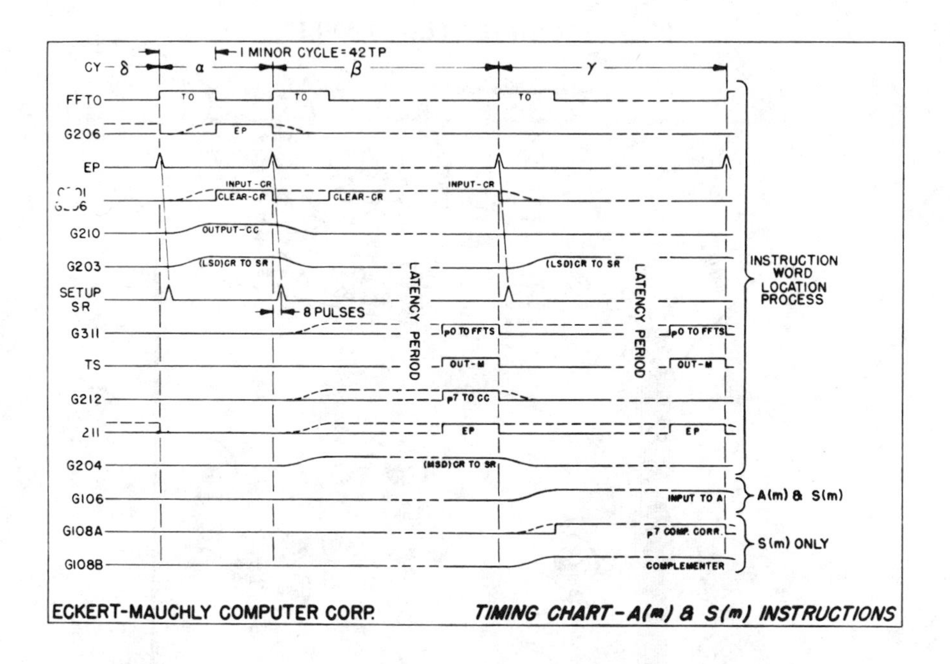

Figure 7.11. Timing chart graphics from the *Final Engineering Report* (Hagley Museum and Archives, Wilmington, DE: Accession 1825, Box 28)

someone else using it. In fact, Chapline was flying blind as to the accuracy of the operational information given to him in the engineers' notebooks. This fact becomes quite evident in a letter written to Northrop headquarters by Dick Baker, an ex-officio "unpaid student" who was the Northrop employee on extended leave acting as liaison on the BINAC project. In his report to Northrop headquarters about the class immediately after it finished on July 17, Baker related a conversation with Weiner, Head of EMCC Engineering:[65]

> As an interesting sidelight on this course, I mention a remark made to me by Mr. Weiner. (I quote his meaning, not his exact words): "I wish you'd sit in on that course, Dick, and check Joe [Chapline], since he has done little work on the machine, and, even though he is writing the instruction manual, he could easily make mistakes not being familiar with the machine itself, but only with the written material available which we both know is very incomplete and, in many instances, unreliable."

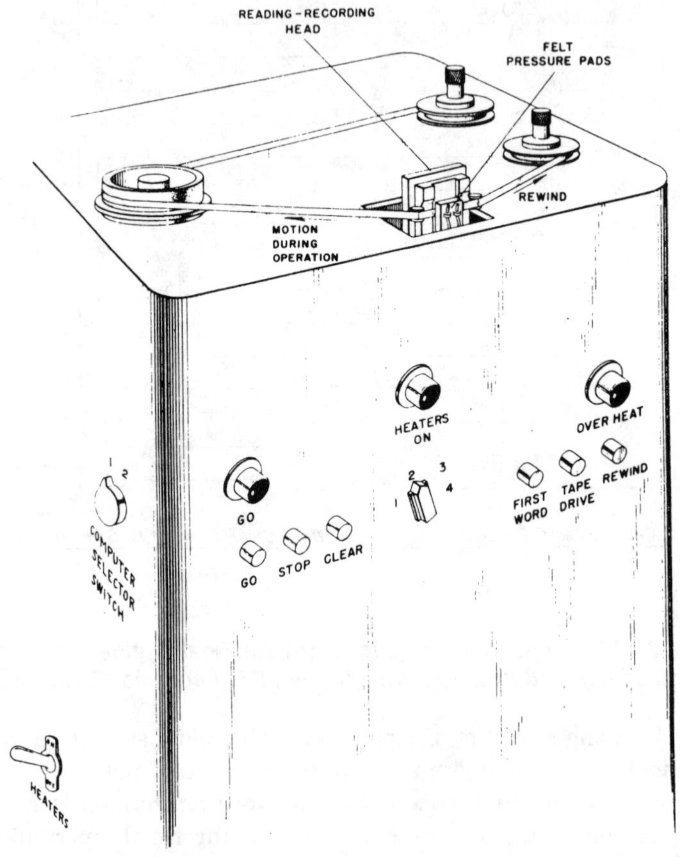

Figure 7.12. Operational graphics of the *Operating and Maintenance Manual for the BINAC* as it was delivered in December 1949

Notice the choice of words—"instruction manual." Either Weiner, Chapline's boss, had the choice of genre quite clear in his mind, or Dick Baker, in recalling and reproducing the conversation, did. Only Chapline was still calling the document a "report." Also notice the observation about how Chapline was gathering information. The raw material of the notebooks given to Chapline were probably complete in their theoretical description of BINAC, but most likely incomplete and sketchy concerning the specifics of the human operation of the computer. How could it be otherwise? Chapline reported that the BINAC was still being debugged as it left the plant for Northrop a month later at the end of August. Perhaps, then, Chapline's choice to include more theory and description, to pattern the *Operating and Maintenance Manual for the BINAC* after the engineering reports, occurred not only because his imagination was constrained by precedent but also because of the preliminary and sketchy operational material given him as raw material.

2.17 Operations

All of the logical elements necessary to the performance of the various instructions have been described in the previous sections. The numerous small block diagrams are actually sections of one main block diagram (D10) which is placed at the end of Section 2. The smaller drawings attempt to preserve the same general appearance as the corresponding parts of D10. The seven registers - CU, CC, CR, A, R, L, and T, are indicated by large Roman letters. The latter six registers are connected together by the HSB system. The HSBA is at the extreme right center of the diagram D10. The memory switch is in the extreme left section below the HSB. The control circuits lie below the HSB and above the grid-like pattern of the FT's.

The purpose of this section is to describe the sequence of events associated with each instruction. Each description of an operation traces the signals on D10 and illustrates their interrelation in time by means of timing charts.

Certain instructions group themselves together. Such grouping is shown by the order from top to bottom of the horizontal lines on the main encoding FT. For example, A(m) and S(m) appear together. The following descriptions preserve the grouping as do the timing charts.

It must be understood that the timing chart patterns are not to be taken as actual electrical signals. They are, in fact, idealized curves forming only a first approximation to the true patterns. They are justifiable, however, as an intermediate step toward an understanding of the computer operation.

In each timing chart the abscissa is time. The main vertical divisions represent minor cycles. When a solid line is labelled, for example, G125, it indicates the condition of the gate at any time. If the solid line is up, it indicates that the gate is open; down, indicates that it is closed. A dotted line indicates that one of the several necessary conditions for opening the gate is fulfilled. Very frequently a gate is conditioned by a FT signal and the TO signal. During TO the FT signal rises to "alert" the gate. Yet the gate will not open until TO ends. The dotted line shows the FT signal as it rises during TO.

Each timing chart shows the gating operation required to remove the instruction word from the memory and how the actual gate operations are timed so as to effect the transfer appropriate to each instruction. Although all instructions (except T(m) and U(m)) are shown in the γ cycle, it is to be understood that any instruction

Figure 7.13. Page describing operations from the *Final Engineering Report* (Hagley Museum and Archives, Wilmington, DE: Accession 1825, Box 28). "The purpose of this section is to describe. . ."

3.5.9 (2)
to
3.5.10

(9) Operate start button. (RCB) CY reads βTO. The LSD half of the word is now set up on the SR lights.

(10) Clear SR (LCB)

(11) Clear C and CY (RCB)

This procedure is useful for examining instruction words. It can not be used for numerical words since there are no lights corresponding to the three digits: sign, p36, and p21.

3.5.10 Typing a Word into a Particular Memory Location without changing CC

When an error in a word in the memory has been discovered, it is often inconvenient to change CC to this value for only one or two words. The following procedure allows any number of words to be typed in but requires considerably more time per word than is required when the CC reading is used for a long series of words.

(1) Prepare converter for typing into memory (3.5.4)

(2) When CY reads βTO, operate SR clear and set up the memory location desired on MSR switches

(3) Type out desired word. When 13th character (0) is struck, word is transferred into specified location. The CC reading is increased one unit and this value will set up in MSR. • If a second word must be inserted into a different memory location, then the SR must be cleared again and the desired memory location set up by operating the MSR switches.

Figure 7.14. Page instructing users about operations from the *Operating and Maintenance Manual.* Notice the explicit link of text and action in the imperative mood step-by-step actions

On August 31, 1949, the BINAC was shipped to Northrop in California, but Chapline continued working on the *Operating and Maintenance Manual* for another four months. The *Operating and Maintenance Manual for the BINAC* was finally delivered to Northrop on December 29, 1949.[66,67]

JD CHAPLINE, TECHNICAL WRITING MANAGER: 1950–1955

The 1954 UNIVAC I Manual Fulfills Northrop's 1949 Directives

Having made his mark in the company with his BINAC work, Chapline was asked to manage the Editorial Department when it worked on the UNIVAC I documentation from 1950 to 1955. During this time Chapline's own individually authored material was increasingly replaced by group-authored material from "the Editorial Department."[68] It was also during these years that the

Editorial Department built on Chapline's BINAC experiences and progressively moved away from prior compromises with precedent—the engineering report genre—and more in the direction of innovations requested by the audience—the automobile instruction manual genre.

For instance, the Preface to the 1954 Manual of Operation for UNIVAC System' announced for the first time a clear understanding of the difference between operational information and engineering design reports:[69]

> The purpose of this manual is to serve as a reference text for personnel learning operational procedures of the UNIVAC System. It was not intended as an encyclopedic reference to the System, and in actuality provides operational information on the central computer only. No prior knowledge of UNIVAC or other large scale digital computers is presumed.

Moreover, in the introduction to "Section 2: Mode of Operation of the UNIVAC," this sensitivity to the difference between operational information and engineering design reports was made even clearer:[70]

> A complete discussion of the operation of the UNIVAC requires a detailed description of the four-stage cycle of operations and each of the forty-three instructions in a manner similar to that contained in Chapter II. This is quite beyond the scope of this book, but a brief description will suffice for operating personnel.

Finally, in the UNIVAC I manual of 1954 Chapline and his colleagues unambiguously followed the 1949 direction from Northrop to "simplify the discussion of the theory behind the maintenance instructions." The audience described by this passage seemed to echo the one described earlier in *Ford Manual: For Owners and Operators of Ford Cars and Trucks* (1915-25). That manual's preface announced it would address: "Owners who in the great majority of cases have little or no practical experience with things mechanical."[71] Thus, not only was the ambiguity of the engineering report genre discarded ("encyclopedic reference"), but so was the assumption of a knowledgeable audience.

The Editorial Department also knew it needed to write more maintenance instructions such as those requested by Northrop in 1949. However, rather than combining operational and maintenance information as had been done in the *Operating and Maintenance Manual for the BINAC*, Chapline and the Department replicated in computer documentation the movement toward specialized manuals being made at the same time by the automobile industry. Thus, the Preface to the *1954 Manual of Operation for UNIVAC System* noted: "Descriptions of the Service Routines, which are of great convenience to the operation of a UNIVAC, are not included as they are already described in another manual."[72]

The formatting innovations of the 1949 *Operating and Maintenance Manual for the BINAC* also continued in the direction of the published automobile instruction manuals and away from the mimeographed technical reports of EDVAC and BINAC. For example, indentation in the 1954 manual now more effectively revealed content parallelism: Compare the increased effectiveness of the 1954 *Manual of Operation for UNIVAC System* in Figure 7.15 with the1949 BINAC in Figure 7.14 or the 1949 EDVAC in Figure 7.3.

MEMORY CLEAR

Next fill the mercury memory with decimal zeros.

1. Depress the Memory Clear Switch, D-3. It will lock down.
2. Set the IOS on ONE OPERATION.
3. Depress and release the Clear C Switch, G-14.1. Note that the PMC Neon, A.2-5.3, will be lit at this time.
4. Hit the Start Bar several times and observe the fifth instruction digit neons in the Static Register change. (Should increase by one with each step.)
5. Set IOS on ONE INSTRUCTION.
6. Hit Start Bar.
7. Lift the Memory Clear Switch to the Inhibit PMC position.
8. Depress the I.F. BIAS CONTROL Switch B.3-4.1, which will light the neon directly above it.

Note at this time, that the PMC Neon will be out, the Static Register will contain all binary zeros, the Cycle Counter will be alpha, and the Time Out and Stop Neons will be lit.

MEMORY CHECK

To see that the memory clear operation was performed successfully, the operator should next let the UNIVAC run on the skip instructions in the memory:

1. Set the IOS on ONE OPERATION.
2. Depress and release the Clear C switch.
3. Set IOS on Normal.
4. Operate Start Bar.

If the memory contains all decimal zeros, the computer will stall -- light the Stall Neon, F-4.3 - after approximately 4½ seconds. The Static Register will contain 01 000, the Cycle Counter will be on beta stage, the Time Out Neons will be lit and the FT Intermediate Checker Neon, B.2-14 will be lit.

(If the memory check operation does not terminate as above, this probably means that the mercury has not as yet reached proper operating temperature. Wait a few minutes, then repeat the Memory Clear operation.)

5. Depress the Stop Switch, F-5.1
6. Set IOS on ONE OPERATION
7. Depress Clear C Switch

Now the UNIVAC is ready to start a problem.

Figure 7.15. Numbered Operational Instructions in the 1954 Remington Rand *Manual of Operation for UNIVAC System* (Hagley Museum and Archives, Wilmington, DE: Accession 1825, Box 372)

Moreover, whereas the 1949 BINAC manual was similar to the ENIAC and EDVAC engineering reports—it was not randomly accessible because there were neither page numbers nor table of content's page numbers—the 1954 *Manual of Operation for UNIVAC System* was more randomly accessible—page numbers appeared for the first time in an initial table of contents just as they had in the automobile instruction manuals. With the inclusion of page numbers, Chapline and his Department also began using typography to indicate heading hierarchy (e.g., large vs. small type size, all uppercase vs. lowercase, centered vs. not centered) rather than heading hierarchy numeral designations (e.g., 1.1, 1.1.1, etc.).

Graphics also played a significantly larger role in the 1954 *Manual of Operation for UNIVAC System* just as they had in the automobile manuals of the 1930s and 1940s; between 50% and 75% of the pages in automobile manuals of that period had illustrations. Only 19% of the pages in the BINAC manual (6 pages) had illustrations, yet 45% in the UNIVAC manual did (59 pages).

Of the 59 illustrations in the UNIVAC manual:

- nine were pictures of the physical units or keyboards or controls
- one was a diagram of the flow of tape through the tape readers
- one was a diagram of the timing of the program in microseconds
- 50 were diagrams and schematics of abstract information such as gating or electronic flow

Along with all these borrowed elements from automobile manuals, the UNIVAC I manual also had an element that illustrated the asymmetric rhetoric of authority. For example, in the 1912 edition of the *Instruction Book for Ford Model T Cars*, the following operator guidance appeared under the heading "Go It Easy:"[73]

> In the flush of enthusiasm, just after receiving your car, remember a new machine should have better care until she "finds herself" than she will need later, when the parts have become better adjusted to each other, limbered up and more thoroughly lubricated by long running. You have more speed at your command than you can safely use on the average roads save under exceptional conditions, and a great deal more than you ought to attempt to use until you have become thoroughly familiar with your machine, and the manipulation of brakes and levers has become practically automatic.

Now, although the operator guidance is not as lengthy as Ford's, Chapline and his staff in the UNIVAC I manual wrote in much the same authoritative manner: "The UNIVAC operator must know what each UNIVAC instruction does in order to perform his duties efficiently."[74] Consider also that the numbered instructions shown in Figure 7.15 enacted the same asymmetric relationship of author and reader as did the sewing machine manuals and the automobile manuals.

With the release of the *Manual of Operation for UNIVAC System* in 1954, Chapline and the Editorial Department finally completed the genre transition from engineering reports for one-of-a-kind computers to instruction manuals for marketing an entire series of computers. The tenor of this whole transition can best be seen in comparing the early draft "Introduction" of the UNIVAC Computer Manual that Chapline probably wrote before EMCC was sold to Remington-Rand with the final published introduction.[75] The earlier draft UNIVAC introduction was still written very much in the format and feel of the compromise with the engineering report genre—the introduction seemed almost tentative in its approach to the audience (see, e.g., page 1):

> 1. Introduction
>
> An electronic computer can be analyzed into many small elements all of which contribute to the total operation. One of the biggest problems in high-speed automatic computing is the input-output equipment. Due to the nature of the problem, the resulting equipment is extremely complex. Yet, it is composed of the same basic elements as are found in the arithmetic and control circuits of the CENTRAL COMPUTER. And, although a naive approach would consider the input-output equipment first, experience had shown that a complete understanding of the inner world of the central part of a computer is the greatest aid in grasping the details of the input-output operations.

The final 1954 introduction, on the other hand, displayed a great deal of self-confidence ("let us consider") along with a new extended use of analogy. These two features gave the final published introduction the self-confident feel of the automobile instruction manuals (see, e.g., page 1):

> SECTION I
>
> Concepts of a Computer
>
> The UNIVAC is a large electronic device designed to perform repetitive clerical or mathematical operations at a high rate of speed and accuracy. In order to make clear the function of each component of the computer and its relationship to other components, let us consider the operation of a payroll clerk.
>
> The clerk has a stack of time cards showing the number of hours a company's employees have worked during the week. She has a second stack of cards which show for each employee certain bits of fundamental information such as the hourly rate of pay, dependents, etc. As a

work aid she uses a calculating machine which performs addition, subtraction, and multiplication for her.

In addition, she herself provides a factor of control or procedure, she selects the time card and hourly rate card for a particular employee. By means of the calculator she computes the man's gross pay, subtracts his income tax, old age insurance tax, etc., and ends up with the man's net pay which he enters in her ledger book.

This payroll process can be thought of in terms of four basic operations: Input, Output, Arithmetic, and Supervision. As shown in figure 1, the time cards and hourly rate cards are the input data, the calculator is the arithmetic unit, and the output is the net pay entered in the ledger. The whole operation is supervised and executed by the clerk.

In taking five years to follow the suggestions of the audience, Chapline and the Editorial Department illustrated just how change in technical communication is evolutionary rather than revolutionary.

Manager Chapline's "Fundamental Experiment" in Staffing

The movement in genera from engineering report to instruction manuals eventually had an effect even on the type of person hired in the Editorial Department. During Chapline's regime from 1950 to 1955, the Editorial Department grew by "a factor of five," adding six writers, two copyreader/checkers, five draftsmen, and assorted typists,[76] and also embarked on a "fundamental experiment" in documentation management and staffing.[77]

In 1949, when Joseph Chapline was first asked to consider technical communication hiring practices, the profession itself was far from being established. For example, even as late as 1956, Reynold Rossi wrote:[78]

> If we find that Technical Writing is not a separate professional occupation . . . then we will recommend that Fordham abandon its attempt at developing a curriculum for technical writing at the university level.

It seems that a great deal of confusion existed as to who should write technical manuals, for Rossi continued:[79]

> We have heard it said, "Give me a man with a solid technical background and I'll make a technical writer of him." Conversely we have heard it said that the important thing is the ability to write clearly and concisely. And finally we have heard "Give me an engineer who can write."

In January 1949, while developing the BINAC documentation, Chapline came down squarely on the side of the first alternative in a letter to Eltgroth, "The best advice seems to be to choose persons of technical ability rather than writing ability."[80] Chapline's decision coincided with what Charles Ressey (1956) later wrote: "In all cases a technical background is a must."[81]

Thus, as the department grew with the UNIVAC I work, Chapline hired people who had degrees in physics, electrical engineering, and mathematics. Soon Chapline found that although he could quiz any of them on the operation of the mechanism they were assigned to document, and although they could easily rattle off accurate information, they began to suffer from what he called "constipation." Without writing ability, Chapline said in his 1989 interview, these technically educated people just could not create the necessary pages from the knowledge they had in their heads. So Chapline took a gamble and hired a nontechnical technical writer named Frank Leary, who was a graduate of Temple University with a degree in journalism.

According to Chapline, his interview with Leary went something like this:

> "Do you know the binary number system?" asked Chapline.
> And Frank answered, "No."
> "Do you know electronics?" asked Chapline.
> "No."

However, what Leary did know were the finer technical points of language usage. Chapline ended the interview saying:

> Frank, a funny thing is going on here. All my objective tests say "No, don't hire this man." But I have a very strong gut reaction that says the opposite. So I'm going to give you a three-month probation, and we'll see if it works.

The very next day things began to happen. In Chapline's words, Leary became a "catalyst." He would wander around the department, coaching the technically proficient members to tell him their ideas. Then Frank would sit down and create pages of text derived from their explanations. When the pages were written, Leary would hand them back to whoever gave him the information. Frequently, they would say, "Frank, this isn't right!" He would retort, "Good, well, what is right?" From this verbal give-and-take, Leary and his technical source would come up with a clearly stated, accurate page.[82] Chapline probably had Frank Leary in mind when he wrote about the collaboration and mutual dependency of the "new technical writer/editor" and his information sources in a 1960 article:[83]

> But he [the scientist] can be helped by the technical editor even if not by writing down his ideas; very possibly the editor, by acting as a rational backboard, can bounce the ideas back to the technical man so that he will see the organization he is looking for; this operation is carried on more conveniently by word of mouth.

Thus, just as the automobile instruction genre was an innovative option for Chapline in designing documentation, so, too, Frank Leary's hiring seemed to epitomize an innovative, new type of "technical writer." Chapline also seemed to think it was Leary's idea to use nontechnical analogies in the UNIVAC manual as illustrated in Figures 7.16 and 7.17 (much like the visual analogies used by Griffiths in the 1850 shipwright treatise).[84]

If Leary was the one who led the Editorial Department to use visual analogies, perhaps he was also the one who led them to use analogy as the basic arrangement of the text in the 1954 *Manual of Operation for UNIVAC System*. As described on page 1 in the "Introduction," Leary, or the Editorial Department under Leary's influence, used greatly simplified versions of a computer to lead the reader gently into more complex descriptions:

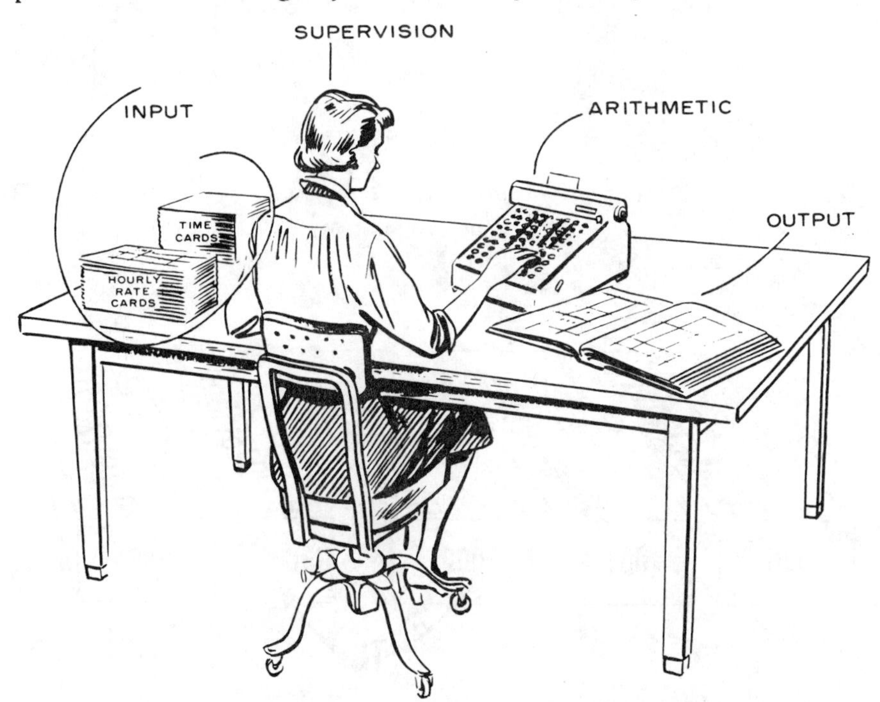

Figure 7.16. Simplified analogy in the 1954 Remington Rand *Manual of Operation for UNIVAC System* (Hagley Museum and Archives, Wilmington, DE: Accession 1825, Box 372)

> Since the logical design of the UNIVAC is exceedingly complex it would be quite outside the scope of this manual. A knowledge of logic is essential to meaningful understanding of the operation of the computer however, and is the reason for describing a simple computer in Chapter II. The elements of UNIVAC are then introduced by analogy with the components of the simpler computer.

More than just chance moved Chapline and his Editorial Department to use both visual and textual analogies in the first manual completed after hiring Frank Leary. Unlike his technically trained colleagues in the Editorial Department and the Liberal Arts trained Chapline who now had been indoctrinated by the EMCC/UNIVAC way of doing things for some three or four years, Leary brought into the EMCC/UNIVAC corporate culture the outside influence and perspective of journalism. Perhaps it was even Leary's personal lack of understanding and background about the electronics of UNIVAC that

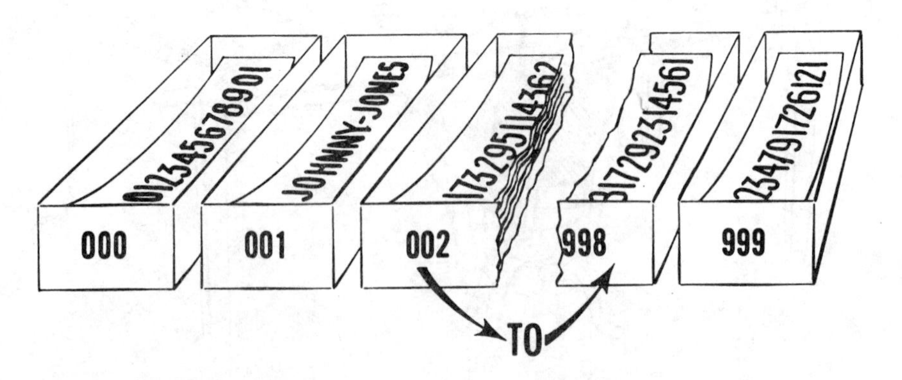

Figure 7.17. Simplified analogy in the 1954 Remington Rand *Manual of Operation for UNIVAC System* (Hagley Museum and Archives, Wilmington, DE: Accession 1825, Box 372)

led him to use analogies for his own understanding. Then, seeing the efficacy of the analogies, Leary simply offered his own method of learning to the manual's audiences—much like Oliver Evans who personally used graphics to understand technology and then used them to explain his technologies to others. Frank Leary's perspectives and journalistic expertise and the 1949 note from Northrop's Frank Bell were influences from outside the EMCC/UNIVAC corporate culture. Perhaps because that corporate culture was still so young, these outside influences quickly permeated it and changed the way EMCC/UNIVAC wrote documentation.

Leary's speedy effect on the Editorial Department convinced Chapline of one thing. His earlier preference for technical people who could write rather than writers who could learn technology was, as he said in his 1989 interview, "a gross error." After Leary, Chapline always tried to hire people who "were canny about language" and to whom he could teach the technology of computers.[85] Eventually, Chapline's philosophy of technical writing management and staffing was so successful in developing his own Editorial Department that it was the reason Chapline left UNIVAC and Remington Rand in 1955. A colleague from Philco, a company also busy developing electronic digital computers in the Philadelphia area, complained to Chapline that he could not get writers nor could he train them. This colleague asked Chapline if he could work his training magic at Philco, and that is exactly what Chapline did from 1955 to 1962. In fact, he managed two different writing groups at Philco.

What is interesting about Chapline's own concept of the "new technical writer" was that he soon began retreating from it. In both published and unpublished articles, Chapline pointed out that the non-technical technical writer could not supplant the knowledge of his or her technological sources. Leary had to collect information from others first before writing his material just as Chapline had done before him with BINAC. Thus, Chapline's "new technical writer" was dependent on his sources for information. Chapline wrote about the interaction of these new nontechnical technical writers and their dependence on information sources in a 1955 unpublished article, "What is a Technical Writer":

> But one of his [the technical writer's] chief problems is the collecting of facts from those skilled in the field and noticing the imbalances and poorly-focused sections. These are the very things that he is there to do and for which the person of talent and genius in the [technical] field needs him. A good technical writer should be able to send the engineer back to his laboratory for further tests when the statement given to the technical writer does not hold logical consistency....Again the technical writer's task is to recognize these distortions and to bring them to the expert's attention for correction.

Five years later, in 1960, Chapline continued to grapple with his staffing "fundamental experiment" and again pointed out the dependency of the nontechnical technical writer on technical information sources. In an article Chapline wrote:[86]

> . . . the technical writer who attempts to communicate for others lives the life of a demimondaine. He is not respected as a writer because he is perhaps only a little more skilled than the scientist in this area, and where science is concerned the technical writer is clearly nowhere near the equal of the scientist.

Chapline's choice of the term *demimondaine* perhaps conveyed more than he intended. *Webster's Third International Dictionary* defines a *demimondaine* as a "group (as within a profession) characterized by dealings of doubtful legality or propriety." (This aspect of the definition follows references in the dictionary to women of ill-repute.) Was Chapline suggesting in his choice of words that his earlier innovative advocacy of nontechnical technical writers with a canniness for language was of doubtful legality or propriety? Could Chapline be undermining his own advocacy by his choice of words, suggesting that such writers were "not respected"? This is indeed a probability, for in his 1989 interview Chapline further characterized the nontechnical technical writer as an "amanuensis"—one who copies or writes from another's dictation. Had Chapline compromised in his technical communication staffing concepts just as he had compromised in his 1949 BINAC manual?

Perhaps Chapline finally escaped from having to choose between technical technical writers and nontechnical technical writers by introducing a third role, the technical editor.[87] In the "The Editorial Function in Scientific Organization," Chapline wrote:[88]

> The technical writer assumes that the less writing a scientist does, the better; the editor says that only a scientist can think and formulate his own ideas in the first instance. The technical editor is a full professional in his own right and lives with the confidence that the scientist is always there to keep him on the technical track. The technical editor has not the technical writer's limitation of knowing only one or a few technical fields, but instead is universal; the technical editor can move from radar to computers to medicine with equal ease, whereas the technical writer cannot.

Perched on the cusp of technical communication precedent and innovation both in the choice of genera and in staffing, and unable to resolve the choices one way or another, Chapline consistently synthesized alternatives and only slowly moved to embrace innovation.

THE MORAL OF THE STORY

There are several lessons to be learned from Joseph Chapline's pioneering work as a technical writer and manager of technical writers. In his BINAC writing experiences, various documentation genera (AN specifications, engineering reports, and automobile instruction manuals) competed for his allegiance. In weighing the efficacy of these various genera, Chapline displayed a wonderful openness to ideas and suggestions from others, while at the same time felt constrained by his own sense of documentation precedents from ENIAC and EDVAC. Eventually, Chapline did participate in the genre change in the UNIVAC I manuals by moving much more surely in the innovative directions requested by the audiences. However, he took some five years to fully embrace these innovations revealing a slow, evolutionary process in his documentation design.

As a manager, Chapline made what he called his "gross error" in hiring technically trained writers instead of people with a "canniness about writing." What he learned from the actions and ideas of his subordinate, Frank Leary, changed his management policies and EMCC/UNIVAC's way of writing computer documentation. Yet, because of the persistent influence of precedent on his management and staffing decisions, Chapline eventually retreated a bit from his innovative hiring practices. Chapline labeled a nontechnical technical writer a "demimondaine" and an "amanuensis," suggesting his own doubts as to the legitimacy or propriety of such writers.

Chapline was a pioneer, not a revolutionary. He repeatedly stepped back from innovation and compromised with precedent. Chapline's difficulty in incorporating the automobile instruction manual as his genre for the BINAC documentation, even though the audience clearly and specifically invited him to this innovative decision, revealed just how conservative and captured by precedent technical communicators can frequently be.

Joseph Chapline was technical communication's Odysseus during his BINAC and UNIVAC adventures. His work was the prototype for many technical communication careers. He endured delays in delivering manuals, handled two or three projects at one time plus teaching, defined his audience over and over again, was mistrusted by engineering and undermined by incomplete and incorrect engineering documentation, lacked accessibility to relevant design engineers, and finally delivered the documentation after much time and effort, only to have some of it shelved, never to see the light of day. Joseph Chapline experienced the frequent frustrating obstacles faced by technical communicators every day.

ENDNOTES

1. Eltgroth to Chapline, 6/21/49, Hagley Museum and Archives, Wilmington, DE: Accession 1825, Box 2.
2. Until 1945 the word "computer" was the term used to describe the operators of differential analyzers and similar computing machines. Only in February 1945 was an effort made to change the definition of a "computer." In a report for the National Defense Research Committee, the following appeared: "By 'calculator' or 'calculating machine' we shall mean a device . . . capable of accepting two numbers A and B, and of forming some or any of the combinations A+B, A-B, A x B, A/B. By 'computer' we shall mean a machine capable of carrying out automatically a succession of operations of this kind and of storing the necessary intermediate results. . . . Human agents will be referred to as 'operators' to distinguish them from 'computers' (machines)." Stibitz, George: February, 1945. "Relay Computers," National Defense Research Committee, Applied Mathematics Panel, AMP Report 171.1R, qtd. in Ceruzzi, Paul E.: "When Computers Were Human." *Annals of the History of Computing* (1991) 13: 237–44.
3. Owens, Larry: "Vannevar Bush and the Differential Analyzer: The Text and Context of an Early Computer." *Technology and Culture* 27(1) January 1986: 63–95 (see page 72, Figure 4); and Aspray, William (ed.): *Computing Before Computers*. Ames, Iowa: Iowa State University Press, 1990: 181–8. See also Bush, Vannevar: "The Differential Analyzer. A New Machine for Solving Differential Equations." *Journal of the Franklin Institute* 212(4) (October 1931). (Bush, of course, is now widely known by technical communicators for his "memex" paper in 1947 that foreshadowed and described "hypertext."). See also Shurkin, Joel: *Engines of the Mind*. New York: W. W. Norton and Co., 1984: 98.
4. Owens, op. cit., 1986: 77.
5. Fisher, Franklin M., McKie, James W., and Mancke, Richard B.: *IBM and the US Data Processing Industry*. New York: Praeger Publishers, 1983: 6.
6. Bashe, Charles M., Johnson, Lyle R., Palmer, John H. and Pugh, Emerson W.: *IBM's Early Computers*. Cambridge, MA: MIT Press (MIT Press Series in the History of Computing), 1986: 26–33. See also Aspray, op. cit., 1990: 215–9.
7. Lundstrom, David E.: *A Few Good Men from UNIVAC*. Cambridge, MA: The MIT Press, 1988: 23—"The Eckert-Mauchly people told the story, believed at the time, that IBM had pretty much ignored the electronic computer, considering it to be a toy for government-funded laboratories. That is, they considered it so until Metropolitan Life Insurance in New York bought one of the early UNIVAC I's. MetLife was a company known

for its tightness with a buck, never spending one dollar unless there could be a two-dollar return. When they bought a $1–1/2 million UNIVAC system, so the story went, there was panic in Poughkeepsie! IBM launched a crash effort to develop their 70X series, which proved to be no great shakes technically but which filled the marketplace gap until IBM could develop something better. Nearly every engineer at UNIVAC understood that *we* were the leaders in electronic computers at that time. The topic of many a lunch time conversation was how we at UNIVAC could maintain and exploit our lead." See also, Watson, Thomas, Jr.: *Father, Son and Company*. New York: Bantam Books, 1990. According to Tom Watson, Jr., head of IBM's first computer projects and later CEO, what drove IBM to computers was the personally felt insult in Eckert and Mauchly's sales claims that magnetic tape, not punch-cards (IBM's mainstay at the time), was the optimal storage device in the future for business, and Eckert & Mauchly was taking business away that had traditionally been IBM's. For example, IBM's Hollerith punch-card machines had been designed for the Census Bureau's work in the 1880s and now the Census Bureau was using a UNIVAC. See also Fishman, Katherine Davis: *The Computer Establishment*. New York: Harper and Row, 1981: 39. Fishman concurs about the defensive role IBM took in regard to magnetic tape. In fact, she speculates that it was the use of magnetic tape on BINAC and later on UNIVAC that most upset the multimillion dollar Hollerith paper punch-card maker and caused it to enter the market in 1952 with the Type 701 which, of course, had paper punch cards as input.

8. Shurkin, op. cit., 1984: 255.
9. IBM did not ignore electronic digital calculators until 1952 when it jumped in with both feet with the Type 701 machine. Although the spotlight of history focuses on EMCC's ENIAC, BINAC, and EDVAC, IBM developed the Selective Sequence Electronic Calculator (SSEC) in 1947 (Bashe, Johnson, Palmer, & Pugh, op. cit., 1986: 47–59). The SSEC was an upgrade of the ASCC technology—it was 250 times faster—but only one was built and according to Bashe et al., "its designers had not extended the frontiers of electronic technology," and "it had surprisingly little influence on the design of subsequent machines" (Bashe, Johnson, Palmer, & Pugh, ibid., 1986: 58). See also Aspray, op. cit., 1990: 244.
10. Bush's first analyzer was build at MIT in the late 1920s. The second and third were built using WPA funds at the Moore School at the University of Pennsylvania.
11. Most notably Shurkin, op. cit., 1984: 139–190.
12. Slater, Robert: *Portraits in Silicon*. Cambridge, MA, 1989: 41–64. It is important to include the word "large" in its historical designation

because some claim that Atanasoff in 1939 and Konrad Zuse in 1941 designed the first electronic digital computers.

13. Stern, Nancy: *From ENIAC to UNIVAC*. Bedford, MA: Digital Press, 1981: 50–1, 133.
14. Shurkin, op. cit., 1984: 75.
15. Shurkin, ibid.: 196.
16. Again one must include the adjective large because history has now shown that the EDSAC invented by Maurice Wilkes at Cambridge University had the first internally stored library of subroutines (Bashe et al., op. cit., 1986: 321). However, much of the work he did in England was inspired by a six-week course of lectures in 1946 at the Moore School, "Theories and Techniques for Design of Electronic Digital Computers," at which Eckert and Mauchly were two of the lecturers (Stern, op. cit., 1981: 94–5).
17. Stern, ibid., 1981: 50–1, 133.
18. The EDVAC project was completed instead by John von Neumann, Herman Goldstine, and Arthur W. Burks. For John Von Neumann's claims regarding his contribution to the stored-program concept, see "First Draft of a Report on EDVAC" (Stern, ibid., 1981: 177–246, "Appendix"), and for Herman Goldstine's claims regarding his contribution to the EDVAC project, see his book, *The Computer: from Pascal to von Neumann*. Princeton, NJ: Princeton University Press, 1972. For a good complete story on EDVAC, see Williams, Michael R.: "The Origins, Uses, and Fate of the EDVAC." *IEEE Annals of the History of Computing* 15(1) 1993: 22–38.
19. NBS acted as a conduit for computer research and development for the Census Bureau (Shurkin, op. cit., 1984: 221) and was quite interested in establishing itself as a "prime center for computer development" (Stern, Nancy: "The BINAC: A Case Study in the History of Technology." *Annals of the History of Computing* 1(1) July 1979: 16).
20. BINAC was not called BINAC because it used binary programming methods—in the 1960s hexadecimal was to become the standard programming number system with the IBM 360 (see Strothman, James E.: "The Ancient History of System/360." *American Heritage of Invention and Technology* 5(3) Winter 1990: 38). Rather it was called BINAC because the final computer actually had two computers linked together so that each could cross-check the other for accuracy. This same technique of dual computers was the secret behind Tandem Computer's multimillion dollar success in the 1970s and 1980s.
21. Stern, op. cit., 1979: 11.
22. The history of BINAC after it left EMCC and arrived at Northrop is somewhat controversial. Saul Rosen states: "The BINAC apparently never worked satisfactorily." ("Electronic Computers: A Historical

Survey." *Computing Surveys* 1(1) March 1969). Stern describes 28 serious deficiencies reported by Northrop (Stern, op. cit., 1979: 13–4). Chapline in his interview observed that one of the reasons why BINAC was perhaps less than satisfactory was because EMCC was not completely finished with the construction and debugging of BINAC even though it was a year late in delivery. Stern in her 1979 article suggests that the Northrop engineers were not entirely objective in their assessment of BINAC because many "had hoped to build their own computer for Northrop and were disgruntled when the BINAC contract was announced" (Stern, op. cit., 1979: 14).

23. Stern, op. cit., 1981: 133.
24. Interestingly, it was General Groves of Manhattan Project fame—user of ENIAC's and EDVAC's calculating powers—who was designated by Remington Rand to act as liaison between the two companies.
25. Rosen, op. cit., 1969.
26. Warner Brothers' Merrie Melodie cartoons had Daffy Duck and Porky Pig in the 1956 "Rocket Squad" cartoon use a player piano as input to a clearly labeled UNIVAC.
27. For more information about the computers used for predicting elections in the 1950s and 1960s, see "Fast Predictors: Computers and the U.S. Presidential Elections." *Annals of the History of Computing* 10(3) 1988: 209–16.
28. Watson, op. cit., 1990.
29. Lundstrom, op. cit., 1988: 23—"Engineers, it was believed, were second-class employees at IBM while all status was enjoyed by marketing. For this reason it was believed that IBM computers were and would continue to be technically inferior because no really first-class creative engineer would want to work there. Nevertheless, the IBM sales force was so well trained and highly motivated that 'they could sell manure in a brown paper sack!' The UNIVAC products, therefore, had to be twice as good as the IBM products to have a chance in the marketplace."
30. Shurkin, op. cit., 1984: 261. Watson, op. cit., 1990. By 1961, 4,000 of the 6,000 computers in existence had been made by IBM.
31. Fishman, op. cit., 1981: 45.
32. Chapline was Mauchly's student at Ursinus College in Pennsylvania when Mauchly taught physics there. In his senior year, Chapline lived on Mauchly's farm, and upon graduation in 1942, Mauchly arranged for him to work on the differential analyzer at the Moore School at the University of Pennsylvania: "I was responsible for the correct operation of the analyzer in the solution of the exterior ballistics equations in fulfillment of a contract between the university and the Aberdeen Proving Ground in the computation of firing tables for the Army Ordinance Department" (Patent Examiner deposition, No. 85, 809, January 21,

1954: 4). While Chapline was working at the Moore School in 1943, he was the one who introduced Mauchly to Herman Goldstine and thus to the military's needs and funding for a computer (Wulforest, Harry: *Breakthrough to the Computer Age*. New York: Charles Scribner's Sons, 1982: 49). Chapline left the Moore School in 1947 and joined Mauchly and Eckert at ECC.

33. Earlier Chapline had written demonstration descriptions for presentation to clients and the manuals for the Army Security Agency's noncomputer device for coding and decoding Army signal traffic. (Chapline to Auerbach, 12/16/47, Hagley Museum and Archives, Wilmington, DE: Accession 1825)
34. Chapline's documentation team in 1948 consisted of himself, Jack Silver, and Bill MacEwan who helped with the technical illustrations. He shared the services of Fran Morello, typist, with three or four executives.
35. An interesting appreciation of Von Neumann's first draft report on EDVAC can be found in Godfrey, Michael D., introduction to the re-edited version of the report in "First Draft of a Report on the EDVAC." *IEEE Annals of the History of Computing* 15(4) 1993: 27–75.
36. Additionally, the EDVAC authors knew their audience intimately because they were mathematicians and scientists much like themselves. Compare this intimate understanding of an audience with the misapprehension of the female audience made by Singer and Howe in the early days of sewing machines when they thought women singularly unsuitable for operating machines.
37. George Eltgroth was ECC's Vice President, General Counsel, and patent attorney, and, as such, he reported directly to Mauchly and Eckert. Eltgroth was not exactly Chapline's superior but more of a collaborator in the development of BINAC documentation: Eltgroth depended on Chapline to help him write the patent applications, and Chapline had to filter all his communications to outside companies through Eltgroth. Because Eltgroth was so prominent in the organization of ECC/EMCC management, there is a wealth of background letters and memos detailing Chapline's experiences from 1947 to 1950. From 1947 to 1950, Chapline was effectively only three-deep in the hierarchy. After 1950, one increasingly loses sight of Chapline and the Editorial Department even though the Editorial Department grew by over 800%. The Department and Chapline got lost as they became buried deeper and deeper beneath growing levels of corporate hierarchy and began reporting to Engineering as only one of many departments. For two years (1951 and 1952) the Editorial Department always appeared as the last item of the monthly Engineering Status Reports until, in 1952, the Editorial Department is dropped altogether from the monthly status reports.

38. Chapline to Eltgroth, 9/9/48, Hagley Museum and Archives, Wilmington, DE: Accession 1825, Box 2.
39. Reisman to Bossi, 11/30/48, Hagley Museum and Archives, Wilmington, DE: Accession 1825, Box 82.
40. Jordanoff Corp. proposal to Eltgroth, 12/9/48, Hagley Museum and Archives, Wilmington, DE: Accession 1825, Box 41.
41. Neither TECHlit nor Jordanoff got the contract. Both proposals for just the UNIVAC manuals were too expensive for a company as seriously under capitalized as EMCC was at that time. TECHlit required $189,500.00, and Jordanoff, for the work of 36 employees, required $194,797.04. There were also concerns that information for the manuals, necessarily highly proprietary, should stay within the company. In January 1949 Chapline submitted a counterproposal to Eltgroth to be allowed to keep the work in-house in the Editorial Department. Chapline suggested he and the Department could do the BINAC and UNIVAC work with the addition of only two more writers (versus the 36 planned for by Jordanoff). Chapline also felt that his current work on the BINAC manuals would "serve as a proving ground for the Editorial Department for its bigger tasks of preparing the UNIVAC book. It is consistent with the type of equipment being manufactured that the accompanying description material should be of high grade" (Chapline to Eltgroth, 1/27/49, Hagley Museum and Archives, Wilmington, DE: Accession 1825, Box 28). In the interview with Chapline he also suggested that Eltgroth never actually intended to award the contracts to outsiders but only used their bids on time and money to develop benchmarks to judge how much internal time and effort the documentation might require.
42. Eltgroth to Gore, 10/27/48, Hagley Museum and Archives, Wilmington, DE: Accession 1825, Box 46.
43. Gore to Eltgroth, 11/10/48, Hagley Museum and Archives, Wilmington, DE: Accession 1825, Box 46.
44. Gore to Eltgroth, 11/10/48, Hagley Museum and Archives, Wilmington, DE: Accession 1825, Box 46.
45. Chapline to Eltgroth, 12/9/48, Hagley Museum and Archives, Wilmington, DE: Accession 1825, Box 46.
46. Eltgroth to Chapline, 12/9/48, Hagley Museum and Archives, Wilmington, DE: Accession 1825, Box 82.
47. Northrop memo, Baker to Bell, 4/28/49 Hagley Museum and Archives, Wilmington, DE: Accession 1825, Box 1.
48. Eltgroth to Chapline, 6/21/49 Hagley Museum and Archives, Wilmington, DE: Accession 1825, Box 2.
49. Eltgroth to Chapline, 6/21/49 Hagley Museum and Archives, Wilmington, DE: Accession 1825, Box 2.

50. Norman, Donald A.: *Things That Make Us Smart: Defending Human Attributes in the Age of the Machine*. Reading, MA: Addison-Wesley Publishing Co., 1993: 133–4.
51. Mendelson transcript, Smithsonian Oral History Collection, #196: 13.
52. EMCC training contract, 7/1/49, Hagley Museum and Archives, Wilmington, DE: Accession 1825, Box 84.
53. The class lasted for two weeks, eight hours a day. In later years, having worked at EMCC until 1955 and later at Philco until 1962, Chapline worked as a technical writing consultant teaching over 200 short courses up and down the East Coast.
54. Typewritten description of training course July 11 to July 15, 1949, Hagley Museum and Archives, Wilmington, DE: Accession 1825, Box 84.
55. Typewritten description of training course July 11–15, 1949, Hagley Museum and Archives, Wilmington, DE: Accession 1825, Box 84.
56. Baker to Ackerlind, 7/17/49, Hagley Museum and Archives, Wilmington, DE: Accession 1825, Box 84.
57. Smithsonian Oral History #196, op. cit.: 49. According to one student, Jerry Mendelson, many of the students were there for the sake of prestige—it looked good to be included in such a cutting-edge technology class—but most were not going to actually use the information in a concrete way. But the class must have worked for Mendelson—who was the only one who was going to be operating the BINAC—because he stated in an interview over 20 years later, "When we came back, I knew that machine inside and out."
58. Handwritten minutes of meeting, 7/49, Hagley Museum and Archives, Wilmington, DE: Accession 1825, Box 4a.
59. Interview with Chapline, November 17, 1989.
60. Hagley Museum and Archives, Wilmington, DE: Accession 1825.
61. The complete documentation for the BINAC, consisting of the *Engineering Report and Operating and Maintenance Manual*, is very much like the contemporary "tiered" documentation approach advocated by Microsoft and AT&T: Different audiences are served with different manuals with different purposes. See Brockmann, R. John: *Writing Better Computer User Documentation, Version 2.0*. New York: John Wiley and Sons, 1990: 50–1.
62. Chapline's genre compromise was well explained by Berkenkotter and Huckin when they described the tension between stability and change at the heart of genre use: "Genres are the intellectual scaffolds on which community-based knowledge is constructed. To be fully effective in this role, genres must be flexible and dynamic, capable of modification according to the rhetorical exigencies of the situation. At the same time, though, they must be stable enough to capture those aspects of situa-

tions that tend to recur. This tension between stability and change lies at the heart of genre use and genre knowledge." (*Genre Knowledge in Disciplinary Communication: Cognition/Culture/Power*. Hillsdale, NJ: Lawrence Erlbaum Associates, Publishers, 1995: 24–5.)

63. Handwritten minutes of meeting, 7/49, Hagley Museum and Archives, Wilmington, DE: Accession 1825, Box 4a.
64. Chapline to Eltgroth, 6/22/49, Hagley Museum and Archives, Wilmington, DE: Accession 1825, Box 83.
65. Baker to Ackerlind letter, 7/17/49, Hagley Museum and Archives, Wilmington, DE: Accession 1825, Box 84. What is fascinating about this observation is how it verified information by Lundstrom (Lundstrom, op. cit., 1988: 20–1), a UNIVAC technician who reminisced about 1956 after Chapline had left for Philco: "This problem in transferring technology, which I have seen demonstrated time and time again, arises from the nature of engineers and engineering. Most engineers hate to write. When absolutely forced to, they will keep their descriptions as cryptic as possible and will plagiarize existing documents wherever they can. A complete engineering design consists of a great stack of engineering drawings, supplemented by a stack of specifications and test procedures. If the developing engineering group has strong discipline and if the engineers are conscientious, this great pile of paper will represent perhaps 90 to 95% of what the receiving group needs to know about the design. Missing will be the subtle tricks of timing, methods of compensating for known weaknesses in components, etc. These are not in writing because no one ever thought of a need to write them down. 'Everybody knows about that.' And, indeed, in the developing organization this may be true: Everybody does know about the tricky timing in the read circuit; it was the topic of hallway conversation for weeks."
66. Resume of Northrop Aircraft, Inc. Contract Relations, Brown to Draper, 2/21/50, Hagley Museum and Archives, Wilmington, DE: Accession 1825, Box 85.
67. However, Chapline did not know that a decision had been made "by Dr. Mauchly and Engineering to withhold the engineering report prepared by Mr. Chapline in that it divulged too much engineering data" (Resume of Northrop Aircraft, Inc. Contract Relations, Brown to Draper, 2/21/50, Hagley Museum and Archives, Wilmington, DE: Accession 1825, Box 85). This was a problem for Northrop because the receipt of this *Engineering Report* was part of their signed contract. Subsequent letters and telegrams from Northrop describe withholding a final payment of $20,000 until the report was delivered, and a legal dispute erupted for two years (Gore to Mauchly telegram, 1/31/50, Hagley Museum and Archives, Wilmington, DE: Accession 1825, Box 85; Gore to Mauchly, 5/1/50, Hagley Museum and Archives, Wilmington, DE: Accession 1825,

Box 86; Rumbles to McNamara, Northrop internal memo, 6/5/51, Hagley Museum and Archives, Wilmington, DE: Accession 1825, Box 86; Eltgroth to Sterck, Bosin and Harr (Northrop's Legal Dept.), 10/6/51, Hagley Museum and Archives, Wilmington, DE: Accession 1825, Box 87; Overton, Lyman, Prince and Vermille Law Firm to Northrop's Schuck, 11/20/51, Hagley Museum and Archives, Wilmington, DE: Accession 1825, Box 87; Eltgroth to Schuck (Northrop's Legal Dept.), 12/20/51, Hagley Museum and Archives, Wilmington, DE: Accession 1825, Box 87).

The reason that EMCC was reticent is suggested by an internal Northrop memo nearly two years later on June 5, 1951, reporting on an investigation into the situation as follows (Rumbles to McNamara, Northrop internal memo, 6/5/51, Hagley Museum and Archives, Wilmington, DE: Accession 1825, Box 86):

> Mr. Johnson, the Controller [of Northrop], told Mr. Ludwig that considerable mystery surrounded the installation of this computer. He said that as far as he knew, the computer was not owned by the Northrop Aircraft Corporation, and he made a cursory examination to verify the fact that no payments had ever been made from Northrop Aircraft Corporation funds. The computer section itself is operated on funds which are allotted periodically, and the source of these funds is not known, either to Mr. Johnson, the Controller, or the operators of the computing section. . . . It might be of interest to note that the two men primarily concerned with the building of the computer have left Northrop [Northrop engineers Wieselman, Sprague, Sarkissian, Steele, Eckdahl, and Reed (three of whom attended Chapline's course) left to found Computer Research Corp. which was later bought by NCR (Smithsonian Computer Oral History Collection #196: Irving Wieselman)], and the computing section is now in charge of an IBM trained engineer. It would appear that IBM, in view of their close relations with this account, may have financed the purchase of the Eckert-Mauchly unit.

Chapline has no knowledge to this day of the withholding of his *Engineering Report*. He has acknowledged, however, a similar situation in which IBM had bootlegged Eckert-Mauchly material on the UNIVAC project. For that project, Chapline was asked to write up some demonstrations about UNIVAC for the Prudential Insurance Company, a "solid IBM customer" (Fishman, op. cit., 1981: 250). Chapline was later surprised to find out that IBM was using "copies" of his UNIVAC demonstrations to give prospective clients demonstrations of IBM's Type 701 computer. No wonder some clients called the Type 701, IBM's UNIVAC!

[This kind of activity on the part of IBM worried NCR, another early computer pioneering company: "Mr. Howard has just telephoned me regarding possible revelations of confidential developments in your Laboratory to representatives of the IBM Company" (Stern, op. cit., 1981: 41)].

68. He also dictated the UNIVAC patent application that was submitted to the Patent Office and continued to write and edit conference papers for UNIVAC engineers and scientists.
69. Hagley Museum and Archives, Wilmington, DE: Accession 1825, Box 372.
70. UNIVAC, *Manual of Operation for UNIVAC System* (1954): 59.
71. Model T manual, *Ford Manual: For Owners and Operators of Ford Cars and Trucks* (1915–25): 2.
72. Even the words now sound like automobile jargon
73. Detroit, MI: Ford Motor Company, Third Edition, March 1912.
74. UNIVAC, ibid., 1954: 62.
75. Hagley Museum and Archives, Wilmington, DE: Accession 1825, Box 372. This document is undated but is labeled at the top with "EMCC Safe file" suggesting that the document was drafted prior to the acquisition of EMCC by Remington-Rand. Certain impressions of the typewriter letters suggest that the same machine typed both the BINAC operating and maintenance manuals. And most important, Chapline indicated the hierarchy of his BINAC manuals (both the *Engineering Report* and the *Operating and Maintenance Manual*) on the top right hand corner of each page (e.g., "1.3," "2. to 2.4.2," etc.), and this UNIVAC manual draft uses the same hierarchy indication method on the top right hand corner of each page ("1.1.1," 1.1.2"). Such heading hierarchy designation totally disappears from the final UNIVAC manual in 1954.
76. It grew by a "factor of five," as Chapline recalled in his interview; if the BINAC had 1,400 vacuum tubes and the UNIVAC was going to have 5,000, the Editorial Department, in order to document it, had to grow by a similar ratio.
77. Interview, 11/17/89, with J. D. Chapline.
78. Rossi, Reynold: "Formal Training Programs." In *Conference Proceedings for the Association of Technical Writers and Editors and the Society of Technical Writers, 1956, New York City* (Vols. 1 and 2) Society for Technical Communication, Arlington, VA: :25–7. (Available on microfiche from UNIVELT Inc.)
79. Rossi, ibid., 1956: 25. The wording of this debate over the qualifications of a technical writer or editor continues to this day. See Samson, Donald C.: "An Editing Project for Teaching Technical Editing." *Technical Communication* 37(3) (Third Quarter 1990): 262–3.
80. Chapline to Eltgroth, 1/27/49, Hagley Museum and Archives, Wilmington, DE: Accession 1825, Box 28.

81. Ressey, Charles: "What Management Expects of Technical Writers." In *Conference Proceedings for the Association of Technical Writers and Editors and the Society of Technical Writers, 1956, New York City* (Vols. 1 and 2) Society for Technical Communication, Arlington, VA: :9. (Available on microfiche from UNIVELT Inc.)
82. This "thinking aloud" during the prewriting stage now comes highly recommended as a prewriting technique. See Dougherty, Barbey N.: *Composing Choices for Writers*. New York: McGraw-Hill, 1985: 25–7.
83. Chapline, op. cit., 1960: 51.
84. With the passage of time away from differential analyzers and women clerks called "computers" (Stern, op. cit., 1981: 14) came the new audience for UNIVAC and a new meaning for "computer." "Computer" came to be seen as a machine, and the former personnel who were called "computers" now became only symbolic analogies to the machine.
85. Perhaps the hiring of Leary was not nearly as innovative as Chapline thought because he himself was hired by EMCC and UNIVAC as the lead writer and manager and his own degree from Ursinus was in history and political science. It was Chapline, a liberal arts major, who was asked by Eckert and Mauchly to write the detailed description for the first computer patent because he understood the whole system better than any one engineer and could convey that sense of the whole to someone else. Did Chapline simply see in Leary a reflection of his own lack of technical expertise and his own canniness about language?
86. "The Editorial Function in Scientific Organizations." *IRE Transactions on Engineering Writing and Speech* (now titled *IEEE Transactions on Professional Communication*). EWS-3 (2) July 1960: 48–53.
87. The British technical communication certifying agency, The London City and Guilds, also discriminates between technical writers and technical editors.
88. Chapline, op. cit., 1960: 50.

Chapter 8

New Computer Technology, New Documentation Methodology? Sidney Lida's Creation of the First IBM Computer Instruction Manual for the Type 701, 1952-3

"First look backward in order to look forward."
—Tom Watson, Sr., President of IBM[1]

Chapter 7 described Chapline's odyssey when he created the first computer instruction manuals for BINAC and UNIVAC I in the late 1940s and early 1950s. Was he alone in compromising between precedent and innovation in using the tradition of instruction manuals bequeathed by the sewing machine and the automobile industries? Or, was there another path to the creation of the earliest computer instruction manuals? To answer these questions, consider the work of Sidney Lida who wrote the documentation for IBM's first electronic computer, the Type 701.

SETTING THE STAGE: THE TECHNOLOGICAL CONTEXT OF THE TYPE 701

When Chapline joined EMCC in 1947, the principals of the company already had designed and built electronic computers. However, when Sidney Lida began writing his documentation at IBM, he could only observe a history of designing and creating electromechanical punched-card calculators. At the time of the Type 701, "most people, insiders and outsiders, considered IBM to be a punched-card company."[2]

Electromechanical punched-card calculators were primarily used by business and government for repetitive, relatively simple calculations such as those used by the Census Bureau. Industry and government had long been heavy users of these electromechanical devices as one can observe from just two of many measures:

- In 1945, the year ENIAC was announced to the public, IBM made $10.9 million net profit after taxes on punched-card electromechanical calculating equipment.[3]
- In 1946, a year after the announcement of ENIAC, 10,000 tons of paper punched cards were used in the United States.[4]

From 1945 to 1951, EMCC pioneered electronic calculators, while, at the same time, IBM continued researching and developing already accepted, electromechanical, calculator technology. For example, in 1944, while Eckert and Mauchly developed ENIAC, IBM donated the Automatic Sequence Controlled Calculator (ASCC) or Mark 1 to the Harvard Mathematics Lab.[5] The Mark 1 was a high point of electromechanical calculator technology. Furthermore, in 1947–48, IBM continued in this line and developed the Selective Sequence Electronic Calculator (SSEC).[6] Although the SSEC was an upgrade of the ASCC technology—it was 250 times faster—"its designers had not extended the frontiers of electronic technology," and "it had surprisingly little influence on the design of subsequent machines."[7]

A number of characteristics namely, flexibility, speed, and reliability, separated the electromechanical punched-card calculators—the ASCCs or SSECs—from the electronic computers—the ENIACs or BINAC. The electromechanical punched-card calculators had their processing operations specifically set up by the wiring of plugboards that looked like old-fashioned telephone switchboards. To change the calculation operations of such a machine meant pulling out a plugboard and replacing or rewiring it. However, to change the operations of the internally stored programming of the BINAC, new instructions could be simply inserted by tape, card, or switch. Thus, the electronic computers were more flexible than the electromechanical calculators in their ability to change calculation operations.

The electromechanical calculators also had physical constraints on the speed of their computations that the computer tracking electrons did not. Thus, the electronic computers were also faster in their calculations. Finally, electromechanical calculators had brushes or photocells to detect holes in the punched cards. In turn these produced an electrical signal activating magnets to open or close relays allowing the calculators to calculate. The problem was, however, that even dust particles could make these relays unreliable.

Yet, users of the electromechanical technology must have wondered if their current low frequency of changing calculation operations warranted the adoption of this new electronic technology. In addition, the users must have also wondered about the advantages of fast computer calculating speeds. Would these expensive computers stand idle as humans attempted to input instructions or assimilate their results? Furthermore, even if humans could keep up with the computers, were there enough mathematicians to develop the programs needed by these super-fast electronic computers? Regarding this problem, Dr. R. F. Clippinger of Aberdeen's Ballistic Research Laboratory[8] observed at the meeting of the American Mathematical Society in 1950:[9]

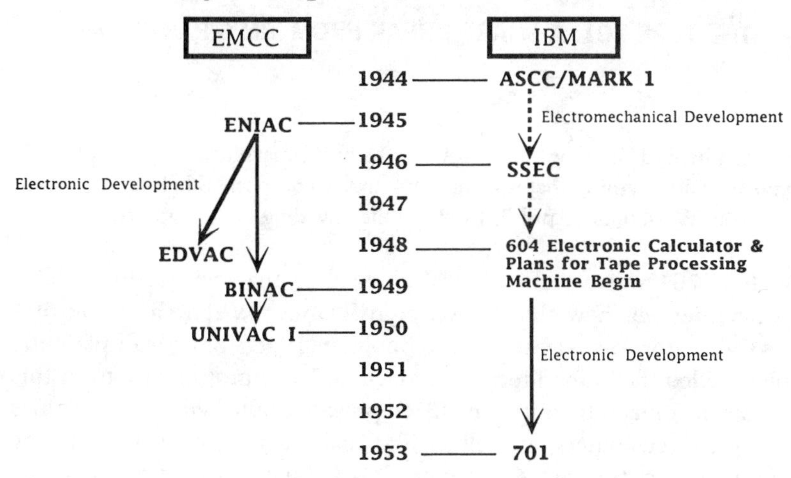

Figure 8.1. The race between EMCC and IBM to create commercial electronic digital computers

> In order to operate the modern computing machine for maximum output, a staff of perhaps twenty mathematicians of varying degrees of training are required. There is currently such a shortage of persons trained for this work, that machines are not working full time.

Finally, even though electromechanical calculators with relays were unreliable, was the electronic computer ENIAC any more reliable when one moth shorting out one circuit could cause the whole system to halt?

Thus, even with the unveiling of IBM's first true electronic computer, the Type 701 in 1953, the focus of IBM production and profits did not shift for quite some time away from electromechanical punched-card calculators. For example, IBM took an important design step toward the Type 701 when it introduced the Electronic Calculating Punch, Type 604, in 1948. The Type 604 allowed greater calculating flexibility because users could perform more than one arithmetic operation during a single pass of the punched cards.[10] Thus, in its electromechanical design and action, the Type 604 prefigured the crucial Type 701 computer flexibility element of "automatically sequenced programs." However, the Type 604 not only prefigured the design of the Type 701, but also succeeded it by selling quite well long after sales and production of the Type 701 had ceased in 1955. In fact, a thousand more of the Type 604s were sold while it was in production until 1958[11] than the combined total of 4,000 for all the first-generation computers produced and sold by all companies up to 1959.[12]

THE TYPE 701: A SHARP BREAK FROM IBM'S PAST?

> The idea behind the new calculator was to build machines that would go far beyond anything that the company had accomplished before.
>
> —Tom Watson, Sr., April 7, 1953, public unveiling of the Type 701[13]

The IBM Type 701 was a technological changeling that developed under a variety of names, each with its own significance. When the Type 604 Electronic Calculator was released to the public in 1948, IBM began planning for a project called the "Tape Processing Machine." The project was given this moniker both to differentiate it from IBM's previous punched-card machines and to suggest to customers, as well as IBM management, how the machine would compete directly with Chapline's EMCC tape-processing machines, BINAC and UNIVAC I.[14] It was especially important for IBM to suggest to customers that it was moving to tape as a storage medium because one of IBM's biggest customers, Metropolitan Life Insurance, told the son of IBM's president, Tom Watson, Jr., that they could no longer afford to rent three floors of a building just to store IBM punched cards. Metropolitan Life Insurance would soon be forced to use EMCC's tape storage devices.[15]

With the outbreak of the Korean War in 1950, the "Tape Processing Machine" on IBM research and development's backburner became a new electronic computer project named the "Defense Calculator." Under that name, the project signified a fulfillment of IBM's pledge to President Truman that it would do whatever was necessary for the war effort. The new name also helped the project sidestep internal IBM opposition to electronic computers by suggesting that the machine was only a wartime special project and not a threat to IBM's primary, peace-time, punched-card product lines.[16] After all, IBM had no prior experience with the electronics of magnetic tape or magnetic drum technologies of data storage, and the few electrical engineers at IBM were dominated by the more numerous and strategically placed mechanical engineers and managers.

Planning for the "Defense Calculator" as a true electronic computer like the BINAC and UNIVAC I began in earnest in January 1951, and its IBM product designation reverted to traditional IBM nomenclature as the "Type 701" with the ceasefire in Korea. It was called the "Type 701" according to Cuthbert Hurd, Director of IBM's Applied Science Division, a special marketing group handling the Type 701: "first to imply that this computer was to be followed by others, and second to distinguish it from the punched-card machinery that had numbers in the 600s."[17]

The first Type 701 was finally shipped in December 1952 to the nuclear bomb test site at Los Alamos Scientific Laboratory in New Mexico and was officially unveiled to the public on April 7, 1953. In 1953, a Type 701 "system" had seven components:[18]

- Type 701 electronic analytical control unit
- Type 706 electrostatic storage unit
- Type 711 punched-card reader
- Type 716 alphabetical printer
- Type 721 punched-card recorder
- Type 726 magnetic-tape reader and recorder
- Type 731 magnetic-drum reader and recorder

Today, the Type 701 would be categorized as a scientific computer; it was designed for calculations of a single, complex problem rather than repetitive, simple calculations, such as the computing of monthly payrolls done by general business computers. General business computing was the goal of UNIVAC, the Universal Numerical Calculator, and of the 701's successors, the 702 in 1955 and 705 in 1956.

Nineteen 701s were produced until April 1955 when the 701 was replaced by the 702. Eight 701s were purchased by aircraft companies, seven by government agencies (three of the seven by Atomic Energy Commission laboratories), two by commercial users for scientific calculations, one by the Rand Corp., and one was kept by IBM for its own use. Yet, beyond these simple statistics and dates was a much more important acme for IBM—when Tom Watson, Jr., later President of IBM, described the Type 701, he defined it as "the machine that carried us into the electronic business."[19]

Technology and methodology separated the Type 701 from anything IBM had ever done before. According to an IBM Confidential "Defense Calculator Preliminary Operator's Reference Manual:"[20]

> In the past the majority of such [scientific and technical] computing has been done on standard or modified punched-card calculators and accounting machines, such as the Type 604 Electronic Calculating Punch. The new machine represents a major step in the development of electronic calculators of greatly increased speed and memory capacity, with an automatically sequenced program and a highly flexible input and output system.

Additionally, in testimony before U.S. Federal Court in the *U.S. vs. IBM* antitrust suit,[21] Cuthbert Hurd described the 701's unique qualities in the following way:[22]

> The technology (electronics) and methodology (stored program and automatic control) for the 701 were sharp breaks from IBM's previous technology (punched cards and relays) and previous methodology (control panels and manual operation).

The Type 701 necessitated the massive hiring of new electrical engineers, rather than the traditional mechanical engineers. From these new hires came corporate officers who later controlled nearly every branch of IBM.[23] The cost of the Type 701 development effort was also a departure from past IBM experiences. It was the most expensive project in all IBM history to that point—$3 million just to develop and design the prototype.[24]

However, the Type 701 was still perceived by both IBM customers and IBM insiders to be a very close kin to previous and contemporaneous electromechanical machines. After all, three of the seven components of the Type 701 system were simply modified electromechanical punched-card machines (the Type 711 punched-card reader, the Type 716 alphabetical printer, and the Type 721 punched-card recorder). The Type 701 was also produced on an assembly-line like previous IBM electromechanical machines, whereas all other first-generation computers, including UNIVAC 1, were either one-of-a-kind machines or built as if they were. The Type 701 also was designed to be easily packaged and shipped like the previous electromechanical calculators. Other first generation computer companies would first assemble their computers in their own plant, then disassemble, crate, uncrate, and reassemble them at customers' sites. Finally, 12 of the 19 customers of the Type 701 had been previous owners of IBM electromechanical devices.

Thus, the Type 701 electronic computer that Sidney Lida was assigned to document was technologically preceded, succeeded, and surrounded by electromechanical punched-card calculators.

SETTING THE STAGE: THE RHETORICAL CONTEXT OF THE TYPE 701

When Chapline was considering the genre of computer documentation manuals for EMCC and UNIVAC, he could draw on only a very short corporate memory—in fact, it was only two years earlier that Eckert and Mauchly had begun ECC and EMCC. Thus, there was little prior experience in writing documentation, and what little experience there was consisted of: articles that Eckert and Mauchly wrote for publication in engineering journals, engineering reports such as those for the EDVAC, and military specifications such as those seen by Eckert when he was working at MIT on radar prior to joining Mauchly. None of these documents were customer or sales support documentation. So although Joseph Chapline was pulled in many directions while making his decision as to the form and design of the computer manuals, there was very little specific corporate documentation precedent to constrain his decisions.

In contrast, Sidney L. Lida, the author of the 1953 *Principles of Operation: Type 701 and Associated Equipment*—the first external IBM computer instruction manual—had quite different and more visible forces contending in his invention process.[25] First, in writing his manual in 1952 and 1953, Lida was working within a relatively long corporate history stretching back to Hollerith's 1896 incorporation of his punched-card tabulator business as the Tabulating Machine Company, which merged in 1911 into a company called C-T-R, the Computing-Tabulating-Recording Company. In 1911, Tom Watson, Sr. was brought from John H. Patterson's National Cash Register Co. to direct CTR, which he renamed the International Business Machine Company in 1924. Thus, the IBM corporate style and documentation standards had had more than 50 years to sink deep roots by the time Lida began his work.

Second, rather than EMCC's documentation forces arising from its academic and military experience, the forces acting on Lida arose from Tom Watson, Sr.'s background as former chief sales manager at National Cash Register under the tutelage of that salesman par excellance, John H. Patterson—the man who, as Chapter 3 described, gave the world the *Silver*

Dollar Demonstration. In fact, Tom Watson, Sr. supposedly often said "Nearly everything I know about building a business comes from Mr. Patterson."[26] Thus, on Mr. Lida's bookshelf was probably a wealth of IBM customer and sales support documentation.

For example, in the late 1920s, IBM published such general presales support brochures as *The Electric Tabulating and Accounting Machine Method and Sales Accounting*. Both of these brochures were designed to inform readers about IBM methodology and convince them to use and purchase the equipment. The mid-1930s showed evidence that IBM continued publication of such material under titles like *IBM Machine Methods of Accounting—The IBM Card and IBM Machine Methods of Accounting—The Preparation and Use of Codes*.[27]

However, in the 1930s and 1940s, IBM also began to publish a type of manual that continued to appear until very late in the 1980s,[28] a post-sales document intended to instruct users about how to operate equipment using specific examples and illustrations just like those in the dramatic oral sales pitch of NCR's *Silver Dollar Demonstration*. These "user documents" were usually named "Principles of Operation." The author of the *IBM Electric Punched Card Accounting Machines, Principles of Operation, Automatic Carriage, Type 921*(1945), explained quite succinctly on the manual's first page the goals and methods of these "Principles of Operation" manuals:[29]

> This manual describes in detail the method of operating the ________ and the principles by which it is controlled to perform the various operations. The basic operations are explained, and a number of typical setups illustrate all phases of automatic ________ and the most commonly used operations. Each of these setups is accompanied by a diagram of control panel wiring and description of the principles by which the ________ and ________ machine are controlled to perform that operation.

Even though these "Principles of Operation" were post-sales documents, they still continued to sell the customer subliminally on the quality of the purchased product and the stature of IBM as a company. The typeset, two-column, fully justified, richly illustrated look of these IBM documents was quite a contrast to the typewritten and carbon-copied manuals for ENIAC and BINAC.[30] The layout and design of the IBM "Principles of Operation" manuals conveyed typographically the same effect that John H. Patterson had taught his Sales Manager, Tom Watson, Sr. to produce with the manners of NCR salesman in the 1890s: "Don't chew gum or tobacco," "Don't tell funny stories," and "Don't put feet on chairs."[31]

Since Mr. Lida's final manual in 1953 for the Type 701 was named the *Principles of Operation*, it is to this type of manual one should look for the 701's rhetorical context.[32] Thus, I examined four "Principles of Operation" manuals in order to understand Lida's rhetorical context:[33]

- *IBM Machine Method of Accounting, Electric Accounting Machines, Type 285 and Type 297* (1936),[34]
- *IBM Electric Punched Card Accounting Machines, Principles of Operation, Automatic Carriage, Type 921* (1945),[35]
- *IBM Electric Punched Card Accounting Machines, Principles of Operation, Alphabetic Collator, Type 89* (1949),[36] and
- *IBM Electric Punched Card Accounting Machines, Principles of Operation, Electronic Calculating Punch, Type 604* (1948).[37]

In deciphering the rhetorical context of the "Principles of Operation" manuals, many of the analytical techniques used in prior chapters are also used here. For example, similar to the analysis of the sewing machine and mower-reaper instruction manuals of the 19th century, I examined these IBM manuals to see:

- how profusely illustrated they were, and how well the text and graphics were coordinated. Knowing this I could judge how well "Principles of Operation" manuals continued the visual communication tradition of American technical communication;
- how easy it was to read the manuals as indicated by the Flesch Reading Ease scale and by the size of paragraphs.

Also, similar to the analysis of Ford and Chevrolet automobile instruction manuals, I examined these IBM manuals to see:

- how well the manuals signaled their hierarchy and structure in headings. Did the heading structure used in the "Principles of Operation" manuals reveal a shallow wide structure with only first- and second-level headings, or did they reveal a deeper structure with headings proceeding to third, fourth, and fifth levels?;
- what kind of reference aids were present in the manuals. Were these manuals intended for random access usage by readers or for sequential cumulative usage?

From this analysis, I derived five signature style elements suggesting that the IBM "Principles of Operation" manuals were pitched to an educated audience as indicated by their sentence-level characteristics and enacted rhetorically in their descriptive—rather than imperative—mood style. These manuals also used a rhetoric of equality between the audience and the author. However, the high reading level of their sentences, and the high education level assumed by the style were counterbalanced by:

- a very profuse style of illustration that often carried the primary message or, at least, was in equal partnership with the text in carrying the message;
- frequent short paragraphs;
- a variety of efforts made to allow readers to access information randomly in the manuals rather than having to read them cover-to-cover.

Each of these signature elements is discussed in turn.

Signature Style Element 1. IBM "Principles of Operation" Manuals Were Well Illustrated and Tightly Linked to the Text Often Through the Use of Call-outs

IBM's "Principles of Operation" manuals during 1930s–1940s were well illustrated as can observed in Figure 8.2. Every manual contained at least one illustration per page, with some having many more. Such frequency of illustration was not nearly as high as in the mower-reaper manuals in the 1910s–1920s that ranged from 120% to 200%. However, it was higher than any of the imperative-mood manuals: the automobile manuals of the 1930s–1940s that ranged only from 50% to 75%, the sewing machine manuals in the 1910s–1920s ranged from 50% to 95%, Chapline's 1949 *Operating and Instruction Manual for the BINAC* had illustrations on fewer than 19% of its pages, and Chapline's 1954 *Manual for the Operation of the UNIVAC System* had illustrations on only 45% of its pages.

Nearly all the illustrations used in the "Principles of Operation" manuals were tightly tied to and explained in the text; they were nearly all correlative just as in the automobile instruction manuals of the 1930s–1940s (Figure 8.3). None of the illustrations seemed to be used solely for the sake of embellishment. In fact, so interwoven were the text and graphics in these "Principles of Operation" manuals that the illustrations and their call-outs often provided the presentation structure for the text. Notice, for example, the priority given to graphics in the 1945 manual: "Each of these setups is accompanied by a diagram of control panel wiring and description of the principles . . . "[38]

Nearly all these illustrations were views of actual physical objects. For example, of the 50 illustrations in the Type 604 "Principles of Operation" manual there were:

- 41 illustrations of the plugboards and their wiring,
- five illustrations of how to fillout the plugboard planning chart,
- two pictures of the physical calculator or its switches,
- one illustration showing the movement of the cards through the calculator, and
- one schematic of the memory assignments.

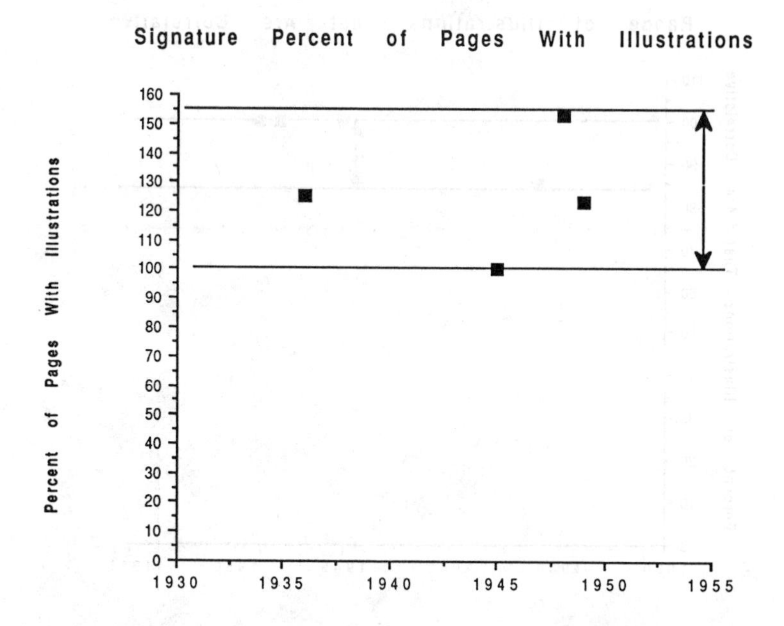

Figure 8.2. The IBM signature style of the 1930s and 1940s ranged between 100% and 153% of pages with illustrations

Thus, the illustrations in the "Principles of Operation" manuals were quantitatively more profuse than those in other industries or other companies, but they were qualitatively like these other companies' illustrations in that they were correlative and illustrations of actual physical objects. For example, only one of the Type 604 manual's 50 illustrations was a conceptual schematic of an abstract idea or concept such as those produced by Griffiths in his 1851 shipwright treatise or Chapline and Leary in the 1954 UNIVAC I manual.

The unusually high *frequency* of illustration in these post-sales manuals may be reflective of the traditional emphasis on sales and marketing in IBM because these "Principles of Operation" manuals were very much like the IBM presales manuals of the 1920s.[39] For example, the *Electric Tabulating and Accounting Machine Method* brochure had 17 separate illustrations in 16 pages, and the *Sales Accounting* brochure had 15 illustrations in 19 pages. Moreover, such an unusually high frequency of illustration may have been part of the "teaching through the eye" tradition bequeathed by NCR's Patterson through Watson to all of IBM:[40]

"Teaching through the eye" is the term most frequently applied to the process [of business]. Mr. Patterson himself often used that phrase in

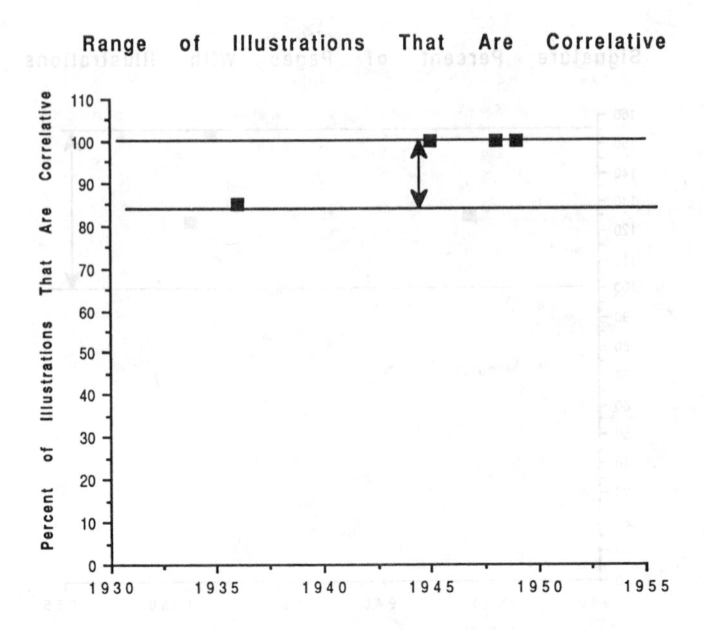

Figure 8.3. The percentage of correlative illustrations of the IBM signature style of the 1930s–1940s ranged between 85% and 100%

> describing it. The term in its literal meaning, however, is hardly broad enough. "Teaching through the eye" in the sense of utilizing pictures and diagrams and graphic symbolism generally to focus the attention and emphasize the point of an argument is only one phase of the system. . . . But in his use of the phrase, Mr. Patterson was not referring exclusively to that. "I have been trying all my life," he said, "first to see for myself, and then to get other people to see with me. To succeed in business it is necessary to make the other man see things as you see them." *Seeing* in this broader sense was the objective. Mr. Patterson "taught through the eye" not merely through charts and diagrams, but by dramatic analogies and objective demonstrations. In the broadest possible sense he was a visualizer.

With their high percentage of illustrations per page, the audiences for the "Principles of Operation" manuals were also given the opportunity to be visualizers.

However, many of the illustrations in the presales brochures were ancillary; often they were simply decorative, whereas the illustrations in the "Principles of Operation" manuals were more directly tied into the mechanical instruction of the text. Perhaps, then, the *content* of illustrations reflected the mechanical world of gears, levers, plugboards, and wires of the dominant engineers in IBM in the 1930s and 1940s, who could reach back to their mechanic predecessors like Oliver Evans whose illustrations were also primarily representations of physical objects.

Signature Style Element 2. IBM "Principles of Operation" Manuals Were Pitched at a "Difficult" Reading Ease Level Appropriate for a Technologically Sophisticated Audience. However the Sentence-level Style Difficulties Were Ameliorated by Frequent Short Paragraphs

When measured by the Flesch Reading Ease scale, IBM's "Principles of Operation" manuals of the 1930s–1940s were "pitched" at a fairly sophisticated audience who could cope with a "Difficult" below "Standard" writing style.

As Figure 8.4 shows, these four manuals had an average score of 41, a score termed by Flesch to be "Difficult." In contrast, only four of the 29 19th century sewing machine and mower-reaper manuals had similar scores; most of those manuals scored above 75, the "Standard" point on the scale.

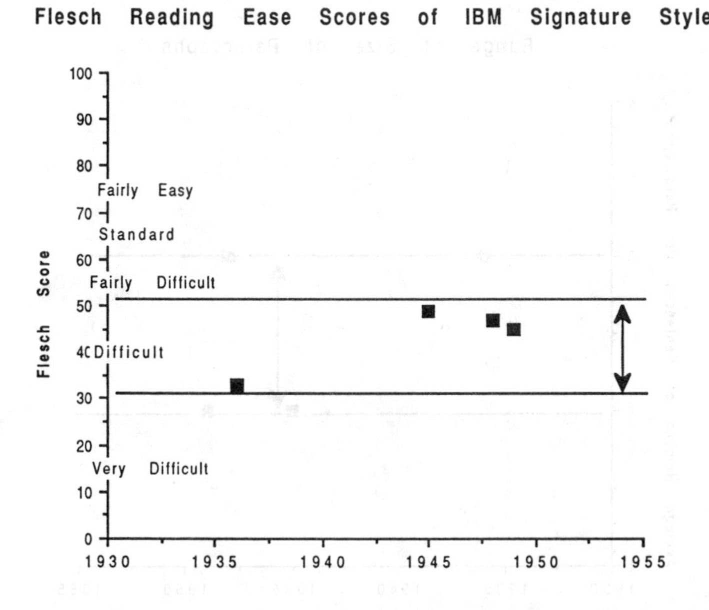

Figure 8.4. The Flesch Reading Ease scores of the IBM signature style of the 1930s –1940s ranged between 33 and 49, Difficult to Fairly Difficult[41]

Nevertheless, as Figure 8.5 illustrates, the size of the paragraphs in this "Difficult" writing style were frequently short with only two or three sentences. This was exactly the size of the paragraphs of the sewing machine and mower-reaper manuals in the 1910s–1920s and suggests that although the individual sentences were "difficult" to read, the short paragraphs attempted to ease that burden.

Signature Style Element 3. IBM "Principles of Operation" Manuals Were Designed For Random Access Reading as Evident by the High Frequency of Headings and Reference Aids. The Manuals Also Displayed a Deep Hierarchical Structure Down to Fourth level Headings

IBM's "Principles of Operation" manuals in the 1930s–1940s made frequent use of headings with sometimes as many as four per page and never fewer than two, Figure 8.6. To put this use of headings in context, the Ford and Chevrolet manuals of the 1930s–1940s had as few as a one heading per page and as many as 2.5.

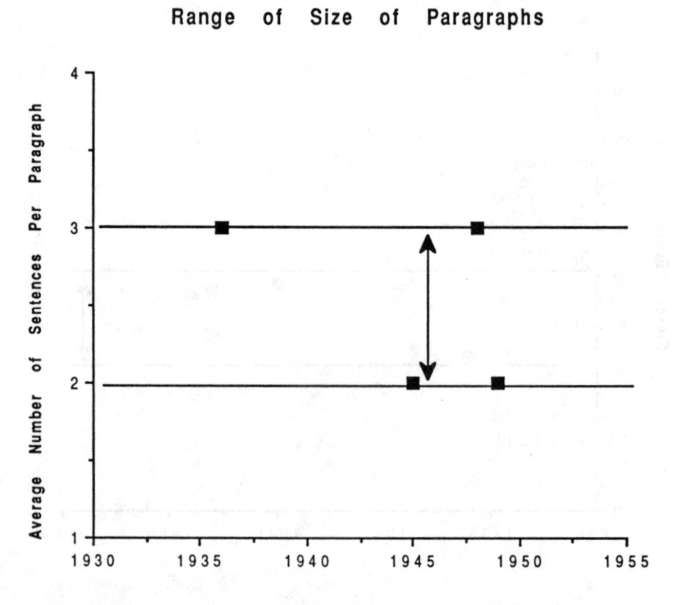

Figure 8.5. The number of sentences in paragraphs of the IBM signature style of the 1930s–1940s ranged between 2 and 3

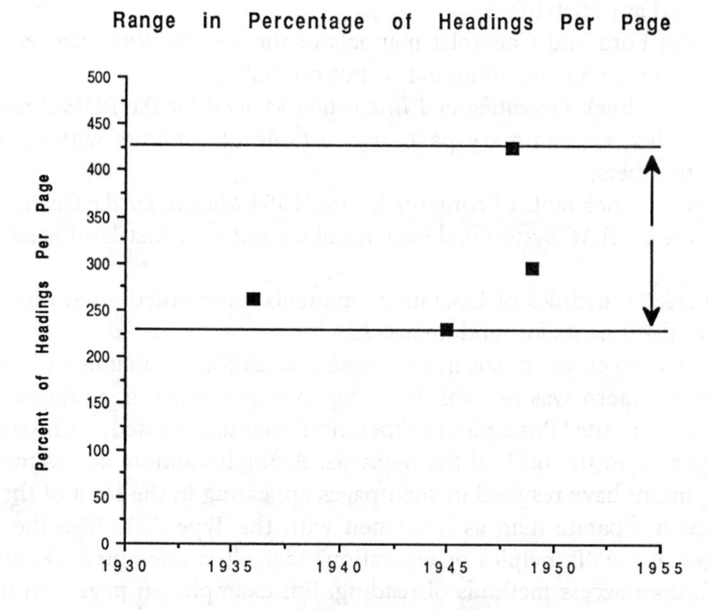

Figure 8.6. The range in headings per page of the IBM signature style of the 1930s and 1940s ranged between 262% and 425%

Table 8.1 Measures of the IBM Signature Style of the 1930s–1940s.

	Number of Pages	Table of Contents?	Index? Page?	Final Summary
1949, Type 89	39	yes	yes	yes
1948, Type 604	71	yes but inconsistent	yes	yes
1945, Type 921	36	yes but inconsistent	no	no
1936, Type 285/297	16	no	no	no

The presence of so many headings, accompanied by tables of contents, indexes, and summary pages of controls or instructions, also indicated that the IBM manuals were being increasingly designed for random access reading (see Table 8.1).[42]

The presence of all these reference aids in the IBM manuals can be compared with the fact that

- only four of the 29 sewing machine and mower-reaper manuals had any such items;
- the Ford and Chevrolet manuals of the 1930s–1940s had either a table of contents or an index, but not both;
- Chapline's *Operating and Instruction Manual for the BINAC* had no index, no summary page, and a table of contents without page numbers;
- Chapline's table of contents for the 1954 *Manual for the Operation of the UNIVAC System* had page numbers but only first-level headings.

Thus, these "Principles of Operation" manuals represented an advance in the design of documents for random access.

However, the break from a model of reading as a cumulative sequential access pattern was not absolute. For example, when a summary page of the contents in the "Principles of Operation" manuals existed, such pages usually appeared in the back of the manuals. A purely random access method of reading might have resulted in such pages appearing in the front of the manuals, or as a separate item as happened with the Type 701. Also the explicit messages of the "Principles of Operation" text often attempted to counteract such random access methods of reading. For example, on page 9 in the Type 604 manual the following appeared:

> The operation of the Type 604 is fully explained in the examples that follow with a detail description of the control panel hubs preceding their use in each example. For a clear understanding of the capabilities of the machine as well as a thorough knowledge of the wiring principles, each example should be *studied in the sequence presented.*

Figure 8.7 illustrates the type of hierarchy displayed in the "Principles of Operation" manuals. All had a fairly deep structure, with three of the four manuals having headings to a fourth level. In contrast,

- of the 15 mower-reaper manuals only two had headings down to three levels, and none down to four levels;
- of the 14 sewing machine manuals, only three had headings down to three levels, and none down to four levels;
- of the six Ford and Chevrolet manuals of the 1930s–1940s, only three had headings down to three levels, and two had headings down to four levels; and
- Chapline's manual for BINAC had many headings down to three levels, but no fourth-level headings, whereas the UNIVAC I manual did have a few headings down to the fourth level.

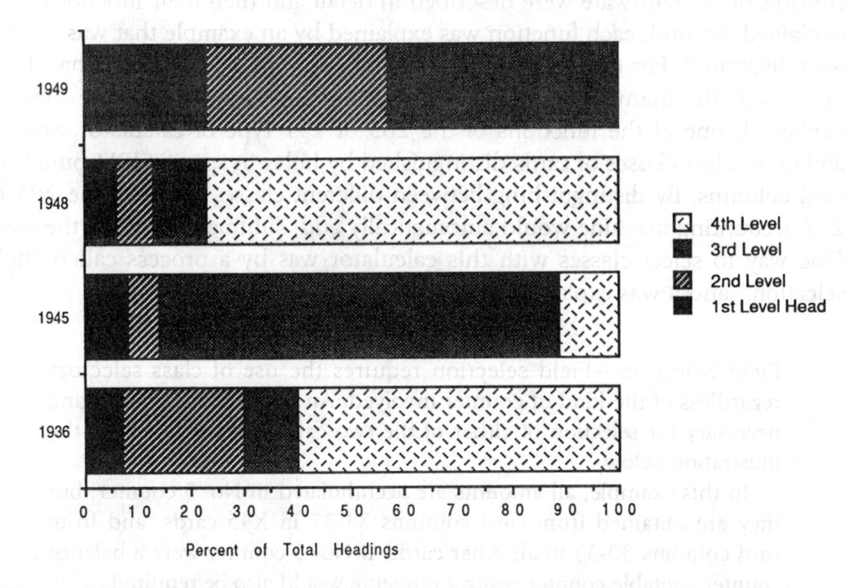

Figure 8.7. Composition of 1-, 2- 3-, and 4-level headings in the IBM signature style of the 1930s–1940s reveals no signature style

However, none of the "Principles of Operation" manuals displayed a similarity in the composition of the headings, and frequently the headings, even second-level headings, were inconsistently displayed in the manual's tables of contents and in its text. For example, sometimes the text headings appeared in the table of contents and sometimes they did not, and the variations seemed to occur from chapter to chapter; that is, some of the headings in a chapter were completely represented and some incompletely represented. This same weakness in execution resulted in the headings on the third- and fourth-levels being typographically displayed very differently from manual to manual and sometimes even within the same manual itself.

Thus, the "Principles of Operation" manuals represented a rather novel effort in creating a randomly accessible deep heading structure as well as an inability to construct such a structure at times.

Signature Style Element 4. IBM "Principles of Operation" Manuals Usually Had Two Parts, a Description of the Hardware and an Explanation of Their Functions. The Explanation of the Hardware Functions was Usually Accompanied by the Use of Illustrated Examples

In addition to the three quantitative elements found in these manuals, two qualitative elements unite the "Principles of Operation" manuals. First, the parts and controls of the hardware were described in detail and then their functions were explained. Second, each function was explained by an example that was usually well illustrated. For example, in the 1936 manual, after the controls had been described, the manual began explaining their functions. Thus, the manual explained, one of the functions of the 285 or 297 type of calculator was the ability to select classes of cards distinguished by holes in specific IBM punched-card columns. By distinguishing between different classes of cards, the 285 or 297 accounting machine would automatically add, subtract, or ignore the card. One way to select classes with this calculator was by a process called "field selection," and it was explained in the manual in the following way:[43]

> Field Selection—Field selection requires the use of class selectors regardless of the kind of counter involved. An example of the plugging necessary for selection of either of the two card fields is shown in the illustration below.
>
> In this example, all amounts are accumulated in No. 5 counter, but they are obtained from card columns 34-37 in X45 cards, and from card columns 30-33 in all other cards. If No. 5 counter were a balance counter, suitable counter control plugging would also be required.

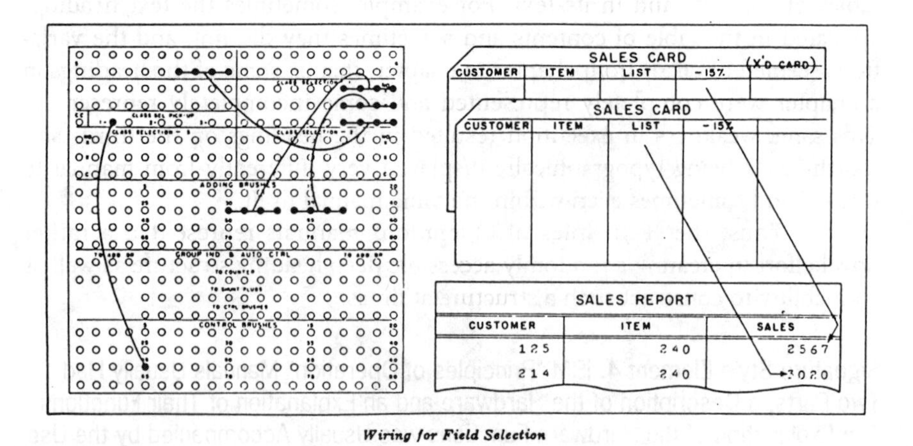

Wiring for Field Selection

Figure 8.8. The illustration accompanying the text example of "Field Selection" from the *IBM Machine Method of Accounting, Electric Accounting Machines, Type 285 and Type 297*(1936)

Notice how even in this very short example, the text was tightly linked to the graphic, and, in fact, the graphic provided the structure for the text.

Signature Style Element 5. An IBM "Principles of Operation" Manual, Unlike Other Instruction Manuals, Did Not Use an Imperative Mood Presentation Style.

One of the characteristics of the instruction manual that seemed so irksome to the 19th century farmers was the imperative mood presentation of information which rhetorically embodied an all-knowing author and an ignorant reader. Such a presentation style was, however, used in the sewing machine manuals and later picked up by the automobile instruction manuals. Chapline was equally worried about such a presentation method and, at first, tempered the suggestion made by Northrop to write his manuals like the automobile ones with the imperative mood style. In his BINAC manuals, Chapline tempered that approach with descriptive elements of presentation like those in engineering reports which, by assuming information on the part of the reader and leaving operational implications ambiguous, represented the author and reader as rhetorical equals. Yet, by the time of the UNIVAC I manuals, Chapline was using an imperative-mood rhetoric more frequently.

The "Principles of Operation" manuals, by and large, tried to enact a rhetorical equality similar to Chapline's BINAC manuals, and, thus, the IBM manuals usually avoided using the imperative mood presentation style. As shown in the earlier "field selection" excerpt, the actions to be taken were described and illustrated, not commanded. The excerpt offered the idea that "Field selection requires the use of class selectors regardless of the kind of counter involved"; it did not say, "Use class selectors to make field selections regardless of the kind of counter involved."

An IBM 1946 form (FORM DE-1-0270-0 (1)) found in the Charles Babbage Institute, Center for the History of Information Processing at the University of Minnesota, outlined how to write instructions for a "manual of procedures": "The purpose of the manual of procedure is to make instructions definite to provide an authoritative reference in answer to questions pertaining to procedures, and to improve administrative control."

Definiteness, authority, and control were all elements of an IBM "manual of procedures" that sound remarkably like the typical sewing machine or automobile instruction manual. However, an IBM "manual of procedures" was not the same as an IBM "Principles of Operation" manual. Remember how in 1945 the author of the *IBM Electric Punched Card Accounting Machines, Principles of Operation, Automatic Carriage, Type 921*, introduced the manual.[44] Description, principles, illustrations, and examples all called for an equality of reader and author because the specifics of actions to be taken were left implied or illustrated but not commanded. Such an approach expected the reader to think and reason out actions and consequences.

Perhaps the high frequency of illustrations also suggested how the "Principles of Operation" manuals were like the type of manual produced for the mower-reapers. The mower-reaper manuals had more illustrations per page than any other group of manuals, and the "Principles of Operation" manuals came in a close second. Both groups of manuals also avoided the imperative-mood rhetoric. How did a visual approach support a descriptive rhetoric of equality?

A graphic is read differently than text; a graphic is read as a super-dense continuous set of symbols, whereas text is read as a discontinuous and discrete symbol system:[45]

> Every mark, every modification, every curve or swelling of a line, every modification of texture or color is loaded with semantic potential. Indeed, a picture . . . might be called a "super-dense" . . . symbol, in that relatively more properties of the symbol are taken into account. . . . A differentiated symbolic system, by contrast, is not dense and continuous, but works by gaps and discontinuities. The most familiar example of such a system is the alphabet, which works (somewhat imperfectly) on the assumption that every character is distinguishable from every other (syntactic differentiation), and each has a compliant that is unique and proper to that character.

If illustrations are "loaded" with semantic potential, it is possible that different individuals can take different ideas and concepts away from an illustration much more freely than they can from text. More alternative interpretations are available from illustrations—even correlated ones such as in the "Principles of Operation"—than from plain text.[46] Moreover, if different interpretations are available, the profuseness of the illustrations support the rest of the efforts in the "Principles of Operation" manuals to leave ideas implied or illustrated but not commanded, again requiring the reader to think and reason out actions and consequences.

Thus, in this last signature element, the IBM "Principles of Operation" manuals reveal themselves as standing in an alternative tradition to the sewing machine, automobile, or UNIVAC instruction manuals. In their rhetorical equality and descriptiveness, the IBM "Principles of Operation" manuals, in fact, appear to display far more kinship to the rhetoric of the mower-reaper manuals than any other type of manual. How then did the *Principles of Operation: Type 701 and Associated Equipment* fit within the traditions of IBM's "Principles of Operation" manuals?

THE WORK OF SIDNEY LIDA IN CREATING THE *PRINCIPLES OF OPERATION: TYPE 701 AND ASSOCIATED EQUIPMENT* (1953)

Chapline's creative process in developing his manuals could be examined rather directly through his paper trail in EMCC–UNIVAC memos, drafts, and notes open to posterity at the Hagley Archives in Wilmington, DE. However, such direct analysis was not possible with Lida's work because the IBM archives in Armonk NY, are closed to the public as well as Lida, unlike Chapline, could not be interviewed. Thus, an indirect method of analysis had to be used.

Chapline had to rely for the source material solely on the engineers' notes and interviews which were notoriously incomplete and unreliable.[47] Lundstrom, a UNIVAC technician, reminisced in 1956 about how information was transferred by the original UNIVAC design engineers:[48]

> This problem in transferring technology, which I have seen demonstrated time and time again, arises from the nature of engineers and engineering. Most engineers hate to write. When absolutely forced to, they will keep their descriptions as cryptic as possible and will plagiarize

Figure 8.9. Sidney Lida, 2nd Row, Center—Enhanced photo from one taken at IBM Customer Education Class, Poughkeepsie, August 1952

> existing documents wherever they can. A complete engineering design consists of a great stack of engineering drawings, supplemented by a stack of specifications and test procedures. If the developing engineering group has strong discipline and if the engineers are conscientious, this great pile of paper will represent perhaps 90 to 95 percent of what the receiving group needs to know about the design. Missing will be the subtle tricks of timing, methods of compensating for known weaknesses in components, etc. These are not in writing because no one ever thought of a need to write them down. "Everybody knows about that." And, indeed, in the developing organization this may be true: Everybody does know about the tricky timing in the read circuit; it was the topic of hallway conversation for weeks.

Thus, Chapline had to rely on these fragmentary sources both because no one wrote much of anything formal in the design stage as well as the BINAC itself was not available for his own personal experimentation and use.

Lida, too, was probably unable to personally use the Type 701 or see it in operation, at best, until he was nearly done with his manual.[49] Furthermore, in 1951, there were also no established procedures at IBM describing how to specify a new computer. In all probability, prior electromechanical punched-card calculators had evolved physically by tinkering from one model to the next so that an established written specification method had previously been unnecessary. Yet, Lida worked with a disciplined design team of electrical engineers who had had management experience, and who could write.[50] Thus, the three principal engineering architects of the Type 701—Nathaniel Rochester, Werner Buchholz, and Morton M. Astrahan—invented a computer specification process themselves at IBM by writing the "Preliminary Operator's Reference Manual."[51] Lida, therefore, was given very solid source material for the creation of his manual in the form of a 135-page preliminary manual.[52] With both this source manual and copies of the Type 701 final manual available to the public at the National Museum of American History, Smithsonian Institution, Division of Computers, Information and Society, indirect conclusions concerning Lida's work can be drawn by comparing and contrasting his source manual and his final manual.[53]

The "Preliminary Operator's Reference Manual"

The "Preliminary Operator's Reference Manual" was written by Rochester et al., who stated their goal for writing both in the manual itself: "Part 2—Organization of the Computer. This part of the manual describes the general organization of the computer from the point of view of the programmer."[54] This goal was restated in a 1983 retrospective on the Type 701 project:[55]

> We wrote a manual that was used by both the engineers who were designing the hardware and the mathematicians who were writing software and application programs. This manual [June 1951] later evolved into the first internal user manual [August 1952]. It described how data were organized, the machine registers, the address system, and all of the instructions. The instructions were described by telling how the contents of the machine registers would change and the other things that would happen when the instruction was executed.

Because Rochester was a co-author of this "Preliminary Operator's Reference Manual," a link between the punched-card technology and documentation, this source material, and Lida's final manual, was quite clear. It became clear when I discovered that at the time of Rochester's hiring by IBM in November 1948, he spent several months becoming acquainted with an electromechanical punched-card machine, the Type 604.[56] The following year he wrote several papers describing an enhanced Type 604 with stored-memory programs—the most important innovation later included in the design of the Type 701.[57,58] Thus, it should come as no surprise that, when Rochester and his two co-authors developed the "Preliminary Operator's Reference Manual," they used the "Principles of Operation" manuals' signature style derived from the Type 604 manual in a number of important ways.

First, the "Preliminary Operator's Reference Manual" followed the usual two-part description–explanation structure of the "Principles of Operation" manuals. For example, pages 11–53 in the "Preliminary Operator's Reference Manual" described the parts of the computer just like the Type 604 "Principles of Operation" manual's pages 5–9, while pages 53–91 of the "Preliminary Operator's Reference Manual" described the functions and use of the computer just as the Type 604 "Principles of Operation" manual's pages 11–66 had done.

Second, just as the Type 604 "Principles of Operation" manual had a summary (pages 67–70), Rochester's "Preliminary Operator's Reference Manual" also had one. However, for the first time, the "Summary of Characteristics" was in the beginning rather than at the end.

Third, this limited distribution, preliminary report had a table of contents with page numbers just like most of the "Principles of Operation" manuals. Thus, the "Preliminary Operator's Reference Manual" was optimized for random access methods of reading just as the earlier "Principles of Operation" manuals had been.

Fourth, just as the "Principles of Operation" manuals had illustrations (e.g., the "Field Selection" example earlier), Rochester illustrated each function of the Type 701 with an example. This can be observed in the "Store Address" instruction section from the "Preliminary Operator's Reference Manual":[59]

STORE ADDRESS STORE A
Code 13
01 101

The rightmost 123 bits of the half-word at the specified memory location are replaced by the 12 bits in positions 6 through 17 in the accumulator. These bit positions are the ones having positional values of 2-6 through 2-17. The remaining 5 bits on the left of the half-word in memory and its sign are not changed.

The contents of the accumulator are not changed by this operation.

STORE A is used only with half-word addresses.

Example.
Accum. +00, 00, 111, 000, 111, 000, 111, 000, 111, 000, 111, 000, 111
Memory - .01, 010, 101, 010, 101, 010
Before
Memory - .01, 010, 000, 111, 000, 111
After

Fifth, and perhaps most important, the rhetoric of the "Preliminary Operator's Reference Manual" created an equality of reader and author because the specifics of the actions to be taken were left implied or illustrated but not commanded in an imperative mood. Thus, like most "Principles of Operation" manuals, the "Preliminary Operator's Reference Manual" appeared—as regards the rhetoric of equality—to be most like the mower-reaper manuals.

Despite these five similar characteristics, the "Preliminary Operator's Reference Manual" did not have all the signature style characteristics of the "Principles of Operation" manuals. Perhaps it did not because this manual was a *confidential* document with only limited distribution to "IBM personnel . . . associated with Laboratory project,"[60] and because the whole project was still only preliminary.[61] Thus, the "Preliminary Operator's Reference Manual" did not have the typographic spit-and-polish of the "Principles of Operation" manuals. For example, it was only typewritten, not fully justified, and had a confusion of layouts with both single and double spacing. Still, unlike the ENIAC, EDVAC, and BINAC *final* manuals, there was more typographic finesse here in the preliminary manual. For example, the ENIAC and EDVAC manuals relied on paragraph numbering alone to signify hierarchy and subordination; Rochester and his co-authors used indentation and paragraph numbering.

Also unlike the earlier "Principles of Operation" manuals, this "Preliminary Operator's Reference Manual" had far fewer illustrations and far fewer headings—less than 10% of the pages had illustrations, and there was only one heading per page rather than the customary two to four in the "Principles of Operation" manuals. Furthermore, in Rochester's manual describing an electronic computer, the content of the illustrations was different. The 21 illustrations were still all correlative, but they were widely separated from their textual discussions, and the majority of the illustrations tried to convey abstract rather than physical information as shown in Table 8.2. Such abstract content was quite different than the standard "Principles of Operation" illustration of actual physical objects.

In general, Lida had an excellent foundation of source material in Rochester's "Preliminary Operator's Reference Manual." No doubt influenced by his close reading of the Type 604 "Principles of Operation" manual for his earlier work, Rochester's manual had many of the characteristics of the "Principles of Operation" manuals: a two-part arrangement, methods of random access reading via a table of contents with page numbers, use of examples, and the rhetoric of equality between reader and author that avoided the imperative mood style.

Nevertheless, because of its limited distribution, the manual lacked the typographic polish typical of the "Principles of Operation" manuals, their wealth of illustrations, and their abundance of headings. Rochester et al. also changed the content of the illustrations to represent the change in technology; the many physical illustrations were replaced by many illustrations of abstract electronic concepts. This change in illustration focus was something quite new in the "Principles of Operation" tradition.

Thus, Lida's work was deeply bound by a long tradition of technology, a long tradition of rhetoric, and extensive source material. Working within these traditions and with this material, Lida did four things:

Table 8.2. Comparing the Content of the Type 604 "Principles of Operation" and the Defense Calculator Preliminary Operator's Manual.

	Pages in Type 604, Principles of Operation	Pages in Defense Calculator Preliminary Operator's Manual
Plugboards and their wiring	41	4
Planning chart	5	4
Physical calculator and switches	2	0
Movement of cards through machine	1	1
Punch card sample	0	1
Microsecond timelines of activity	0	5
Diagrams of purely abstract information, e.g., information paths, memory registers	1	6

- First, he sought to both improve the readability of the text and to make changes to accommodate a shift in audience.
- Second, he maintained a number of "Principles of Operation" elements already present in Rochester et al.'s source manual.
- Third, he introduced a number of elements from the "Principles of Operation" manuals absent from the source manual.
- Fourth, he changed a number of signature style elements.

Each of these four aspects of Lida's work is discussed in turn .

Improving the Readability of the Text and Making Changes to Accommodate a Shift in the Audience

In 1951 and 1952, Rochester and his co-authors thought they were writing their "Preliminary Operator's Reference Manual" for an audience of "engineers who were designing the hardware," "mathematicians who were writing software and application programs," and "programmer[s]." However, in 1953, the only audience Lida explicitly mentioned in his manual was the following: "This section is intended primarily to give the relative beginner an understanding of some of the basic techniques of programming."[62]

Consequently, Lida must have felt he needed to reorient Rochester's material deleting irrelevant material, shortening the source manual by nearly 20% or 29 pages, and adding new material. The deleted pages primarily consisted of technical material in Rochester et al.'s appendices:

- "Appendix C. Basic Input-Output Subprogram,"
- "Appendix E. Data for Time Estimates,"
- "Appendix F. Programming during Input-Output Time," and
- "Appendix G. Rearranging Words Stored by READ BACKWARD."

In place of this material more appropriate for the Type 701 design engineers, Lida added appendices that a relative beginner programmer would need in order to handle the novel[63] processing methods of the Type 701:

- "Appendix A. Binary and Octal Number System,"
- "Appendix B. Table of Powers of 2," and
- "Appendix C. Octal-Decimal Integer Conversion Table."

In addition to adding and deleting material from Rochester's source manual, Lida reorganized the material for his new audience. For example, Rochester's engineers and mathematicians may have needed to know that a particular type of Type 701 magnetic storage was either "main" or "auxiliary," but the relative

Table 8.3. Lida's Manual Compared to his Source Material.

	Audience Addressed	Number of Pages	Table of Contents?	Index?	Summary Page? Where?
Rochester's Preliminary Operator's Reference Manual 1951–52	engineers , mathematicians, programmers	135	yes	no	yes/in front
Lida's Principles of Operation: Type 701 and Associated Equipment 1953	relative beginners	109	yes	no	yes/in back also included Programmer's Quick Reference Card

beginners addressed by Lida either did not need to know this or would not be able to distinguish such esoteric characteristics of storage. After all, the electromechanical technology within which all IBM customers had operated had never previously had such electronic distinctions. Hence, Lida evidently took topics that were widely separated both by page and levels in the heading hierarchy such as:

- "2. Electrostatic Storage,"
- "4. Address System for Electrostatic Memory,"
- "8.2 Magnetic Tapes," and
- "8.3 Magnetic Drums"

and regrouped them in parallel ways under new parallel-level headings, "Storage," "Address System," and "Input-Output Components" (see Figure 8.10). Such a reorganization not only increased the consistency of the manual's organization, but it also overcame the former esoteric basis of the organization: the fact that some storage was "main" rather than "auxiliary."

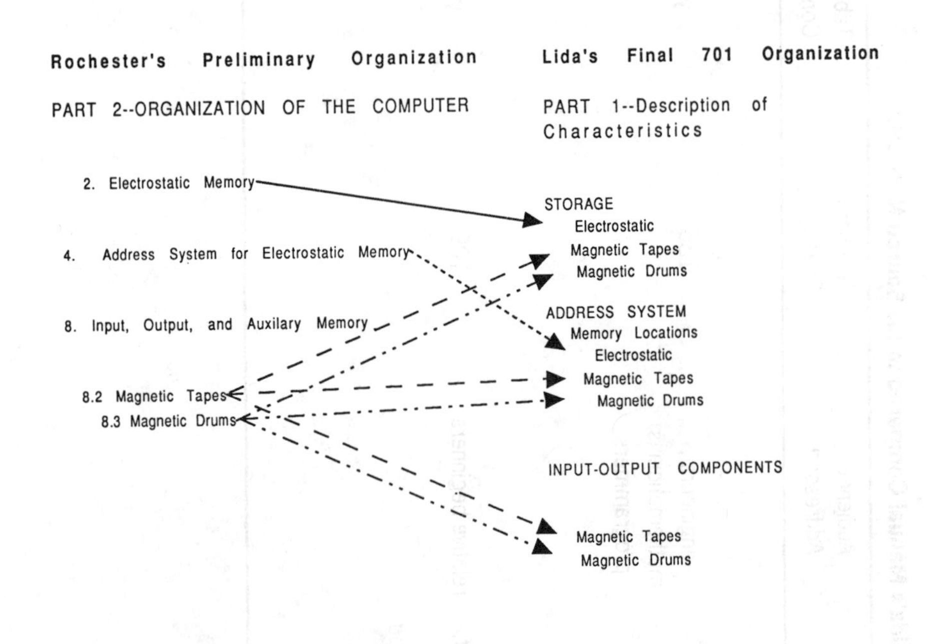

Figure 8.10. Lida's reorganization of the Type 701 materials

Lida not only deleted, added, and reorganized the material, but also evidently rewrote the material to simplify it. This can be observed by comparing the first description of the "electrostatic memory" (what might today be called "RAM") from Rochester's "Preliminary Report" with Lida's final published description.

In Rochester's "Preliminary Report," information on the electrostatic memory was in two locations: "2. Electrostatic Storage" and "4. Address System for Electrostatic Memory":[64]

> 2. Electrostatic Memory
>
> The high-speed, electrostatic memory has a capacity of 2048 full-words (36 bits per word). These are stored on 36 sets of cathode-ray tubes. Since any full-word can be split into two half-words, this allows the total number of words which can be stored to be increased. If the entire memory is assigned to half-words, the capacity becomes 4096 words.
>
> 4. Address System for Electrostatic Memory
>
> Different addresses are assigned to distinguish memory locations used for half-words from those used for full-words:
>
> (a) The 4096 possible memory locations for half-words are addressed by the positive numbers 0000 to 4095 (in decimal notation).
>
> (b) The 2048 possible memory locations for full-words are given by the negative even numbers -0000 to -4094. If an instruction refers to a full-word memory address, the entire instruction behaves as a negative number.
>
> The relation between these addresses is as follows:
>
> If -2n is the address of a full-word memory location, then the same location can also be split into the left half-word, address +2n, and the right half-word, address +(2n+1). n here is any positive integer less than 2048.
>
> Example:
>
> 110100010011001000 010011000111101010
>
> Full-word, Address: -2962
>
> or
>
> 1.10100010011001000 0.10011000111101010
>
> Left Half-Word, Address: +2962 Right Half-Word, Address: +2963
>
> The manner in which the addresses are used to read information from the electrostatic memory and to store information in memory is illustrated in Fig. 3 and Fig. 4 (a), (b).

Lida took these two sections and produced the following in 1953:[65]

ELECTROSTATIC

The heart of the machine is the electrostatic storage unit, through which all information to and from all other components of the machine must pass. Electrostatic storage consists of a bank of cathode-ray tubes. Information is stored on the screen of each tube through the presence or absence of charged spots at certain locations on the screen. In this way, a certain number of binary digits (or "bits") may be stored on each tube. One electrostatic storage unit can accommodate 1024 full words or 2048 half words. However, two such units may be used to produce a maximum storage of 2048 full words or 4096 half words. It is assumed in what follows that maximum electrostatic storage has been provided for this installation.

Principal advantages of electrostatic storage over other types is [sic] the very small time necessary to extract information from any given location and send it to the computing unit and the fact that the programmer has random access to any electrostatic storage location. Information is lost when the power is turned off.

Electrostatic

The 2048 different locations for full words in electrostatic storage are identified by the negative even integers from -0000 to -4094. The 4096 possible locations for half-words in electrostatic storage are distinguished by the positive integers +0000 to +4095. The relation between full and half-word addresses is as follows: if -2n refers to a full-word location, then +2n identifies the left half-word, and +(2n+1) the right half-word, into which the full-word location may be split.

For example, if the full-word address is -1962, then the left half-word address is +1962 and refers to the sign position and positions 1 to 17 of the full word. The right half-word address is +1963 and refers to the sign position and positions 18 to 35 of the full word location, position 18 being the sign position of the right half word (Figure 1). If a full word is to be obtained from or supplied to electrostatic storage and, through error, a negative odd address is given (e.g. -1963), the result will be the same as if the next lower (in absolute value) negative even address (-1962) were given.

All the sections from either the "Preliminary Operator's Reference Manual" or the final manual have examples, a characteristic of the "Principles of Operation" manuals. Moreover, Rochester used binary numbers as examples (e.g., 110100010011001000) probably because such numbers accurately represented the Type 701's novel binary processing system. However, in using binary numbers, Rochester immediately limited his audience to those with knowledge of such novel number processing. Lida, on the other hand, did not trust that his audience would be so knowledgeable in the binary number system—remember the new appendix Lida included, "Appendix A. Binary and Octal Number System"—and so Lida used the more widely known decimal numbers in his examples (e.g., +1962 and -1962).

In addition to all these changes because Lida sensed a shift in his audience, he also had the opportunity as a professional technical writer—Rochester and his co-authors were not—to rewrite the information in a more effective way. For example, in the earlier excerpts, Rochester et al. began by first discussing "full words" and then moved to "half-words," but in the second section they reversed their order:

> 2. Electrostatic Memory
>
> The high-speed, electrostatic memory has a capacity of 2048 **full-words** (36 bits per word). These are stored on 36 sets of cathode-ray tubes. Since any **full-word** can be split into two **half-words**, this allows the total number of words which can be stored to be increased. If the entire memory is assigned to **half-words**, the capacity becomes 4096 words.
>
> ---
>
> 4. Address System for Electrostatic Memory
>
> Different addresses are assigned to distinguish memory locations used for **half-words** from those used for *full-words*:
>
> (a) The 4096 possible memory locations for **half-words** are addressed by the positive numbers 0000 to 4095 (in decimal notation).
>
> (b) The 2048 possible memory locations for **full-words** are given by the negative even numbers -0000 to -4094. If an instruction refers to a **full-word** memory address, the entire instruction behaves as a negative number.

Lida was more consistent. In the excerpt below, he began his description of the Electrostatic Memory by first discussing "full words" and then moved to "half-words," and he did so consistently:

> [O]ne electrostatic storage unit can accommodate 1024 full words or 2048 half words. However, two such units may be used to produce a maximum storage of 2048 full words or 4096 half words . . .
>
> ---
>
> *Electrostatic*
>
> The 2048 different locations for **full words** in electrostatic storage are identified by the negative even integers from -0000 to -4094. The 4096 possible locations for **half-words** in electrostatic storage are distinguished by the positive integers +0000 to +4095. The relation between **full** and **half-word** addresses is as follows: if -2n refers to a **full-word** location, then +2n identifies the left **half-word**, and +(2n+1) the right **half-word**, into which the **full-word** location may be split.

By remaining consistent in his presentation of material, Lida improved on the ability of readers to follow the train of thought in his material.

Lida also established a more even flow to his material by carefully planting references back to previous material in each new sentence. In the excerpt above, Rochester et al. failed to establish any pattern or focus in the subjects of their first sentences:

- The high-speed, electrostatic memory has a capacity of 2048 full-words (36 bits per word).
- These are stored on 36 sets of cathode-ray tubes.
- Since any full-word can be split into two half-words . . .

However, in his first sentence, Lida wrote: "The 2048 different locations for full words in electrostatic storage are identified by the negative even integers from -0000 to -4094." His second sentence referred back to the first and gave the reader some familiar ground in the word and sentence placement: "The 4096 possible locations for half-words in electrostatic storage are distinguished by the positive integers +0000 to +4095."

Lida's revision was a good example, on a small scale, of "contextual editing" or the creation of a predictable pattern to improve flow and readability in technical communication.[66] Lida's revision also exemplified what has been described as maintaining the "topic focus" in discourse.[67]

Lida also emphasized *what* the user should do and *why* the user of the electrostatic memory should do these activities, while at the same time deemphasizing Rochester's discussion of the *how*. This shift in emphasis can quite easily be seen by looking again at the two discussions of the electrostatic memory. Lida deleted explanations of how things were achieved in memory by omitting Rochester's words:

- (36 bits per word),
- These are stored on 36 sets of cathode-ray tubes,
- Different addresses are assigned to distinguish memory locations used for half-words from those used for full-words,
- The manner in which the addresses are used to read information from the electrostatic memory and to store information in memory is illustrated in . . .

Lida also added new material to explain why customers should care about these facts:

> Principal advantages of electrostatic storage over other types is [sic] the very small time necessary to extract information from any given location and send it to the computing unit and the fact that the programmer has random access to any electrostatic storage location.

This change in emphasis was perhaps warranted by his new target audience of "relative beginners," but perhaps it was also warranted by the traditional sales emphasis of IBM. In deleting these specific words, Lida was carrying out a policy taught long ago to his IBM boss, Tom Watson, Sr., by his former boss at NCR, John Patterson. Patterson's turn-of-the-century *Sales Manual* had an interesting characteristic similar to Lida's editing efforts:[68]

> It is worthy of special notice that this exposition of [cash] register functions in the [Sales] Manual was not a technical, mechanical exposition, and at no time did Mr. Patterson equip the salesmen with knowledge of the mechanical construction and operation of the machine. They were taught *what* the machine did, and *why* it did it; but *how* the machinery worked was outside their province [emphasis added].

Lida's shift in emphasis to the what and why is even more evident in a comparison of the figures referenced by the two electrostatic excerpts and shown in Figures 8.11 and 8.12.

Lida deleted everything visually designating "bits," "zeroes," and distinctions among memory "registers," "locations," and "accumulators." They were beyond the needs of his audience and, in deleting them, his figure became elegantly simple.

Maintaining "Principles of Operation" Elements Already Present

Lida explicitly acknowledged the link of his Type 701 "Principles of Operation" manual with others, and he assumed that readers were equally familiar with the content of such manuals. For example, in the introduction to his section on control panel wiring, Lida announced that he would describe only the modifications to the three electromechanical devices in the Type 701 system (the Type 711 punched-card reader, the Type 716 alphabetical printer, and the Type 721 punched-card recorder). Lida continued: "Questions on how standard features operate can be answered by referring to the appropriate principles of operation manual."[69] Later on the same page, Lida suggested: "It is important when reading this section that familiarity with wiring of the standard machines, of which the 701's are modifications, is assumed."[70]

Not only did he explicitly link his manual with previous "Principles of Operation" manuals, but Lida like Rochester et al. in the "Preliminary Operator's Reference Manual" offered the reader a table of contents with page numbers down to the fourth level. Also, following the example of the summary pages in the Type 604 (1948), Type 89 (1949), and the "Preliminary Operator's Reference Manual," Lida also included summary pages and added, in what may have been a first, a quick reference card called the "Programmer's

-49-

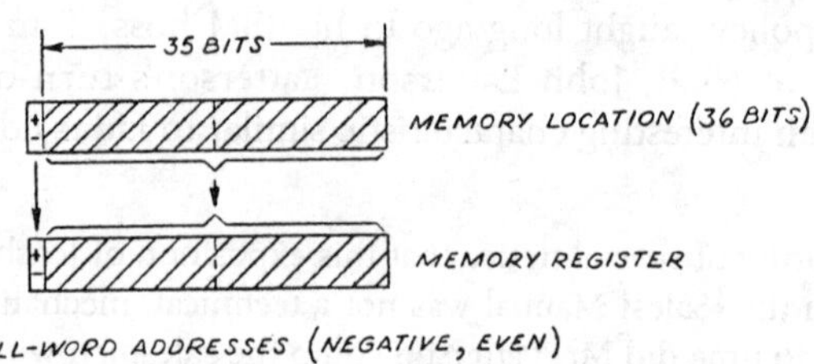

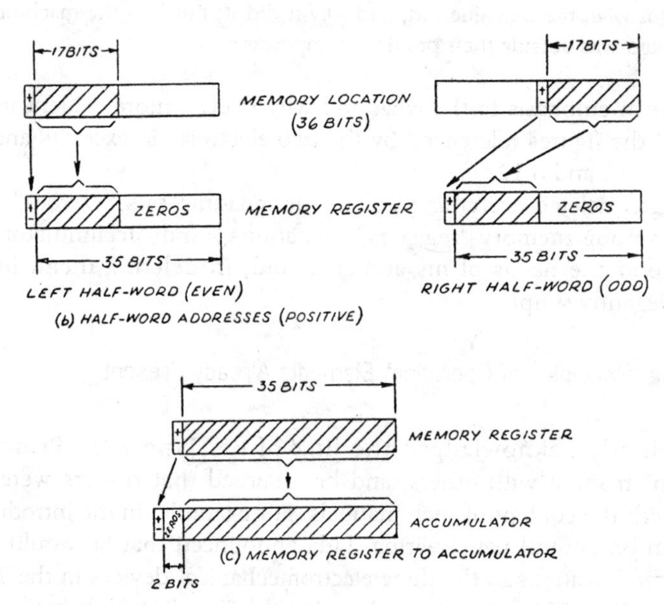

Figure 8.11. Rochester's illustration explaining electrostatic memory is much more complex than Lida's. (Notice the material on bits and zeroes)

Card" (Figure 8.13). In this way, Lida and Rochester et al. maintained the "Principles of Operation" tradition of including the means by which selective reading could occur.

However, in an almost identical fashion to the Type 604 "Principles of Operation" manual, Lida toned down the possibilities of a purely random way of reading. Remember how the Type 604 manual introduced examples:[71]

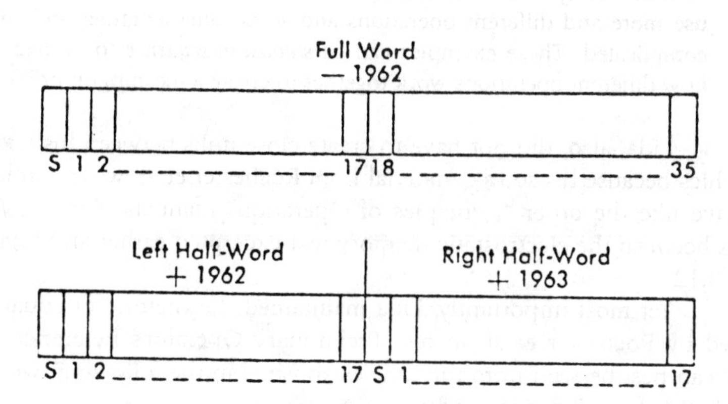

Figure 8.12. Lida's illustration explaining electrostatic memory is much simpler than Rochester's. Notice how Lida implies the presence of 35 Bits

OPERATION CODE OCT	(DEC)		OCT	(DEC)	
00	(00)	STOP	20	(16)	MPY
01	(01)	TR	21	(17)	MPY R
02	(02)	TR OV	22	(18)	DIV
03	(03)	TR +	23	(19)	ROUND
04	(04)	TR 0	24	(20)	L LEFT
05	(05)	SUB	25	(21)	L RIGHT
06	(06)	R SUB	26	(22)	ACC LT
07	(07)	SUB ABS	27	(23)	ACC RT
10	(08)	NO OP	30	(24)	READ
11	(09)	ADD	31	(25)	READ B
12	(10)	R ADD	32	(26)	WRITE
13	(11)	ADD ABS	33	(27)	WR EOF
14	(12)	STORE	34	(28)	REWIND
15	(13)	STORE A	35	(29)	SET DR
16	(14)	STORE MQ	36	(30)	SENSE
17	(15)	LOAD MQ	37	(31)	COPY

CONVERSION			
40	(32)	64	(52)
41	(33)	65	(53)
42	(34)	66	(54)
43	(35)	67	(55)
44	(36)	70	(56)
45	(37)	71	(57)
46	(38)	72	(58)
47	(39)	73	(59)
50	(40)	74	(60)
51	(41)	75	(61)
52	(42)	76	(62)
53	(43)	77	(63)
54	(44)	100	(64)
55	(45)	101	(65)
56	(46)	102	(66)
57	(47)	103	[illegible]
60	(48)	104	[illegible]
61	(49)	105	[illegible]
62	(50)	106	[illegible]
63	(51)	107	(71)
		110	(72)

	NO ZONE	12	11	0
ZONE ONLY		.	-	0
1	1	A	J	
2	2	B	K	S
3	3	C	L	T
4	4	D	M	U
5	5	E	N	V
6	6	F	O	W
7	7	G	P	X
8	8	H	Q	Y
9	9	I	R	Z
8-3	+	.	$	,
8-4	-	□		°

T H I N K

Figure 8.13. The front of the 701's quick reference card in 1953, the Programmer's Card

The operation of the Type 604 is fully explained in the examples that follow with a detail description of the control panel hubs preceding their use in each example. For a clear understanding of the capabilities of the machine as well as a thorough knowledge of the wiring principles, each example should be *studied in the sequence presented.*

In the last chapter of his manual, "Examples," Lida wrote in almost identical words:[72]

> This section gives several examples of programming that progressively use more and different operations and so become lengthier and more complicated. These examples can be *studied in sequence* to understand how different operations work together to create a meaningful program.

Lida, also, did not have to create close links between his text and his graphics because his source material from Rochester et al. was completely correlative like the other "Principles of Operation" manuals. Look again at the links between the electrostatic memory text excerpted earlier and Figures 8.11 and 8.12.

Yet most importantly, Lida maintained the rhetoric of equality established by Rochester et al. in his "Preliminary Operator's Reference Manual." This can best be seen from the tone established in the following two excerpts. The first is from Lida's manual:[73]

> BINARY INPUT
>
> Programs that transmit binary information into electrostatic storage are presented in the following sections. These programs were chosen because they:
>
> 1. Illustrate many programming techniques.
> 2. Are relatively simple.
> 3. Are useful programs that may be used by many installations.
> 4. Have been completely checked out on a 701 installation.
>
> These programs do not necessarily represent the most efficient method of programming and, in addition, programs may be written that are specifically adapted to the problems of a particular installation.

Notice that the use of the passive voice, like that in the objective-impersonal style of patents, remove the individual and focus attention on the object. Both Lida and his readers looked at the object from an equal distance and with equal authority. For example, Lida suggested that the reader could even improve on what is offered: "These programs do not necessarily represent the most efficient method of programming . . . " Compare Lida's example of the "Principles of Operation" rhetoric of equality with the following 1956 manual also describing the Type 701:[74]

> II. Programming the Problem
>
> A. Storage Assignment
>
> 1. Plan a rough storage assignment of auxiliary storage.
> 2. Assign Region 2 storage.
>
> B. Design Load Sheets
>
> 1. Discuss with the engineer the form and arrangement of the input data.

2. Form a rough draft of load sheets.
3. Have coordinator and machine room leadman check load sheets.
4. Make vellum load sheets (or have them drawn by artist). Always try bluelines of load sheets for a few production runs before having printed ones made.
5. Obtain machine room leadman's final approval of completed load sheets.

This second example was produced at Douglas Aircraft Company by two programmers, C. L. Baker and Bob Ruthrauff, in order to overcome the lack of documentation available prior to the delivery of a Type 701 to Douglas on May 23, 1953.[75] (The Computing Engineering Group at Douglas felt they needed to train and prepare for the arrival of the Type 701, so they proceeded to write their own manual that evolved as a document from 1951 to 1956.) Many of the documents available to Lida when he wrote his manual were also available to Baker and Ruthrauff at the same time they wrote Douglas's manual:[76]

- The Type 604 was being used at Douglas prior to the arrival of the Type 701 and thus its "Principles of Operation" manual must have been available at the site,
- Lectures from an IBM representative covered in detail the 32 operation codes described in "Part 3" of Rochester's "Preliminary Operator's Reference Manual."[77]

However, these two programmers, in contrast to Lida and the "Principles of Operation" manuals, were clearly operating within the tradition of imperative-mood instruction manuals in which the author was clearly directing and dictating to a novice audience. This can be seen most readily in the introductory material preceding the Douglas excerpt above:[78]

> The success of the application of these methods and procedures has been recognized throughout the industry. This fact *dictates* that the material specified must be followed if a high standard of quality is to persist. On the other hand, it is recognized that progress depends upon the flexibility of procedures. Recommendations for improvement are invited; but *deviation from the established practices should not be made without proper authorization.* [emphasis added]

Baker, one of the co-authors of the Douglas Aircraft manual, pointed out in a 1983 recollection that the tone set in these excerpts was largely derived from John Lowe, who was[79]

> the driving managerial force behind the establishment, codification, and enforcement of a rigid set of rules, procedures, and standards for all phases of design, specification, coding, and testing by the programmer, and of production running by the machine-room staff.

The relationship of author and reader in the Douglas manual's imperative-mood rhetoric can be observed no better than in this description of how Lowe, who set the tone for this manual, physically treated his programmer subordinates:[80]

> We learned to respect his intolerance for sloppy-looking paperwork of any kind . . . but all were terrified of his randomly timed twice- or thrice-daily '"cruises" through the machine room. Woe unto anyone who had left working materials of any kind strewn carelessly on equipment, keypunches, or tables—or who left a deck of [punched] cards unlabeled, unbanded, or unblocked! All of these were swept indiscriminately into the nearest trash barrel. The loss might cost a day's or a week's worth of work, but no matter; the lesson had to be learned. Some of us still flinch when we hear footsteps behind us in a computer room and instinctively look for stray papers.

Placing his manual clearly with the "Principles of Operation" rhetoric of equality, Sidney Lida took the same content available to Lowe, however, Lida chose to convey it within a different relationship with his audience rather than Lowe's use of terror and woe on a flinching audience.

Introducing a Number of Elements from the "Principles of Operation" that were Absent from the Rochester Source Material.

Lida's manual, unlike Rochester et al.'s successfully duplicated the typographic finish of all the previous "Principles of Operation" manuals; that is, the manual was typeset and fully justified in two columns. Also, most "Principles of Operation" manuals not only ensured the correlation of text and graphics by verbal call-outs but also by visual juxtaposition of text and graphic on the same or opposite pages. Rochester's typographical finesse in this area was not as fine as Lida's; for example, Rochester's references to Fig. 3 and 4 in the electrostatic memory excerpt earlier sent readers some 30 pages ahead, whereas Lida's reference to Fig. 1 simply directed the reader to the opposite page.

Lida also brought Rochester's source material into more of a characteristic balance of headings according to the signature style of the "Principles of Operation" (see Figure 8.14). Lida evidently broke up Rochester's headings into finer units, thus creating more 4th-level heads than Rochester (45% versus 10%) but also fewer 3rd-level headings (15% versus 50%). In changing the complexion of the hierarchical levels, Lida achieved a pyramiding of headings (few 1st-level headings, slightly more 2nd-level headings, even more third level headings, and the most numerous 4th-level headings) much more characteristic of the "Principles of Operation" signature style as typified by the Type 604 manual or the 285/297. In other words, Lida ensured that the lower level headings would always be more numerous than the higher level headings.

Changing a Number of Signature Style Elements in the "Principles of Operation" Manuals

Because Lida's version of Rochester's material usually either employed or enhanced the "Principles of Operation" signature elements present in the text, it was all the more unusual when Lida did not or could not. Hence, it was surprising to observe that there were dramatically fewer illustrations in the Type 701 documents written by either Rochester or Lida (Figure 8.15). Lida did increase the percentage of pages with illustrations from Rochester's 10% to 50%, but even, such a percentage is less than half the number of the lowest range of illustrations in any other "Principles of Operation" manuals. What

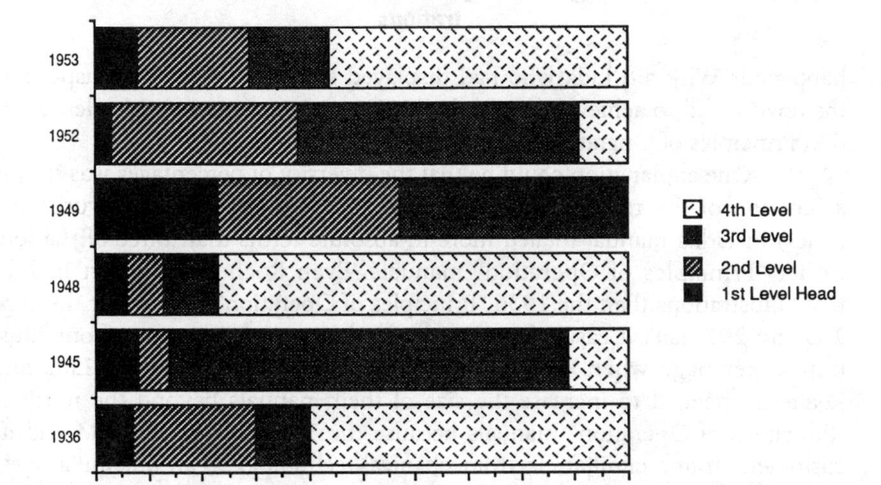

Figure 8.14. The preliminary and final versions of the Type 701 manual compared to the IBM signature style of the 1930s-1940s reveals identical deep structure down to fourth-level headings

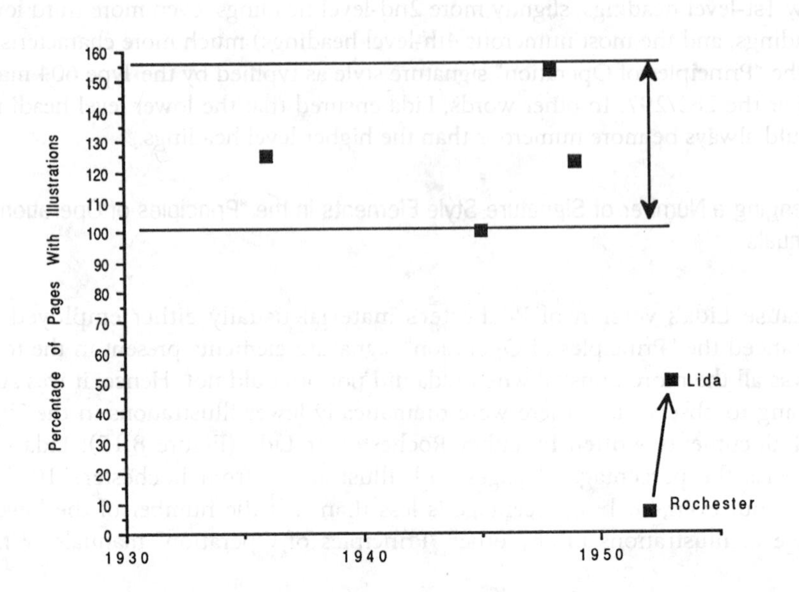

Figure 8.15. The preliminary and final versions of the Type 701 manual compared to the IBM signature style of the 1930s-1940s reveals fewer illustrations

happened? Why did Lida who had displayed such finesse in other aspects of the rewrite, fail to achieve completely this important signature style element of the "Principles of Operation" manuals?

One explanation could be that the diversity of percentages was simply a function of the relative size of manuals. The 54 illustrations (figures and tables) in Lida's manual totaled more in absolute terms than three of the four earlier "Principles of Operation" manuals. Even Rochester's report had 15 more illustrations than the 1936 "Principles of Operation" manual for the Type 285 and 297. Isn't it easier for an author to maintain an average of one illustration per page when there are only 30 pages or so? However, Lida and Rochester needed to increase the size of their manuals beyond the norm of "Principles of Operation" manuals because the Type 701 moved IBM and its customers from a familiar electromechanical technology to an unfamiliar electronic technology. Nevertheless, if there was more to explain and describe, why weren't there, accordingly, more illustrations especially since visual communication seemed to play such a large role in the "Principles of Operation"? Remember, for example, the priority given graphics in the 1945 manual: "Each of these setups is accompanied by a diagram of control panel wiring and description of the principles . . ."

Perhaps the explanation for the lack of illustrations was the very transformation that warranted the manuals' increase in size, the transformation from an electromechanical technology to an electronic technology. Compare the example used in the 1936 Type 285 and Type 297 manual, Figure 8.8, with the examples used by Rochester and Lida, Figures 8.11 and 8.12. Figure 8.8 was a representation of a physical plugboard and IBM cards, whereas Figures 8.11 and 8.12 were conceptual models of ideas. Perhaps as memory and instructions moved from the mechanical engineer's physical world of plugboards to a series of electrostatic bits with 0s and 1s, there was simply less that could be visually represented or, at least, was thought so by Lida and Rochester, the first authors to deal with this subject matter in the "Principles of Operation." Wasn't Lida being challenged by IBM's new electronic technology to make the same visual transition that John W. Griffiths had had to make some 100 years earlier, as Chapter 2 described, in the case of the half-hull, lift model or as Frank Leary had to make contemporaneously at UNIVAC, as Chapter 7 described?

John W. Griffiths's fellow apprentice, Lauchlan McKay, showed illustrations of half-hull, lift models representing actual wooden models in order to increase the rhetorical acceptability of his book, the *Practical Shipbuilder* (1839). McKay wanted to show the common, established way of nautical design. Griffiths, however, wanted to introduce a more scientific analysis of ship design especially called for by new shipbuilding materials such as wrought iron and because of the increasing internal complexity of the ship caused by the placement of steam power inside the hull. Consequently, John W. Griffiths used the half-hull, lift-model in new ways in his 1851 *Treatise on Marine and Naval Architecture* to convey abstract scientific ideas. Griffiths used the illustration of the half-hull lift-models as analogies of ideas rather than as representations of real things. Chapter 2 suggested that Griffiths was able to make this visual transition by his off-hand observation of the English dilettante's Beaufoy's scientific apparatus picture that looked remarkably like the carved half-hull lift-models. Griffiths, thus, had to reach outside the normal sources of American nautical design to make this visual communication breakthrough in his *Treatise*.

Similarly, consider the visual and textual analogies introduced by the "new" nontechnical technical writer, Frank Leary, in Chapline's UNIVAC Editorial Department as described in the previous chapter. For example, the draft introduction of the UNIVAC I manual probably authored by Chapline working within the established engineering context of EMCC was not nearly as effective as the 1954 introduction with its analogies to the payroll clerk that was authored after Leary's arrival. Leary may have been able to make this breakthrough in UNIVAC documentation or, at least, to aid it because his degree was in journalism and not engineering and he had only very recently become part of the EMCC–UNIVAC Editorial Department. Thus, Leary was able to use information and communication techniques outside the UNIVAC Editorial Department to make his visual metaphor breakthrough.

Sidney Lida's background and tenure at IBM are unknown. Yet, Lida must have had an established, successful track record at IBM to have been entrusted with the task of authoring the manual for such a crucial new product. However, such a track record could only have been achieved within the mechanical engineering world of punched-card calculators. Thus, the success of his authorial insight in an earlier technology may, in fact, have contributed to his authorial blindness in this new technology. Perhaps Lida was too accustomed to illustrations having a one-to-one relationship with physical reality that he was unable to use illustrations in new ways as analogies for abstractions. Accordingly, whereas Griffiths and Leary used illustrations as analogies for electronic information, Lida ignored this possibility. Lida ignored this possibility of illustration by replicating his own and the "Principles of Operation" tried-and-true methods of having illustrations with a one-to-one relationship with reality: Nearly 50% of Lida's 701 illustrations are just pictures of the planning form with various blocks filled out (see Table 8.4). In all earlier "Principles of Operation" manuals, illustrations of the planning forms were usually a very minor item, but here Lida took such illustrations and made them the predominate ones so as to achieve the "Principles of Operation" manuals' typical profuseness of illustration. Lida made the planning forms his predominate illustration because they alone could give him both profuseness as well as one-to-one representations of reality just as had McKay's illustrations of half-hull, lift-models. Both Lida and McKay hoped to build rhetorical creditability by accurately representing reality rather than imaging reality at one remove as metaphors and analogies.

Figure 8.16 reveals how Lida compensated for the loss of visual communication abilities. Unable to use his two hands—text and pictures—to communicate with his audience, Lida dramatically increased the reading ease of his words as measured by the Flesch Reading Ease scale just as dramatically as he decreased his use of illustrations. Not only was Lida's manual nearly 50% higher on the scale than any other "Principles of Operation" manual, but Rochester's—the amateur's—was not too far behind. With their scores in the 70s, Lida and Rochester achieved in a single bound what had been characteristic of the mower-reaper manuals 75 years earlier.

However, it seems that by combining the information in Figure 8.16 with that of paragraph size in Figure 8.17, the two may have tended to cancel out each other. In other words, on a sentence level (Figure 8.16), Lida and Rochester dramatically improved the reading ease, but by increasing the size of the paragraphs (Figure 8.17), they gave the reader more at a single gulp, thus diminishing their texts' reading ease.

Whereas Lida acted uncharacteristically in his employment of illustrations and in his style of writing, Lida and Rochester also both had dramatically

Table 8.4. Comparing the Illustration Content of the *Type 604, "Principles of Operation"*, the *Defense Calculator Preliminary Operator's Manual*, and Lida's *Type 701 and Associated Equipment: Principles of Operation.*

	Pages in *Type 604, Principles of Operation*	Pages in *Defense Calculator Preliminary Operator's Manual*	Pages in *Type 701 and Associated Equipment: Principles of Operation*
Plugboards and their wiring	41	4	4
Planning chart	5	4	21
Physical calculator and switches	2	0	1
Movement of cards through machine) (all nearly identical	1	1	2
Punch card sample	0	1	1
Microsecond timelines of activity	0	5	6
Diagrams of purely abstract information, e.g., information paths, memory registers	1	6	8

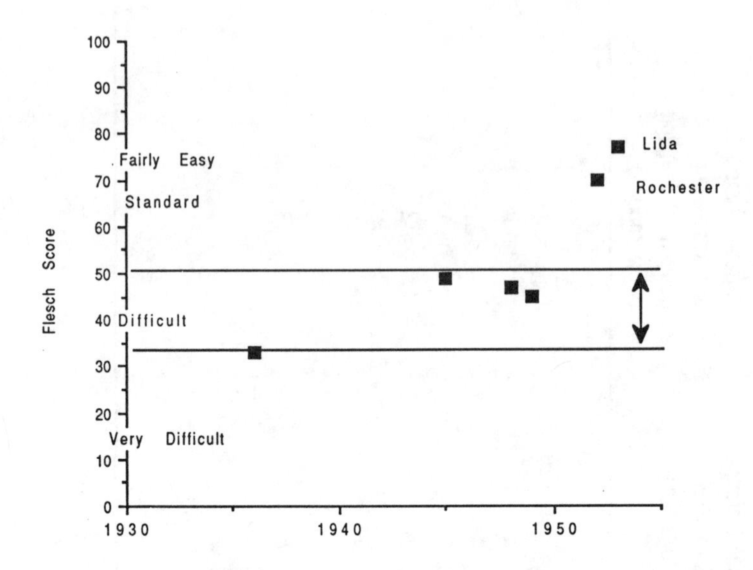

Figure 8.16. The preliminary and final versions of the Type 701 manual compared to the IBM signature style of the 1930s-1940s reveals significantly higher scores on the Flesch Reading Ease scale

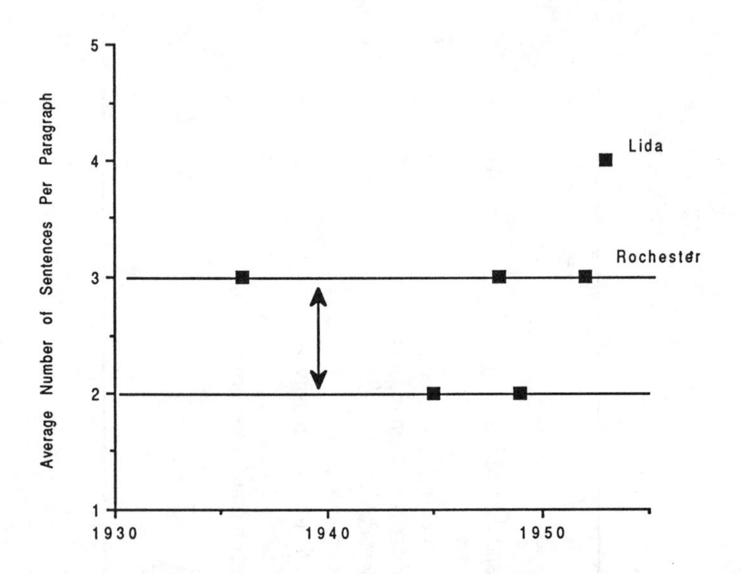

Figure 8.17. The preliminary and final versions of the Type 701 manual compared to the IBM signature style of the 1930s-1940s reveals significantly higher scores on the Flesch Reading Ease scale

fewer headings as is shown in Figure 8.18. This also could have just been a function of size; perhaps it was easier for a 39-page document to have three headings per page, but can such a pace be maintained for a manual nearly 280% larger?

Perhaps another reason why there were fewer headings per page is implied by Figure 8.17, which reveals larger paragraphs; with fewer, larger paragraphs, there would be fewer headings per page. Figures 8.17 and 8.18 suggest that both Rochester and Lida moved to larger bite-sizes of information perhaps because the complexity of their novel electronic material eluded the formerly easy, short segmentation of text. It could have been that the concepts were so large and complex that it just took more space for Rochester and Lida to discuss them. Certainly it took more to explain "electrostatic memory" and "electrostatic addresses" than it did to explain "field selection" punches in a card. Could it also have been that the novelty of the ephemeral, electronic information that had earlier denied them ease of packaging in illustrations also denied them ease of organization?

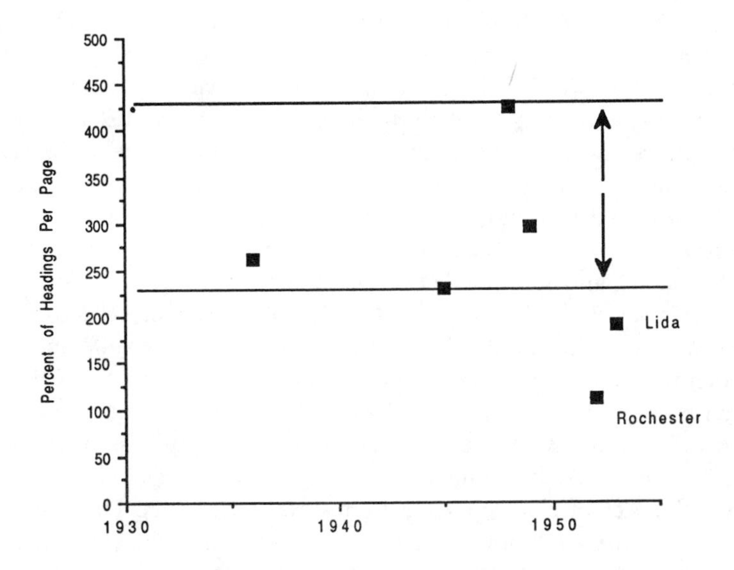

Figure 8.18. The preliminary and final versions of the Type 701 manual compared to the IBM signature style of the 1930s-1940s reveals many fewer headings

LIDA'S ODYSSEY COMPARED TO CHAPLINE'S

Both Lida and Chapline struggled with precedent and innovation. Fortunately after more than 40 years, it is still possible to observe their journeys. In Chapline's case, his letters, memos, rough drafts, and interview directly described his authorship odyssey. With Lida it was different. Only by inference, can one today reassemble part of the bookshelf beside his desk, observe how he transformed his original source materials, and suggest the motivations that led him in various directions. In his odyssey, Lida performed well as a technical writer. His work in altering Rochester's source material to meet the needs of a new audience as well as to ease its comprehension would today still be the marks of a technical communication craftsman.

However, a successfully completed voyage reveals as much about the sea as it does the sailor, or a finished manual as much about the corporation as the writer. EMCC—later the UNIVAC division of Remington-Rand—was a young company in existence only two years prior to Chapline's work. The boundaries of such a young company's corporate culture and the definitions of its particular methods of communication and documentation were still permeable. True, the predisposition of EMCC was to use the methods and organization of familiar technical reports arising from Eckert and Mauchly's prior publications. This predisposition, of course, resulted in Chapline's 1949 two-part BINAC project that had both an *Engineering Report* as well as the *Operating and Maintenance Manual*. Yet, the influence of outsiders—such as Frank Bell from Northrop Aircraft with his 1949 letter, and Frank Leary the journalist hired in 1950—was able to break through the boundaries of the EMCC–UNIVAC corporate culture and combine to influence the creation of the 1954 *Manual of Operation for UNIVAC System*. Bell and Leary were able to combine and predispose Chapline and his Editorial Department to break the internal mold of engineering documentation and reshape it into the tradition of the earlier sewing machine and automobile instruction manuals.

However, the boundaries of IBM's corporate culture and the definitions of its particular molds of communication were quite different. IBM and its earlier manifestations had been in existence 56 years, and the "Principles of Operation" as an IBM genre was older in 1950 than Chapline himself. Hence, the boundaries of IBM were relatively more solid, and the documentation forms more impervious to outside extratextual influences of change as well as inside influences of change arising from novel technology. Lida worked very hard to make the 701's novel electronic content fit into the electromechanical genre of the "Principles of Operation" manual. When Lida could, he simply continued elements of the "Principles of Operation" from his source material. When "Principles of Operation" elements were not present in the Rochester source material, such as the customary large number of illustrations, Lida added them. Moreover, Lida added illustrations even when they did not quite fit; he chose to take a formerly minor type of illustration in the electromechanical manuals, pictures of actual planning forms, and dramatically increased their number in the Type 701 manual when the possibility of other pictures of actual physical objects decreased. By hook or by crook, Lida worked to take the abstract world of electrical engineering and rework it into the type of illustrations appropriate to the earlier IBM world of mechanical engineering. In doing so, Lida missed an opportunity to make the illustrations of the "Principles of Operation" into analogies of electrical engineering in the way that Frank Leary had done over at UNIVAC or as John W. Griffiths had 100 years earlier with half-hull lift-models.

Perhaps it was too much to expect that Lida single-handedly could create a technical communication revolution at IBM. After all, it had taken nearly five years for a variety of outside influences to permeate UNIVAC and to lead to a conceptual innovation in documentation. Furthermore, at the time, Lida's choice of the customary "Principles of Operation" framework for his documentation must have been well received because 12 of the 19 owners of Type 701s had also been owners of earlier electromechanical calculators and 3 of the 7 components of the Type 701 system were simply modified electromechanical machines. If only one could sample Sidney Lida's later work, say from 1958 or 1959 with the later 700 series computers, one may have been able to see how Lida could have worked to bring the "Principles of Operation" manuals into the new world of electronic computers. He may have been able to bring the "Principles of Operation" manuals to the new world of electronic computers once he had come to understand he stood in the midst of revolutionary technical change. However, after his work on the Type 701 manual, documentation specifically attributed to Sidney Lida disappeared, and then he too disappeared from published history.

IBM continued to produce "Principles of Operation" manuals for another 30 years. Along the way they became known as "red books" forming part of a suite of documents (including a user guide, a technical reference manual, and a planning guide) accompanying each new IBM computer. One IBM employee with over 30 years of documentation experience with the company[81] recalled that a 1967 "Principles of Operation" manual for a model of the IBM 360 mainframe was the epitome of all that a "Principles of Operation" manual offered. Nevertheless, by the late 1970s, IBM had discovered that few customers wanted a "red book," so the assignment of their authorship was passed onto writers of lesser ability. Finally, in the late 1980s, IBM dropped the "Principles of Operation" from its suite of documents.

Yet, just at the time that the "Principles of Operation" manual disappeared in the late 1980s, the two traditions of instruction manuals—the sewing machine/automobile/UNIVAC asymmetric tradition and the mower-reaper/"Principles of Operation" symmetric tradition—were made explicit in

IBM. In 1982 IBM issued an Information Development Guideline, *Designing Task-oriented Libraries for Programming Products*, and, two years later in 1984, John Carroll at the T. J. Watson Research Center published his first article, "Minimalist Design,"[82] which, in 1990, led to his book, *The Nurnberg Funnel: Designing Minimalist Instruction for Practical Computer Skill*.[83] This guideline and book asked for far more than simple comprehension from their audiences; together, they grappled for the soul of technical communication. It is to this final explicit crossroads of the American instruction manual traditions that the last chapter of this book now turns.

ENDNOTES

1. Qtd. in Hurd, Cuthbert C.: "Epilogue." *Annals of the History of Computing* 3(2) April 1981: 219.
2. Rochester, Nathaniel: "The 701 Project as Seen by Its Chief Architect." *Annals of the History of Computing* 5(2) April 1983: 115.
3. Shurkin, op. cit., 1984: 255.
4. War Department Bureau of Public Relations, "History of Development of Computing Devices," 2/ 16/46. (Material used in connection with the public announcement of ENIAC.) Hagley Museum and Archives, Wilmington, DE: Accession 1825, Box 47: 2.
5. Bashe, Charles M., Johnson, Lyle R., Palmer, John H. and Pugh, Emerson W.: *IBM's Early Computers*. Cambridge, MA: MIT Press, MIT Press Series in the History of Computing, 1986: 26–33; Aspray, William. (ed.): *Computing Before Computers*. Ames: Iowa State University Press, 1990: 215–9.
6. Bashe et al., op. cit., 1986: 47–59.
7. Bashe et al., op. cit., 1986: 58. See also Aspray, 1990, op. cit.: 244.
8. Clippinger was perhaps not as off base then as he sounds now. He was personally involved with Von Neumann in developing the floating-point mathematical concepts behind EDVAC. See Williams, Michael R.: "The Origins, Uses, and Fate of the EDVAC." *IEEE Annals of the History of Computing* 15(1) 1993: 26.
9. Clippinger, R. F.: "Mathematical Requirements for the Personnel of a Computing Laboratory," American Mathematical Monthly 57 (1950): 439, qtd. in Ceruzzi, Paul: "An Unforeseen Revolution: Computers and Expectations, 1935–1985." In (Corn Joseph J., ed.) *Imagining Tomorrow: History, Technology, and the American Future*. Cambridge, MA: MIT Press, 1986: 194.
10. McClelland, W. F. and Pendery, D. W.: "701 Installation in the West." *Annals of the History of Computing* 5(2) April 1983: 167.

11. Aspray, op. cit., 1990: 149. Cortada tells this story in his book (*Before the Computer: IBM, NCR, Burroughs, and Remington Rand and the Industry They Created, 1865–1956*. Princeton, NJ: Princeton University Press, 1993: 323): "As an IBM salesman in the Data Processing Division in 1979, I found an intact data center belonging to Worthington Pump Corporation at a foundry in New Jersey that was completely stocked with punched card equipment from the 1940s and 1950s, proof that even decades later, computers had not fully displaced all tabulating technology."
12. Aspray, op. cit., 1990: 150.
13. "IBM President Describes 701 Development." *Annals of the History of Computing* 3(2) April 1981.
14. Bashe et al., op. cit., 1986: 130–1. See also, Watson, Thomas, Jr.: *Father, Son and Company*. New York: Bantam Books, 1990. According to Tom Watson, Jr., head of IBM's first computer projects and later CEO, one of the things that drove IBM to develop computers was the personally felt insult in Eckert and Mauchly's sales claims that magnetic tape, not punchcards (IBM's mainstay at the time), was the optimal storage device in the future for business. See also Fishman, Katherine Davis: *The Computer Establishment*. New York: Harper and Row, 1981: 39. Katherine Fishman concurred about the defensive role IBM took in regard to magnetic tape. In fact, Fishman speculated that it was the use of magnetic tape on BINAC and later on UNIVAC that most upset the multimillion dollar Hollerith paper punch card maker and caused IBM to enter the market in 1952 with the 701 which, of course, had paper punch cards as input.
15. Watson, Thomas, Jr., op. cit., 1990.
16. Hurd, Cuthbert C.: "Early IBM Computers: Edited Testimony." [testimony given January 3–February 1, 1979, U.S. Federal Court, Southern New York District, U.S. vs. IBM antitrust suit]. *Annals of the History of Computing* 3(2) April 1981: 164–5, 168. Interestingly, the final two names for the computer that were so cleverly chosen for different specific reasons get conjoined in the source material used by Lida, the "Preliminary Operator's Reference Manual." The cover for the June 1951 version of the "Preliminary Operator's Reference Manual" announced that the book was the "Defense Calculator Preliminary Operator's Reference Manual," but when the same material was expanded and rereleased in August 1952, it was called the "Type 701, Electronic Data Processing Machines, Preliminary Operator's Reference Manual." National Museum of American History, Smithsonian Institution, Computers, Information and Society: holding #1977, 3001 (June 1951 version and August 1952 version).
17. Hurd: "Early IBM Computers," op. cit., 1981: 168.
18. Bashe et al., op. cit., 1986: 161.

19. Hurd, Cuthbert C.: "Prologue." *Annals of the History of Computing* 5(2) April 1983: 110.
20. *Defense Calculator Preliminary Operator's Reference Manual*: 8.
21. Watson, Thomas, Jr., op. cit., 1990: 376–389. The suit was filed during the Johnson Administration in 1969 and charged IBM with monopolistic practices in the computer industry and sought to break IBM into a set of smaller companies. The suit was settled in 1981 during the Reagan Administration. (Tom Watson, Sr. had been sentenced to a jail term under similar charges when he was part of the National Cash Register Co. and working for John Patterson in 1903. Similar charges had been filed by the Federal government against IBM in 1935 and 1952.)
22. Hurd: "Early IBM Computers," op. cit., 1981: 168.
23. Hurd, Cuthbert C.: "Epilogues." *Annals of the History of Computing* 3(2) April 1981: 218.
24. Watson, Thomas, Jr., op. cit., 1990: 204.
25. Hurd, Cuthbert C.: "Sales and Customer Installation." *Annals of the History of Computing* 5(2) April 1983: 163. A letter from Mr. Hurd to this author dated 9/21/90 explained that Mr. Lida has last been heard of working in Oklahoma on bovine artificial insemination. And there is some ambiguity as to the exact title of the manual. In the copy of the manual obtained from the National Museum of American History, Smithsonian Institution, Computers, Information and Society Division, the title page did not have the name, "Principles of Operation." However, Cuthbert Hurd, who both directed the 701 project and edited the special issue of the *Annals of the History of Computing*, 5(2), which described the 701 and all the related engineering and design activity, named the manual twice in the issue with "Principles of Operation" (page 161 and 164). Hence, that name will be used in this chapter.
26. Watson, op. cit., 1990: 13.
27. Theoretical manuals of the 1920s and 1930s also included *Sales Accounting* (1928) and *IBM Machine Methods of Accounting–The Preparation and Use of Codes* (1936). (These are available from the Charles Babbage Institute, Center for the History of Information Processing, University of Minnesota, 103 Walter Library, 117 Pleasant Street SE, Minneapolis, MN 55455.)
28. An Internet search using Worldcat turned up dozens of IBM Principles of Operations manuals from the 1960s–1980s, with the most recent citation including the *Principles of Operation for the IBM 370* (1989).
29. *IBM Electric Punched Card Accounting Machines, Principles of Operation, Automatic Carriage, Type 921*: 1.
30. In fact, some of the early presales brochures produced by Chapline for the UNIVAC I were typewritten (1948 and 1951).

31. Johnson, Roy W. and Lynch, Russell W.: *The Sales Strategy of John H. Patterson* (Sales Leaders Series). Chicago and New York: The Dartnell Corp., 1932: 195.
32. It is unclear because of their lack of industry universality to term these "Principles of Operation" instruction manuals a "genre," "subgenre," or simply a "company format." See JoAnne Yates and Wanda J. Orlinkowski: "Genres of Organizational Communication: A Structurational Approach to Studying Communication and Media." *Academy of Management Review* 17(2) (1992): 303–4.
33. The search for manuals included communications with the Charles Babbage Institute, Center for the History of Information Processing, University Of Minnesota, 103 Walter Library, 117 Pleasant Street SE, Minneapolis, MN 55455, and IBM Archives, Old Orchard Road, Armonk, NY 10504.
34. Obtained from National Museum of American History, Smithsonian Institution, Computers, Information and Society, IBM Accounting Machines, Box 304349, Document stamped L26585.
35. Obtained from Morris Library open stacks, University of Delaware.
36. Obtained from National Museum of American History, Smithsonian Institution, Computers, Information and Society, IBM Accounting Machines, Box 304349, Document stamped L26582.
37. Obtained from National Museum of American History, Smithsonian Institution, Computers, Information and Society: holding #1977, 3001 (x), Nat Rochester's signed-out copy. Rochester was the 701's Chief Architect and co-author of the preliminary documentation.
38. *IBM Electric Punched Card Accounting Machines, Principles of Operation, Automatic Carriage, Type 921*: 1.
39. Lundstrom, David E.: *A Few Good Men from UNIVAC*. Cambridge, MA: The MIT Press, 1988: 23—"Engineers, it was believed, were second-class employees at IBM while all status was enjoyed by marketing...The IBM sales force was so well trained and highly motivated that 'they could sell manure in a brown paper sack!'"
40. Ibid., 1932: 125–6.
41. The Flesch Reading Scores were based on the following information:

	Average Sentence Length in Words*	Number of Syllables Per 100 Words
1936	32	167
1945	26	156
1948–49	23	162
1949	22	165

*Average of five sentences chosen randomly in the manual.

42. The progressive presence in the "Principles of Operation" manuals of tables of contents, indexes, and summary pages of controls or instructions may also have reflected the growing number of pages in such manuals. For example, if a "Principles of Operation" manual had as few as 16 pages as did the 1936 example, perhaps the author thought that such items were unnecessary. By that same token, when the 1948–49 manual grew to 71 pages, perhaps the very same author would have now seen a good reason to include such items.
43. *IBM Machine Method of Accounting, Electric Accounting Machines, Type 285 and Type 297*, 1936: 9.
44. *IBM Electric Punched Card Accounting Machines, Principles of Operation, Automatic Carriage, Type 921*: 1.
45. Mitchell, W. J. T.: *Iconology: Image, Text, Ideology*. Chicago: University of Chicago Press, 1986: 67–8.
46. The variety of interpretations offered by illustrations in comparison with text is very much like the "affordances" of use and understanding in different technologies discussed by Donald Norman in *Things That Make Us Smart: Defending Human Attributes in the Age of the Machine*. Reading, MA: Addison-Wesley Publishing Co., 1993: 105–113.
47. Chapline to Eltgroth, 6/22/49, Hagley Museum and Archives, Wilmington, DE: Accession 1825, Box 83—"It has been learned that only very slight use (incidental only) can be made of the engineers concerned. Therefore, the resulting report will be based only on those facts recorded in the notebooks and then only on those which form a complete story or which can be made complete by a question or two put to the appropriate engineer." See also, Baker to Ackerlind letter, 7/17/49, Hagley Museum and Archives, Wilmington, DE: Accession 1825, Box 84—"As an interesting sidelight on this course, I mention a remark made to me by Mr. Weiner. (I quote his meaning, not his exact words): 'I wish you'd sit in on that course, Dick, and check Joe [Chapline], since he has done little work on the machine, and, even though he is writing the instructions manual, he could easily make mistakes not being familiar with the machine itself, but only with the written material available which we both know is very incomplete and, in many instances, unreliable'."
48. Lundstrom, op. cit., 1988: 20–1.
49. Haddad, Jerrier A. (co-engineer of the 701 with Rochester); "701 Recollections." *Annals of the History of Computing* 5(2) April 1983: 124—"Late in the development process it became obvious that the first machine out of manufacturing would not be ready until very late in 1952." However, in the 1953 manual, Lida wrote: "These programs were chosen because they [h]ave been completely checked out on a 701 installation." These words suggest that Lida indeed saw a working Type

701 during the creation of the manual. However, if Haddad was correct in saying that the machine would not be out until late 1952, and, if IBM shipped Lida's manual when the Type 701 was released in April 1953, that would give Lida only a very small window of opportunity to work with a functioning Type 701 to help write his manual.

50. Like EMCC's Eckert, Rochester had also been a wartime employee of the MIT Radiation Lab where radar research and development had taken place. Eckert went directly from MIT to the Moore School to join Mauchly in designing UNIVAC computers—he never left an academic situation. Rochester, chief architect on the 701 and co-author of this "Preliminary Operator's Manual," did have corporate experience with Sylvania Electric Products Co. as an Engineering Manager (Bashe et al., op. cit., 1986: 108–9). Perhaps the additional corporate experience as well as the influence of IBM documentation precedents led him to write manuals less like the academic engineering reports of Eckert at EMCC.

51. Rochester, op. cit., 1983: 116. Rochester's claim to co-authorship received some physical verification by the presence of his penciled-in name at the top of the manual's introductory memorandum in the Smithsonian's copy of this manual. Other claims of authorship came from William F. McClelland (head of a New York City Mathematics Committee that was a rival to Rochester's Committee in Poughkeepsie) who claimed "to have [written] the first 701 customer operating manual, and taught the first 701 class" (McClelland, W. F. and Pendery, D. W.: "701 Installation in the West." *Annals of the History of Computing* 5(2) April 1983: 168). However, in "Sales and Customer Installation" (*Annals of the History of Computing* 5(2) April 1983: 163), Cuthbert C. Hurd stated that John Mayhew wrote the second version.

52. Rochester, op. cit., 1983: 116. The "Preliminary Report on the IBM Electronic Data Processing Machine" can be viewed at National Museum of American History, Smithsonian Institution, Computers, Information and Society: holding #1977, 3001 (June 1951 version and August 1952 version). Personal correspondence with Cuthbert Hurd (10/90) verified that this was Lida's working source material. However, there are in the archives two versions of the manual. The versions are identical in all respects except that the August 1952 version has an additional eight appendices of 25 total pages and separate manuals for the card reader, card punch, and printer. This analysis focuses on the later, expanded edition.

53. The method described here examining Lida's editing of his written source materials to reveal information about the Lida's rhetorical situation has been used fruitfully for a long time in the area of scriptural analysis and has been called redaction criticism. See Harrington, Daniel J.: *Interpreting the New Testament: A Practical Guide*. Collegeville, MN: Liturgical Press, 1990: 96–107.

54. "Preliminary Operator's Reference Manual:" 11.
55. Rochester, op. cit., 1983: 116.
56. Bashe et al., op. cit., 1986: 108–9.
57. Bashe et al., op. cit., 1986: 108–9.
58. Also, both copies of the Type 604 manual and the "Preliminary Operator's Reference Manual" collected from the holdings of the Computers, Information and Society Division of the Smithsonian Institution appeared to be personal copies of Mr. Rochester with his name penciled in at the top of a number of pages in each manual.
59. Preliminary Operator's Reference Manual:" 68.
60. In fact, lack of advance documentation for the 701 caused Douglas Aircraft Co. to write their own manual (Baker, C. L.: "The 701 at Douglas, Santa Monica." *Annals of the History of Computing* 5(2) April 1983: 188), and an IBM salesman, R. Blair Smith, to have the following adventure: "Manuals for the 701 were scarce, and Paul Armer was upset because I could not get more than one copy for Rand [Corp.]. Armer wanted to use the 701 at the Technical Computing Bureau and needed several more copies. Meantime, I was trying to sell him a 701. Rand certainly appeared to have plenty of UNIVAC manuals. Every manager I visited at Rand had a UNIVAC manual on his desk. Then I learned that every one of the new IBM Applied Science employees had a 701 manual. I stormed into their office one day, introduced myself as a senior salesman, told them I had an urgent need for three manuals, confiscated the manuals, and took them to Rand. Later, when Rand ordered a 701, Paul Armer invited me to his office for a little celebration. When everyone had come and coffee was served, Armer said, 'Blair do you recall seeing UNIVAC manuals in the various offices here?' I said, 'Sure, everyone had one.' He said, 'When you couldn't get us more than one 701 manual, we pulled a trick on you. We had only one UNIVAC manual. However, when the front desk announced your arrival and someone came to escort you, we sent the UNIVAC manual ahead to the office you were going to visit.'" Smith, R. Blair: "The IBM 701–Marketing and Customer Relations." *Annals of the History of Computing* 5(2) April 1983: 170–1.
61. Baker, C. L.: "The 701 at Douglas, Santa Monica." *Annals of the History of Computing* 5(2) April 1983: 187.
62. *Principles of Operation: Type 701 and Associated Equipment (1953)*: 72.
63. Novel that is for IBM customers; BINAC and UNIVAC I had already used such a number system in their calculation processes.
64. Preliminary Operator's Reference Manual," 11, 13.
65. *Principles of Operation: Type 701 and Associated Equipment* (1953): 13, 15.
66. Mathes, J. C. and Dwight W. Stevenson: *Designing Technical Reports*. Indianapolis, IN: Bobbs-Merrill, 1976.

67. Faigley, Lester and Stephen P. Witte: "Topical Focus in Technical Writing." In (Anderson, Paul V., Brockmann R. John, and Miller, Carolyn R., eds.) *New Essays in Technical and Scientific Communication: Research, Theory, Practice*. Farmingdale, NY: Baywood Publishing Co., Inc., 1983: 59–68.
68. Johnson and Lynch, op. cit., 1932: 198.
69. *Principles of Operation: Type 701 and Associated Equipment (1953)*: 55.
70. *Principles of Operation: Type 701 and Associated Equipment (1953)*: 55.
71. *Type 604 Principles of Operation*: 9.
72. *Principles of Operation: Type 701 and Associated Equipment (1953)*: 86.
73. *Principles of Operation: Type 701 and Associated Equipment (1953)*: 78.
74. Douglas Aircraft Company: "Computing Engineering Manual, 1956." *Annals of the History of Computing* 5(2) April 1983: 194.
75. Baker, op. cit., 1983: 188.
76. Ibid.: 188.
77. Ibid.: 188. Baker states in the article that he had no record of Rochester's "Preliminary" manual being available at the representative's lecture in December 1951, but the presentation was so "extremely detailed" that a 1/4/52 Douglas document by Jack Middlekauff captured all 32 operational codes in Part 3 of the "Preliminary" manual. In May 1952, Middlekauff updated the Douglas manual and specifically referenced Rochester's "Preliminary" manual.
78. Ibid.: 191.
79. Ibid.: 190.
80. Ibid.: 191—"Those who have worked for John Lowe will quickly recognize the tone that is set here. His three rubrics—neatness, timeliness, accuracy—are engraved indelibly in all our minds, and, lest we forget, are hammered home on the very next page of the manual."
81. Personal conversation with Jeff Hibbard, former president of the Society for Technical Communication, 10/5/94.
82. Carroll, John: "Minimalist Design." *Datamation* 30(18) 1984: 125–36.
83. The Nurnberg Funnel: *Designing Minimalist Instruction for Practical Computer Skill*. Cambridge, MA: MIT Press, 1990.

Part Two

Using a Historical Perspective in Technical Communication

"France in Vietnam, 1954, and the US in Vietnam, 1965—A Useful Analogy?"

—June 1965 memorandum sent to President Johnson by White House National Security Assistant, McGeorge Bundy[1]

In the previous eight chapters I sketched out three themes in the history of American technical communication:

- the varying weave of text and graphics and the importance of visual communication;
- the influence of genres and their birth and development in response to both internal and external pressures; and
- the role of the individual technical communicator coping with technological innovation within the constraints of genres.

I sketched out these themes using stories of both the famous and the forgotten, the quantitative analysis of dozens of patents and instruction manuals, and a nearly day-by-day examination of the development process of individual manuals. Surely, these chapters have shown American technical communicators that their history stretches back far. However, there is much that remains to be explored.

For example, from the 1840s, the time just prior to the introduction of the sewing machine and mower-reaper instruction manuals, there were "rule" booklets with instructions for workers on how to operate the new technology of railroad lines,[2] as well as books with instructions for travelers along the Oregon Trail.[3] For example, Johnson and Winter's *Route Across the Rocky Mountains, with a Description of Oregon and California; Their Geographical Features, Their Resources, Soil, Climate, Productions, &c, &c.* (1846) well represents the visual communication theme in American technical communication with its final "Bill of the Route" (see Figure P2.1). Thus, can analysis artifacts such as these trace the separation in America of narration and instruction manuals to a point earlier than the 1859 Grover and Baker's *A Home Scene; Mr. Aston's First Evening with Grover and Baker's Celebrated Family Sewing Machine Containing Directions for Using*? Johnson and Winter, for example, offer their "Instructions to Emigrants" only after 176 pages of narration.

An analysis of radio or wireless instruction manuals of the 1920s and 1930s will certainly yield interesting information on how authors first conveyed intangible electronic information to the general American public in comparison to the automobile manuals of the time in which authors sought to convey physical mechanical information. For example, two previously unacknowledged authors pioneered techniques of communicating both intangible electricity to the general American public as well as the physical mechanical workings of automobiles. Milton Blake Sleeper wrote dozens of manuals about radio and the wireless from 1917 to 1950,[5] while at the same time Victor Alfred Pagé wrote hundreds of books and articles about automobiles, as well as planes, tractors, motorcycles, and power boats, at about the same time from 1906 to 1950. His biographer called Pagé "the foremost technical writer of his day," yet how many of heard of him or studied his treasury of technical writing?[6]

Moreover, a comparison of the 1946 ASCC's documentation (Chapter 8)[7] with the 1953–57 documentation for the C.S.I.R.O. Mark 1, the first Australian computer,[8] could also yield interesting insights on how varied genres as well as varied cultures of technology affected computer instruction in its very earliest days. For example, a manual was used for the American machine, whereas a series of scholarly articles was used for the Australian machine. It may also be interesting to compare the first known computer documentation authored by a woman—Grace Hopper co-authored the ASCC manual—with that for similar technology authored by men—Trevor Pearcey and G. W. Hill co-authored the C.S.I.R.O. Mark 1 documentation.

However, if the field of technical communication is instrumental communication, communication that gets things accomplished, so must its history.[9] Simply recording technical communication accomplishments of the

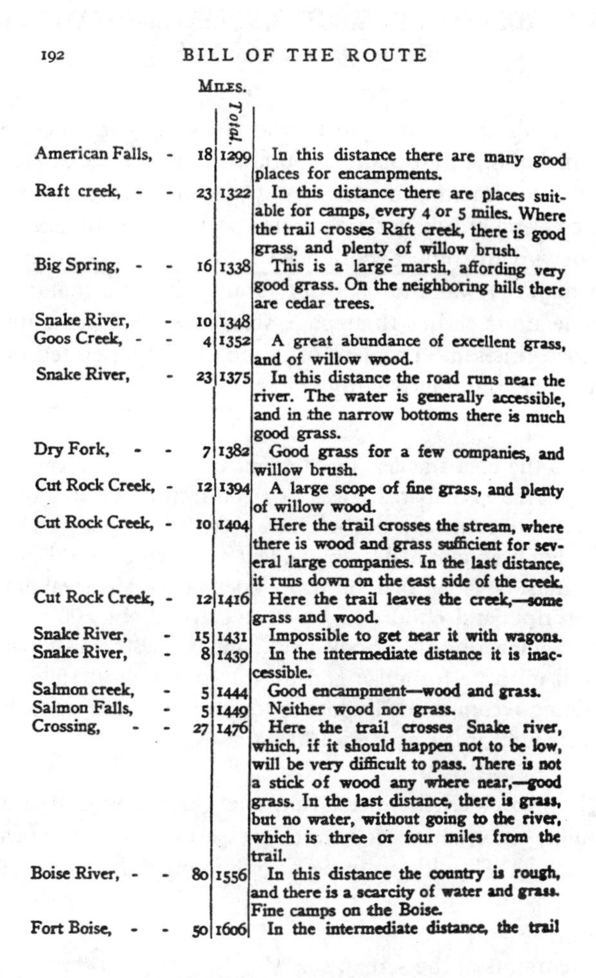

192 BILL OF THE ROUTE

	Miles.	Total.	
American Falls, -	18	1299	In this distance there are many good places for encampments.
Raft creek, - -	23	1322	In this distance there are places suitable for camps, every 4 or 5 miles. Where the trail crosses Raft creek, there is good grass, and plenty of willow brush.
Big Spring, - -	16	1338	This is a large marsh, affording very good grass. On the neighboring hills there are cedar trees.
Snake River, -	10	1348	
Goos Creek, - -	4	1352	A great abundance of excellent grass, and of willow wood.
Snake River, -	23	1375	In this distance the road runs near the river. The water is generally accessible, and in the narrow bottoms there is much good grass.
Dry Fork, - -	7	1382	Good grass for a few companies, and willow brush.
Cut Rock Creek, -	12	1394	A large scope of fine grass, and plenty of willow wood.
Cut Rock Creek, -	10	1404	Here the trail crosses the stream, where there is wood and grass sufficient for several large companies. In the last distance, it runs down on the east side of the creek.
Cut Rock Creek, -	12	1416	Here the trail leaves the creek,—some grass and wood.
Snake River, -	15	1431	Impossible to get near it with wagons.
Snake River, -	8	1439	In the intermediate distance it is inaccessible.
Salmon creek, -	5	1444	Good encampment—wood and grass.
Salmon Falls, -	5	1449	Neither wood nor grass.
Crossing, - -	27	1476	Here the trail crosses Snake river, which, if it should happen not to be low, will be very difficult to pass. There is not a stick of wood any where near,—good grass. In the last distance, there is grass, but no water, without going to the river, which is three or four miles from the trail.
Boise River, - -	80	1556	In this distance the country is rough, and there is a scarcity of water and grass. Fine camps on the Boise.
Fort Boise, - -	50	1606	In the intermediate distance, the trail

Figure P2.1. Visual communication in Johnson and Winter's "Bill of Route" for the Oregon Trail (1846)[4]

past does not help the profession get things accomplished in the here and now. Thus, Part Two and this last single chapter demonstrate how history may be used to gain valuable insights into contemporary technical communication issues.

History is tricky—one can never absolutely say that a past event caused a present event. So, one cannot look to history for absolute causes of present problems. Yet, one can look to history for help in solving present problems by using historical analogies. If, for instance, one was presented with an option from the past that is very similar to a present option, one might examine the past outcome of that option to inform the present decision.

ANALOGOUS PROBLEMS: COMPUTER SCREEN LIMITATIONS AND THE COST OF TELEGRAMS

Consider the possibilities of using history as an analogy when choosing how to group or "chunk" online information. Gomoll et al. found despite the fact that more and more documentation was being placed online, in-house company standards for creating online documentation were rare.[10] In Gomoll et al.'s follow-up interviews with nine experienced online writers, an area of concern stressed by a majority was the need for "chunking" information into screen-size stand-alone units rather than page-size, interdependent units.[11] When Horton discussed this same problem of "chunking," he pointed out the severe limits of screen-size, stand-alone units:[12]

> Not only is the area smaller, the amount of detailed information that can be displayed in that area is less. . . . Contrast the 1250-dot-per-inch typeset paper document with the typical 70-100-dot-per-inch computer screen. For even minimal legibility, text and graphics on the computer screen must be larger than on paper documents. Medieval illuminated manuscripts and children's books average about 200 words per page. An engineering handbook may contain 1500 words per page, along with intricate formulae and diagrams. A typical Help display, incorporating recommended white space and text large enough to be legible, may hold as few as 50 to 100 words.

Furthermore, Girill noted that the generally agreed-on need for "chunking" online information does not necessarily resolve the basis on which chunks are created.[13] Should grouping or online information or "chunking" be based on

- the constraints of the screen size,[14]
- the content of the information (functional units),[15] or
- a combination of size, fixed number of lines for a "chunk", and content, a "chunks" would be centered around keywords?[16,17]

Contemporary online writers, however, have neither been the first ones to face severe technological constraints on their communication techniques nor the first to search for a solution.

Consider, for example, late 19th- and early 20th-century American journalists facing the costs of telegrams. When newspaper reporters began to take advantage of the telegraph in the late 1840s, the usual cost of sending 10 words 100 miles was 25¢, and the cost for sending a message further than 100 miles was, on average, 2.5¢ per word[18] (compare this cost to 3¢ for a prepaid letter to be sent up to 3,000 miles[19]). When the Civil War ended, the cost escalated to 3.5¢ per word. As a result, newspapers paid $521,509 for telegraph services for the year 1866.[20] Telegram costs, of course, did not cease escalating in 1866, with the result that the cost of telegrams for newspapers on a single night could reach astronomical figures. For example, telegrams wired by reporters during the Tunney-Dempsey fight in September 1926 rang up a single night's tab of $500,000.[21]

Many sought to contain such communication costs by reducing the number of words wired through a process of condensing or telescoping the material, thus making the full meaning of the message dependent on information elsewhere, for example, on the prior knowledge of the receiver. Thus "Majority Rule" could be telescoped by the sender to "Maj" and yet still be understood by the receiver who would use his or her prior knowledge to translate it back out to its full meaning, "majority rule." This example was used in 1844 when Samuel Morse and his partner Alfred Vail operated the Washington-Baltimore line. Morse told Vail to leave out the "the's" and to try and make three letters represent words of three or four syllables or even entire phrases:[22]

> Condense your language more, leave out "the" whenever you can. The beginning of a long common word will generally be sufficient—if not I can easily ask you to repeat the whole. For example, "Butler made communication in favor of majority rule" could be telescoped into "Butler made com in fav of maj"—the word "rule" or similar words are unnecessary to repeat when the subject has been considered already.

This custom of telescoping continued as the newspaper business grew.[23] Finally, Walter P. Phillips, who for many years alternately managed the Associated and United Presses, went through the entire English vocabulary and telescoped many long words or phrases to create the Phillips Code.[24] A few examples of the Phillips Code and its translated full meanings follow:

Phillips Code Telescoped Word	Translated Full Meaning
fapib	filed a petition in bankruptcy
gx	great excitement
pwf	powerful
potus	President of the United States
pnr	prisoner
pmy	prominently
pskv	prospective
scotus	Supreme Court of the United States

Thus, in the Phillips Code, the following telegram dispatch:[25]

> The Supreme Court of the United States today confirmed the sentence of five years in the penitentiary imposed on J. Smith, a banker of Squamus, Oregon, convicted of stealing $228,000 of state funds.

would be telescoped and sent like this:

> Washn 27—t scotus tdy cmfd t stc o fv ys d pen impsd on j smith, a bnkr of squamus ore cnvctd o stealg sx 228tnd o sta fnds.

Another example of telescoping was Ernest Hemingway's famous dispatch:[26]

> KEMAL INSWARDS UNBURNED SMYRNA GUILTY GREEKS.

which when translated by his editors at the Hearst International News Service was printed in full as:

> Mustapha Kemal, in an exclusive interview today with the correspondent of the International News Service, denied vehemently that the Turkish forces had any part in the burning of Smyrna. The city, Kemal stated, was set afire by incendiaries, the troops of the Greek rear guard, before the first Turkish patrols entered the city.

After many, subsequent experiences with this "cabelese" (telescoping of language with its almost poetic density of meaning), Hemingway became fixated, telling anyone who would listen to "read the cabelese, only the cabelese. Isn't it a great language?"[27] Thus, the constraints of a technology changed the way in which journalists wrote: More specifically, a commentator in 1877 "gave the telegraph credit for the compactness of modern journalism":[28]

> The telegraph required short, pithy paragraphs, he wrote: it was responsible for the breaks and new headings that made a newspaper page look more like a checkerboard.

USING A HISTORICAL ANALOGY TO CREATE A SOLUTION

Just as the late 19th- and early 20th-century journalists changed how they wrote in response to the constraints of their medium of communication, so too will online writers—if history can be used as an analogy. Just as a journalist telescoped his or her telegrams, an online writer can telescope large pieces of information through the use of hypertext links. These links can automatically translate short "chunks" of information for readers by taking them to other screens of information that flesh out and expand their full understanding of the original short "chunk".

For example, in the hypertext screen taken from *The Writer's Pocket Almanack* (1988) (Figure P2.2), the word "Ovid" is short but powerful enough as a link to translate to other related items on Ovid or to other pieces related to what Ovid said. Another example on the screen is "Great Moments In Writing," which serves as both an icon and as a telescoped phrase that is small enough to fit on the buttons to the right of the screen but powerful enough to translate to fuller, longer pieces of information elsewhere.

Finally, just as the 1877 commentator noted how the technology of telegraphs affected specific ways in which journalists wrote, so too, a 1989 commentator noticed how the technology of hypertext began to affect specific ways in which contemporary technical communicators wrote:[30]

> The writers in the group were very positive about using the hypertext technology for writing technical manuals. To paraphrase from a discussion about modularity [for instance]: 'Lots of technical writers have tried for modularity; having it enforced by the environment is very different.

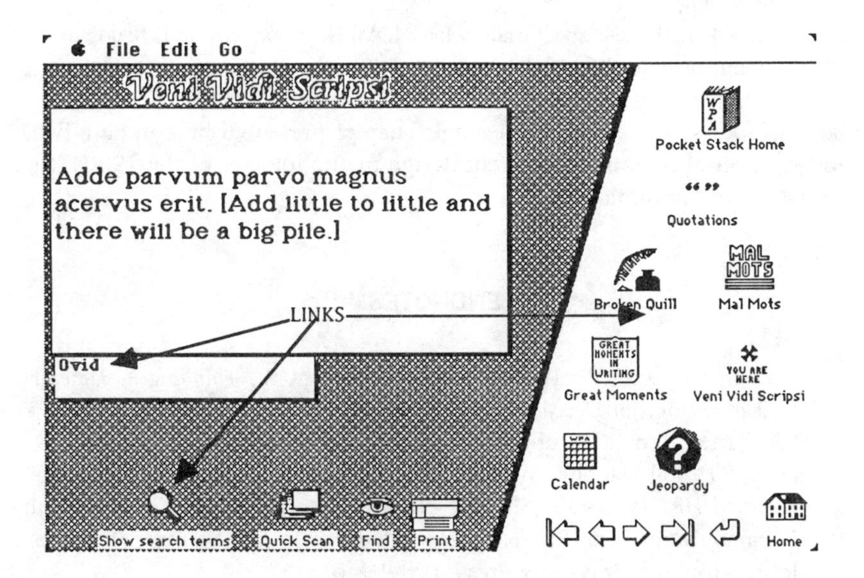

Figure P2.2. A screen with links from *The Writer's Pocket Almanack* (Horton and Brockmann, 1988)[29]

It helps you isolate units and pushes you in the direction of thinking things through logically. . . .' Basically the writers felt that being able to see the modular structure of their [hypertext] documents as they were writing contributed significantly to the ease of doing their job.

HOW TO DEVELOP HISTORICAL ANALOGIES

One of the best uses of a history of American technical communication is the creation of analogies on which to make decisions.[31] Of course, there will always be the need to unearth and describe the past as accurately as possible. However, in a profession that gets things done, the payoff for technical communicators is how to make decisions based on a perspective offered by historical analogies. Neustadt and May (1986) offered the following mini-method for using historical analogies:[32]

> Step 1. Separate what is KNOWN NOW from what is UNCLEAR and both from what is PRESUMED.
> Step 2. Do the same for all relevant THENs.
> Step 3. Compare THEN with NOW for likenesses and differences (skip if no analogues).
> Step 4. Articulate specifically what NOW is of concern and, if possible, commensurate objectives.

I use this four-step method in the single chapter presented here in Part Two, a consideration of two IBM document design methodologies of the 1980s: task-orientation and minimalism.

ENDNOTES

1. Qtd. in Neustadt, Richard E. and May, Ernest R.: *Thinking in Time: The Uses of History for Decision-Makers*. New York: The Free Press, 1986: 75.
2. See Bartky, Ian R.: "Running on Time." *Railroad History*. Autumn 1988 (159): 19–34; Jacobs, Warren: "Early Rules and the Standard Code." *Railroad History*. 1939 (50): 29–55; and Yates, JoAnne: *Control through Communication: The Rise of System in American Management*. Baltimore: Johns Hopkins University Press, 1989: 4–6.
3. See Gilbert, Bil: "Pioneers Made A Lasting Impression on Their Way West." *Smithsonian*. May 1994 25(2): 40–53, and Johnson, Overton, and Winter, William H.: *Route Across the Rocky Mountains, with a Description*

of Oregon and California; Their Geographical Features, Their Resources, Soil, Climate, Productions, &c, &c. Lafayette, IN: John B. Semans Printers, 1846 (reprinted in facsimile by Cannon, Carl L., Princeton, NJ: Princeton University Press, 1932.)

4. Consider how Johnson and Winter's 1846 "Bill of the Route" was like medieval pilgrimage maps: "The first medieval maps included only the rectilinear marking out of itineraries (Performative indications chiefly concerning pilgrimages), along with the stops one was to make (cities which one was to pass through, spend the night in, pray at, etc.) and distances calculated in hours or in days, that is, in terms of the time it would take to cover them on foot. Each of these maps was a memorandum prescribing actions." De Certeau, Michel: *The Practice of Everyday Life* (translated by Steven T. Rendall). Berkeley, CA: University of California Press, 1984: 120–1. Qtd. in Barton, Ben F. and Barton, Marthalee S: "Ideology and the Map: Toward a Postmodern Visual Design Practice." In (Sims, Brenda, ed.). *Studies in Technical Communication*. Denton, TX: University of North Texas, 1991.
5. Search the Internet resources of WorldCat to find the exact citations of the dozens of manuals and books written by Milton Blake Sleeper.
6. Derato, Frank C.: Victor W. Pagé: *Automotive and Aviation Pioneer.* Norwalk, CT: Cranbury Publications, 1991: vii. See also Brockmann, R. John: "Victor W. Pagé's Early Twentieth Century Automotive and Aviation Books: "Practical Books for Practical Men." *Journal of Business and Technical Communication*. 1996 10(3): 285-305.
7. Hopper, Grace, et al. (Staff of the Computation Laboratory): *A Manual of Operation for the Automatic Sequence Controlled Calculator*. Cambridge, MA: Harvard University Press, 1946.
8. Pearcey, T. and Hill, G. W.: "Programme Design for the C.S.I.R.O. Mark 1 Computer, I. Computer Conventions." *Australian Journal of Physics*. 1953 6(3): 316–34; Pearcey, T. and Hill, G. W.: "Programme Design for the C.S.I.R.O. Mark 1 Computer, II. Programming Techniques." *Australian Journal of Physics*. 1953 6(3): 335–56; Pearcey, T. and Hill, G. W.: "Programme Design for the C.S.I.R.O. Mark 1 Computer, III. Adaptation of Routines for Elaborate Arithmetical Operations." *Australian Journal of Physics*. 1954 7(3): 485–504; Hill, G. W. and Pearcey, T.: "On Starting Routines for the C.S.I.R.O. Mark 1 Computer." *Australian Journal of Physics*. 1955 8(3): 412–16; Hill, G. W. and Pearcey, T.: "Programme Design for the C.S.I.R.O. Mark 1 Computer, IV. Automatic Programming by Simple Compiler Techniques." *Australian Journal of Physics*. 1957 10(1): 137–161.
9. Killingsworth, M. Jimmie and Gilbertson, Michael K.: *Signs, Genres, and Communities in Technical Communication*. Amityville, NY: Baywood Publishing Company, Inc., 1992: 102: 213.

10. Gomoll, K., Gomoll, T., Duffy, T., Palmer, J., and Aaron, A.: "Writing Online Information: Expert Strategies (Interviews with Professional Writers)." (Communication Design Center Technical Report Number 37, 1988). Pittsburgh, PA: Carnegie-Mellon University, Communications Design Center, 1988: 4.
11. Gomoll et al., ibid., 1988: 9.
12. Horton, William: "Writing Online Documentation" "The Wired Word" column. *Technical Communication.* 36(2) 1989: 152–3.
13. Girill, T. R.: *On-line Documentation: A Practical Workshop.* (Syllabus and Training Materials). UCLA Division of Continuing Education, June 5–6, 1986.
14. Girill, T. R.: "Display Units for Online Passage Retrieval: A Comparative Analysis." In *Proceedings of the 31st International Technical Communication Conference, Seattle, Washington, April 29–May 2, 1984.* (Washington, DC: Society for Technical Communication, 1984): ATA-87–ATA-90; see also Ridgway, Lenore S.: "Information on Display Screens." In *Proceedings of the 30th International Technical Communication Conference, St. Louis, Missouri, May 1-4, 1983* (Washington, DC: Society for Technical Communication, 1983): ATA-18–ATA-21.
15. Girill, op. cit., 1984; see also Price, Lynne A.: "THUMB: An Interactive Tool for Accessing and Maintaining Text." *IEEE Transactions on Systems, Man, and Cybernetics.* SMC-12, March–April, 1982: 155–161.
16. Kehler, T. P. and Barnes, M.: "Design for an Online Consultative System." *Association for Educational Data Systems Journal.* 14, 1981: 113–127.
17. Sander, Linda K. and Rischard, Susan I.: "Automating the Newspaper Library with BASIS." *Proceedings of the 45th American Society for Information Annual Meeting, Columbus, Ohio, October 17–21, 1982.* White Plains, NY: Knowledge Industry Publications, Inc. 1982: 262–264.
18. Thompson, Robert L.: *Wiring a Continent: The History of the Telegraph Industry in the United States, 1832–66.* Princeton, NJ: Princeton University Press, 1947: 44.
19. Yates, op. cit., 1989: 23.
20. Lee, Alfred McClung: *The Daily Newspaper in America: The Evolution of a Social Instrument.* New York: Macmillan Co., 1937: 509.
21. Lee, ibid., 1937: 551.
22. Harlow, Alvin F.: *Old Wires and New Waves: The History of the Telegraph, Telephone, and Wireless.* New York: D. Appleton-Century Co., 1936: 199.
23. Lubar, Steven: *Infoculture: The Smithsonian Book of Information Age* Inventions. Boston: Houghton Mifflin Company, 1993: 91—"There were many published codes. The *ABC Telegraphic Code* listed 25,000 business phrases that could be sent with a one-word code. *Jacob's Friend-to-Friend Cable Code* encoded vacationer's greetings. A cotton purchaser using *Shepperson's Cotton Trade Cipher Code* could send one word to describe

the kind and quality of cotton he bought. The American Code Company even sold a book of '401,000,000 pronounceable words,' so that customers could make their own codes. If the first letter of a code word was mistransmitted, you could still find it in the *Terminal Index*, 'Invaluable for quickly correcting mutilated code words.' And, if, even with codes, telegraph costs were too high, you could use *Pieron's Code Condenser*, which reduced two code words to one." See also Yates, op. cit., 1989: 221—"Its second, and perhaps most important, goal was to reduce telegraphing costs. Before even sending the new code book to the company, [Elliot S.] Rice [duPont's Chicago agent] announced to Francis G. [du Pont] that he had tested the code on a real telegram and 'it works like a charm, cutting down the number of words nearly one half and simplifying very materially.' [July 3, 1888] After the code had been put into use by the company, he commented, 'I am pleased to note that the code is working to your satisfaction, and it seems to me that it is saving a great expense' [July 19, 1888]." Also see Yates, ibid., 1989: 225.

24. Harlow, ibid., 1936: 199–200.
25. Harlow, ibid., 1936: 200.
26. Cowley, Malcolm: "Introduction." In Ernest Hemingway: *The Sun Also Rises*. New York: Charles Scribner's Sons, 1962: x.
27. Steffens, Lincoln: *The Autobiography of Lincoln Steffens*. New York: Harcourt, Brace and Co., Inc., 1931: 834.
28. Qtd. in Lubar, op. cit., 1993: 25.
29. Brockmann, R. John. and Horton, William: *The Writer's Pocket Almanack* (paper and hypertext software versions). Santa Clara, CA, Info Books, 1988.
30. Walker, Janet H.: "Hypertext and Technical Writers." In *Proceedings of the 36th International Technical Communication Conference, Chicago, Illinois, May 14–17, 1989*. Washington, DC: Society for Technical Communication, 1989: RT-178.
31. The method of historical analogy was quite nicely summarized in Hartman, Harold and Robinson, Patricia C.: "From Quills to Keyboards: The Parallel Challenges of the Renaissance and the Information Age." In *Proceedings of the 37th International Technical Communication Conference, May 20–23, 1990, Santa Clara, California*. Washington, DC: Society for Technical Communication, 1990: RT-122–RT-124.
32. Neustadt and May: op. cit., 1986, Appendix A: 273.

Chapter 9

A Historical Perspective on Technical Communication Paradigms, IBM's Task Orientation, and Minimalism[1]

> We will win, and you will lose. You cannot do anything about it, because your failure is an internal disease. Your companies are based on Taylor's principles. Worse, your heads are Taylorized, too.
>
> —Kososuke Matsushita, chairperson of the Matsushita Electric Company, 1988[2]

DEFINE THE IMMEDIATE SITUATION AND THE DECISION MAKERS' PROBLEMS—IBM IN THE 1940S AND 1950S VERSUS IBM OF THE 1980S

The IBM of the late 1940s and early 1950s, the IBM of Sidney Lida's *Principles of Operation: Type 701 and Associated Equipment* manual, was not the IBM of the 1980s. Under Tom Watson, Sr., the IBM of the late 1940s and early 1950s had remained close to the individual customer because most of IBM's machines were punched-card calculators and typewriters—stand-alone, relatively low-cost equipment that required minimal technical expertise. For example, in IBM's typewriter business during that period, staying close to the customer meant:

> [S]alesmen in the 1940s and 1950s personally took each electric typewriter out of its wooden crate, installed it at a customer's office, and instructed the user on how it operated. The salesmen spent so much time hand-holding that they could usually sell only a dozen or so of the imposing devices a month.[3]

How minimal that minimal level of technical expertise was can be gauged by a 1942 book intended to coach readers on the Civil Service examination for the position of "Calculating Machine Operator." The book described the duties of the typical punched-card operator in the following way:[4]

> The duties of the Junior Calculating Machine Operator consist of operating a calculating machine in the solution of problems of average difficulty involving the arithmetic processes of addition, subtraction, multiplication, and division.

Moreover, the technical operating requirements of IBM equipment were so minimal that an *IBM Operator's Guide* written at the same time as Lida's 701 manual suggested that learning how to operate an IBM electric punched-card machine "is no more difficult that learning how to drive an automobile."[5]

However, under Tom Watson, Jr. and his successors, IBM drew its revenue and prestige from a new source—large, expensive, mainframe computers, such as the IBM 360 and 370. The creation and marketing of such large, expensive mainframe computers gave IBM a 70% share of the computer market and moved the company to the top-10 list of America's biggest companies.[6] Yet, these large, expensive mainframe computers also moved the focus of IBM documentation and training to the computer specialist working in the climate-controlled, glass-enclosed computer room. Thus, only a decade and a half after suggesting IBM equipment was "no more difficult than learning how to drive an automobile,"[7] the author of the *IBM System/360 Principles of Operation* manual assumed much more prior knowledge of the audience:[8]

> The reader is assumed to have a basic knowledge of data processing systems and to have read the IBM System/360 System Summary, Form A22-6810 which describes the system briefly and discusses the input/output devices available.

Furthermore, in the *IBM System/370 Principles of Operation* manual, the author also assumed a similar rich background of the reader:[9]

> The information in this publication is provided principally for use by assembler-language programmers. . . . It assumes the user has a basic knowledge of data-processing systems.

The office managers or workers who were not computer specialists were only left to input "problems" to the specialists who had to translate such input into the arcane languages of FORTRAN and ASSEMBLER. The typical office worker might also receive the printout from the specialists with all the numbers or design problems neatly typed in rows, but he or she did not interact directly with the mainframe computer.

Because the specialists who did interact directly with the mainframe computers were relatively few in number and had training in computer languages and operations, they received special treatment in support and training. Focus and emphasis on the specialists worked when the computer systems cost hundreds of thousands of dollars and when the number of specialists was so few.[10]

Yet, on August 12, 1981, IBM reentered the world of stand-alone, relatively low-cost systems used by nonspecialist office workers when it introduced the IBM PC.[11] The following January, in *Time*'s famous man of the year issue that named the personal computer as their man-of-the-year, *Time* also noted that IBM had[12]

> launched itself in a new direction by marketing a small, low-cost personal computer. The creamy white PC (for personal computer), introduced in August 1981, has set a standard of excellence for the industry . . . With PCs now selling at a brisker rate than ever, the marketplace apparently agreed that IBM had built the Cadillac of the 1982 class.

With the advent of personal computers, IBM and all its competitors were presented with a huge population of computer users to educate at a time when the money to pay for training was rapidly diminishing.[13] As Charles E. Pankenier, Director of Communications for IBM's Entry Systems Division, noted, computers were becoming less high-tech, scientific machines and more like retail products that the consumer could "make the tradeoffs on price, on delivery, on levels of support and on the countless other factors that may be important in an individual purchase decision."[14]

In a nutshell, IBM and all its competitors were re-presented with an audience similar to that of Sidney's Lida's in the 1950s and the audiences of punched-card machines in the 1930s and 1940s—an audience of relative beginners to whom IBM suggested it was all "as easy as driving a car." Yet in the 1980s, the audience was probably tens of thousands times larger, with even less a minimal background in the "arithmetic processes of addition, subtraction, multiplication, and division" than the most junior of punched-card calculator operators in the 1940s.

DEFINE THE IMMEDIATE DECISION MAKER'S SOLUTION—*PUBLICATION GUIDELINES, DESIGNING TASK-ORIENTED LIBRARIES FOR PROGRAMMING PRODUCTS*

By 1980, the profession of technical communication had grown in the United States. There were dozens of colleges and universities offering classes and degrees in technical communication. There were numerous textbooks and handbooks, and annual conferences of such groups as the Society for Technical Communication and the IEEE Professional Communication Society drew thousands of participants. The result of all this was that the profession of technical communication had become more self-conscious of both its techniques and its products. The re-presentation of an old problem—an audience of relative beginners who wanted it all "as easy as driving a car —combined with the new professional technical communication self-consciousness led to the release of *Publications Guidelines: Designing Task-Oriented Libraries for Programming Products*[15] and an almost evangelical fervor of its advocates.

Prior to the *Publications Guidelines* of 1982, there had been other professional technical communication design theories in the 1960s and 1970s: Information Mapping™,[16] "playscript,"[17] and STOP.[18] However, only two or three articles on these design theories appeared in the professional journals or in the conference proceedings; facts about Information Mapping™ and "playscript" remained largely within proprietary seminars, and information on STOP was mostly known only by those in Hughes Aircraft where it had been created. However, between 1982 and 1988 there was an unprecedented avalanche of conference presentations on task-orientation design principles and on *Publications Guidelines*—at least 16 conference presentations were given by IBM employees on task orientation between 1982 and 1990.[19] Soon the principles of this "Z" document intended for IBM internal eyes only[20] passed into the mainstream of technical communication in such textbooks as Brockmann (1986).[21] Moreover, as early as 1987, the task-oriented design principles had also passed from the medium of paper manuals to that of online documentation.[22] The principles passed so quickly into the technical communication mainstream that by the time of Hackos's 1994 textbook, *Managing Your Documentation Projects*,[23] the author never cited IBM nor the *Publications Guidelines* as sources even though she spent 16 pages discussing various aspects of task-orientation design principles. Hence, by 1994, IBM's task-orientation design principles had become an unquestioned, integral element of technical communication's document design paradigm.[24]

How Was Task-Oriented Writing Defined?

Task-oriented design principles suggested that the writer tell readers how to perfotm a task with the software or hardware rather than inform readers about the design of the software, "the descriptions of the internal workings of the product and descriptions of the tools that the product provides, such as commands, macro instructions, and utility programs."[25] The reason for the "product-oriented" style of writing was because:[26]

> In the past, most products were designed to provide tools for solving business problems rather than to solve the problems directly. . . . The users of these products were technical people: programmers, mathematicians, and scientists. . . . Users of software documentation were highly skilled computer users. There were not many software products used at any one time. . . . They knew, better than product developers, the best way to use the products in their environment. They did not want procedural information; they wanted maximum flexibility. Today, more and more special-purpose products are being produced. The goal is to solve business problems, to provide tools to let the user solve the problems. Computer software is being used by people with less data processing background. . . . The users of the products will be accountants, secretaries, managers, and others that have no interest in dealing with the technical aspects of computers. . . . The objective of a reader using a manual is to find out how to do something; the most efficient way for a writer to meet the reader's objective is to tell how to do it.

Additionally, task orientation was both a method of analysis and a way of organizing information. As a way of analysis, one advocate, Annette Bradford, described it in 1988 in terms very similar to the electrical engineering design of AND gates and NAND gates. Bradford suggested creating a matrix with a:[27]

- *Performer*: the person (or persons) taking the action to accomplish the goal.
- *Starting state*: where the performer begins (a thing, a set of things, a state of being, or a condition),
- *Action*: the mental or physical activities or operations.
- *Ending point* (or Goal): a thing or a state of being that indicates completion of the task.

Bradford continued:[28]

> Testing here is not just a validation procedure; if testing proves that your task analysis was 100% correct the first time, then that's truly exceptional. Task analysis is most likely a highly iterative process, performed several times and fine tuned continually.

Information for such a task-analysis matrix could be gleaned by surveys, interviews, direct observation, protocol analysis, and by simulations and prototypes.[29]

To complement this method of analysis, as a way of organizing information, the *Publications Guidelines* suggested that

- the documentation should be limited to task-required information and to a simple and direct writing style;[30] "The body of the manual was designed to be as procedural (cookbook) as possible."[31]
- the documentation should minimize the number of cross-references to other sections of the manual.
- the documentation should have headings that suggest action rather than descriptions such as "Entering and Exiting XYZ," "Moving through XYZ Panels," "Using the Commands," and "Getting Help."[32]

The distinction between task-oriented design principles and its predecessor, software product-oriented principles, was illustrated with examples such as those in Figure 9.1.

Software-Product

There are two ways to enter the XYZ program. One way is to call it from <situation a>. The other way is to call it from <situation B>. The method you use depends on the way XYZ was customized when it was installed. If you have any questions about which method to use, see the person responsible for installing XYZ.

Task- Oriented

There are two ways to enter the XYZ program.
If you begin in <situation A>, you must do this:
COMMAND TO CALL FROM SITUATION A.
If you begin in <situation B>, you must do this:
COMMAND TO CALL FROM SITUATION B
If you don't know whether you are starting from <A> or <B>, ask your system programmer or other person who is responsible for installing XYZ.

Figure 9.1. Software product-oriented versus task-oriented documentation[33]

An application of task-oriented design principles to the IBM System 36 planning information resulted in all the documents being packaged in a binder labeled *What To Do Before Your Computer Arrives*, and inside were eight booklets. The first was entitled *Your Guide to Planning*, which offered overviews for the major planning tasks, a road map charting a path though the other seven booklets. The other booklets were titled *General Planning Activities, Preparing a Place for Your Computer, Planning to Get Your Computer Up, Planning for Data Communications, Planning for System Configuration, Planning for System Security, and Planning to Receive Your Computer.*[34]

According to the advocates of these design principles, the result of using a task orientation is that it[35]

> [m]akes few demands on the reader and reduces user errors. The writer defines the step-by-step instructions for doing everything that must be done with a product. . . . The reader need understand the product only to the extent that it affects the task.

Yet, there was a price for this type of analysis and organization; task orientation required more time in planning because more time was spent on examining the users and their environment rather than allowing previously existing patterns of organization (software structure, menu order, or user titles) to organize the information. This design orientation also required more pages because the task approach made explicit what was only implicit in previous design orientations. This cost in time was explored by Odescalchi (1986).[36] Her study compared the productivity of users with a software-product oriented manual and others with a task-oriented manual, and her results are shown in Table 9.1.

Table 9.1. Software-Internals Oriented Manual versus Task-Oriented Manual.

Software-internals oriented manual

- Dissatisfaction rate 79% higher
- Failure rate 310% higher
- Error rate 480% higher
- Ability to follow 88% lower

Task-oriented manual

- Productivity 41% higher
- Amount of time required to produce 42% higher
- Number of words 33% higher
- Number of pages 87% higher

IBM's pride in task orientation could be measured by the avalanche of presentations their employees gave at professional meetings in the 1980s. Many at IBM even felt they had invented task orientation with the publication of the *Publications Guidelines*. Such a claim was made explicit by Bethke et al. in their *IBM Systems Journal* article (1981) that first announced task orientation to the company: "Task-orientation is a new approach and contrasts with other approaches that have been used to design libraries of programming publications."[37] This was emphasized again by Mitchell (1984) when she described it to a convention of technical communicators: "In 1980 an IBM corporate wide task force designed a guideline that introduced a new concept to technical communication—the concept of task analysis."[38]

However, not all the IBM presenters thought they had invented task orientation. For example, Simpson (1986)[39] suggested that precedent could be found in military manuals such as the *Guidebook for the Development of Army Training Literature*.[40] Like the IBM style guide book, the military Guidebook drew a distinction between topic- and performance-oriented writing. Compare Figure 9.1, for example, to the first quote from Army standard operating procedures that was used to illustrate what writers of these procedures were *not* supposed to do:

> TOPIC-ORIENTED WRITING
> CBR: THE LOCAL ALARM
>
> The local alarm (warning) is given by any person recognizing or suspecting the presence of a CBR [Chemical, Biological, Radiological Warfare] hazard. Unit SOP's must provide for the rapid transmission of the warning to all elements of the unit and to adjacent units. Brevity codes should be used where feasible. Suspicion of the presence of a chemical hazard is reported to the unit commander for confirmation. It is important to avoid false alarms and to prevent unnecessary transmission of alarms to unaffected areas. Consistent with the mission and circumstances of the unit, the alarm will be given by the use of any device that produces an audible sound that cannot be easily confused with other sounds encountered in combat. Examples of suitable devices for local alarms are empty shell cases, bells, metal triangles, vehicle horns, and iron pipes or rails. The unit SOP should specify the devices to be used, locations of the devices in the unit area, and procedures to be followed. As a supplement to the audible (sound) alarms or to replace them when the tactical situation does not permit their use, certain visual signals are used to give emergency warning of a CBR hazard or attack. These visual signals consist of donning the protective masks and protective equipment, followed by an agitated action to call attention to this fact. In the event of a chemical agent attack, there is a danger of breathing in the agent if the vocal warning is given before masking. The individual suspecting or recognizing this attack will mask first and then

give the alarm. The vocal alarm for chemical agent attack will be "SPRAY" for a spray attack, and "GAS" for an agent delivered by any other means. The vocal warning is intended for those individuals in the immediate vicinity of the person recognizing the attack. The vocal alarm does not take the place of the sound alarm or the visual signal to alert a unit of a chemical attack.

This second text block example was, like the IBM style guide, shows how the material could be rewritten using a "procedure-oriented" style:

PERFORMANCE-ORIENTED WRITING

CBR: THE LOCAL ALARM

The local alarm or warning is given by any person who knows, or thinks that a CBR hazard is present.

Unit Commanders Responsibility for the Local Alarm Procedure

The Unit Commander will prepare a Unit SOP describing the procedures to be followed in giving the local alarm. These procedures must provide for:

a. A vocal alarm to warn people who are near to the person who gives the alarm;

b. A sound alarm to warn people in the Unit's area—

(1) This alarm should produce a sound that will not be confused with other sounds of combat. Examples of objects that might be used are empty shell cases, bells, metal triangles, vehicle horns, and iron pipes or rails.

(2) The Unit SOP should identify the object to be used, where it is located, and how it is to be used to give the alarm;

c. A visual signal to be used in addition to the sound alarm, or, in place of the sound alarm when silence is necessary;

d. A way for members of the Unit to quickly report a suspected CBR hazard to the Commander for confirmation;

e. Rapid communication of the warning to other nearby units, making use of brevity codes if at all possible.

How to Give the Local Alarm

In case of Chemical attack, use these steps to give the local alarm:

a. Put your mask on first;

b. Give a vocal alarm—

If spray attack, say "SPRAY"

For all other kinds, say "GAS"

c. Give the sound alarm, visual signal, or both as directed in your Unit SOP

d. Pass the warning to the Unit Commander as directed in your Unit SOP.

Like the IBM software task-oriented instructions, the military's performance-oriented instructions had no cross references; they were written in the order in which the user would use the instructions, and, most importantly, for discipline and to maintain the chain of command, planning was separated from action.[41]

Yet, beyond the precedent of the military manuals, even a brief search of communication practices at other companies would have uncovered the Bell System's *Standards for Task Oriented Practices* (TOP) (1977).[42] Similarly, from the history explored in the earlier chapters of this book, it is quite obvious that IBM's task-orientation style guidelines of 1980 were simply the most contemporary avatar of the asymmetric rhetorical arrangements used by:

- the sewing machine manuals of the 19th century,
- the automobile industry manuals from the 1920s onward,
- the 1954 *Manual of Operation for UNIVAC I System*,
- Douglas Aircraft's "Computing Engineering Manual, 1956" for the IBM Type 704 computer, and even
- IBM's own "Operator's Guides" such as the *IBM Operator's Guide* of 1951/5 that in almost identical fashion to the guides of the 1980s described its purpose as:[43]

> to help you to understand how several IBM Electric Accounting Machines operate and to serve as a guide during practice sessions. Study of the sections Operating Pointers and How to Operate the Machine will do much to avoid the confusion that often is the result of a trial-and-error approach. After instruction, practice, and study of each machine, evaluate your understanding of the machine by answering the review questions without reference to the text. Once you learn to answer the questions in your own way, you will approach each machine with confidence, because you will know what you are doing and why. Do not be discouraged if at first you cannot remember some of the terms used to describe the various features of the machines. The important thing at first is to learn where they may be found on the machine and when and how to use them.

However, beyond simply knowing that in 1980 IBM's task orientation was a continuation of a particular type of rhetorical arrangement repeatedly used in the military and, in fact, throughout American technical communication history, it is important to know there was a historical analogy of a much different type with much greater significance. This other historical analogy—Frederick Winslow Taylor's theory of "scientific management"[44]—explicitly put the asymmetric, imperative mood tradition of task-oriented design principles into a philosophical framework.

DEFINE THE SITUATION THEN AND THE DECISION MAKER'S PROBLEMS—DECENTRALIZED CONTROL OF THE AMERICAN FACTORY, 1880S

> With this Desk a man absolutely has *no excuse for slovenly habits* in the disposal of his numerous papers, and the man of method may here realize that pleasure and comfort which is only to be attained in the verification of the maxim, "*a place for everything, and everything in its place.*"
>
> The operator having arranged and classified his books, papers, etc., seats himself for business at the writing table, and realizes at once that he is "*master of the situation.*" Every portion of his desk is accessible without change of position, and all immediately before the eye. Here he discovers that perfect system and order can be attained, confusion avoided, time saved, vexations spared, dispatch in the transaction of business facilitated, and peace of mind promoted to the daily routine of business.
>
> —Advertisement by the Wooton Desk Manufacturing Company, circa 1880.[45]

With the rise of large factories in the United States with dozens or even hundreds of workers and before the rise of middle management in the 20th century, work was handled on an individual basis. Initially, when the factory was small, the manufacturers themselves could direct activities on the shop floor as well as handle the minor demands of marketing and finance. Thus, early management was largely centralized in one person, and directions and orders were carried out in face-to-face conversation.[46] Yet, when factories grew, manufacturers soon discovered that their time and attention were being increasingly consumed by marketing and finance and their ability to direct the workers personally was diminishing. Hence, many manufacturers hired foremen as their surrogates to hire, fire, set wages, supervise, and direct workers.[47] Thus, management became decentralized. One of the primary ways in which these foremen established wages as well as decided who to keep and who to fire in this decentralized arrangement was by the rates and wages of the piecework system. Furthermore, the manner in which these foremen developed these rates was largely by rule of thumb:[48]

> We take all but the very slow . . . and get the average [time] of the lot, Then we deduct about ten per cent for loafing and go to some factory in our town that has a piece price on that work. If no one in our town has a piece price we compare it with factories in other towns, and if we are not much too low or too high we put it in.

Not only was the rate established by rule of thumb, but if workers failed to maintain the set rate, foremen would employ "driving"—authoritarian rule and physical compulsion—to ensure that they did meet the rates.[49] The result of such ad hoc, decentralized management was a cycle of continuous labor strife and unrest in America in the second half of the 19th century.

This cycle of strife and unrest drew the attention of the manufacturers who were often mechanician-entrepreneurs fascinated with machines and the predictability of physical laws. These mechanician-entrepreneurs longed to see the entire manufacturing process run as a smoothly as Evans's fully automated flour mill. This could only happen, the mechanician-entrepreneurs reasoned, if the manufacturing process was handled more efficiently with more knowledge and forethought.[50] Within this context of management problems and aspirations for solutions, Frederick Winslow Taylor's "scientific management" technique of task analysis and instruction cards was born.

DEFINE THE DECISION MAKER'S SOLUTION THEN—SCIENTIFIC MANAGEMENT

Taylor wrote that "task work" was invented at the Bethlehem Steel Company in the 1890s, and in his 1903 handbook, *Shop Management*, he suggested that "tasks" were[51]

> [a] clearly defined task laid out before him. This task should not in the least degree be vague nor indefinite, but should be circumscribed carefully and completely, and should not be easy to accomplish....All orders must be given to the men in detail in writing....

In a later book which more fully defined "tasks," Taylor modestly undercut any claim to their invention:[52]

> Perhaps the most prominent single element in modern scientific management is the task idea. The work of every workman is fully planned out by the management at least one day in advance, and each man receives in most cases complete written instructions, describing in detail the task which he is to accomplish, as well as the means to be used in doing the work. . . . These tasks are carefully planned, so that both good and careful work are called for in their performance. . . . Scientific management consists very largely in preparing for and carrying out these tasks.

> There is absolutely nothing new in the task idea. Each one of us will remember that in his own case this idea was applied with good results in his schoolboy days. No efficient teacher would think of giving a class of students an indefinite lesson to learn. Each day a definite, clear-cut task is set by the teacher before each scholar, stating that he must learn just so much of the subject; and it is only by this means that proper, systematic progress can be made by the students.

Taylor institutionalized the concept of "tasks" with a way of organizing writing, "instruction cards," and with a method of analysis, time and motion studies. In Frederick A. Parkhurst's book, *Applied Methods of Scientific Management* (1917),[53] Parkhurst gave an example of the instruction card (Figure 9.2) theorized by Taylor but actually used by Parkhurst at the Ferracute Machine Company in Bridgeton, NJ. As in the IBM task-oriented instructions, there was an emphasis on doing rather than on describing in these 25 steps; there were no cross references, and it was written in the order in which the user would have performed the instructions (see Figure 9.2):[54]

1. Preparation.
2. 0 first raise.
3. Set jig by first raise and clamp.
4. Place raise in jig & clamp.
5. Swing boxes to position.
6. Start machine.
7. Change feed.
8. BO [Bore out]
9. Start machine and raise head.

The key to developing these tasks and writing an "instruction card" was [55]

> [t]he deliberate gathering in on the part of those in management's side of all the great mass of traditional knowledge, which in the past has been in the heads of the workmen, and in the physical skill and knack of the workmen, which he has acquired through years of experience.

At the Bethlehem works, for instance, Taylor assigned his subordinates such accoutrements of science as a stopwatch and slide rule to study laborers shoveling coal:[56]

> On March 13 Gillespie and Wolle selected ten of the "very best men" and ordered them to load a car "at their maximum speed." Working at that rate, each man loaded the equivalent of seventy-five tons per day, nearly six times the previous average of thirteen tons. The men, howev-

158 APPLIED METHODS OF SCIENTIFIC MANAGEMENT

INSTRUCTION CARD — Form F.A.P. 36a — FERRACUTE MACHINE CO.

NOTE: All time is expressed in Hours and Decimals.

INSTRUCTION 10.
1 SHEETS, SHEET 1 DATE Oct. 19, 1909

ARTICLE OR JOB RM93A
OPERATION BO and RM

Quantity 20 | Material I. C. | Hand or Machine 3B | Sketch or Drawing 10177

TOTAL LOT TIME 45 | TOTAL PIECE TIME .35 | BONUS CHART 10

SUB OPERATION	DETAIL INSTRUCTIONS	TOOL	CUT	FEED	SPEED	PIECE TIME	LOT TIME
1	Preparation						200
2	I. O. first ram						050
3	Set jig by first ram & clamp						10
4	Place ram in jig & clamp					020	
5	Swing borer to position					007	
6	Start machine					001	
7	Change feed					001	
8	BO		.156"	.047"	G5	103	
9	Stop machine & raise head					002	
10	Swing reamer to position					007	
11	Start machine					001	
12	BO		.059"	.047"	G5	103	
13	Stop machine & raise head					002	
14	Swing reamer shank to position					007	
15	Couple reamer to shank					009	
16	Start machine					001	
17	Change feed					001	
18	RM		.035"	.293"	G5	021	
19	Stop machine & raise head					002	
20	Uncouple reamer from shank					008	
21	Unset ram					010	
22	Place ram on floor					010	
23	Clean out jig					002	
24	DM and clean machine and						
25	change work order						100
26							
27	Add 10% to piece time					032	
28							
29							
30							
31							
32							
	Carried Forward, or TOTAL					350	450

WHEN MACHINE CANNOT BE RUN AS SPECIFIED, GANG BOSS MUST REPORT AT ONCE TO P. Meyers.
INSTRUCTION 10.

FIG. 30.—Instruction card for shop use. Original is 9×12 ins. These are issued from the tool room in the form of blueprints.

Figure 9.2. Sample instruction card of the 1910s with a written task

> er, were exhausted after filling only one car. Additional observations confirmed the conclusions that seventy-five tons per day was the theoretical limit. . . . Gillespie and Wolle took their findings to Taylor on March 15 . . . [and] he set a piece price . . .

When photography became more accessible, Frank and Lillian Gilbreth transformed these early scientific accoutrements of scientific management into their more precise photographic time-and-motion studies.[57] If these instruction cards were specific and the studies were done scientifically, then, Taylor reasoned, "[All] men could do, workers and manager alike, was to accommodate themselves to its scientific imperatives."[58]

Instruction cards, time-and-motion studies, and scientific imperatives all effectively removed power from the capricious foremen in a decentralized management arrangement and recentralized it in the hands of enlightened, corporate management. Of course, there were other elements of Taylor's scientific management system including:

- attacking the disorganization of the old style factory by systematically reorganizing tool rooms,
- employing an early method of ergonomics by redesigning and standardizing tools to fit the task and the material, and, most importantly,
- establishing differential bonus rates as incentives for able and cooperative workers.[59]

Nevertheless, it was largely the new methods of communication—the instruction card rather than the idiosyncratic instructions of the foreman and the substitution of time-and-motion studies for a rule of thumb—that together gave recentralized management a way to instruct and measure the new hires flooding into American factories at the turn of the century:[60]

> The instruction card can be put to wide and varied use. It is to the art of management what the drawing is to engineering, and, like the latter, should vary in size and form according to the amount and variety of the information which it is to convey. In some cases it should consist of a pencil memorandum on a small piece of paper which will be sent directly to the man requiring the instructions, while in others it will be in the form of several pages of typewritten material, properly varnished and mounted, and issued under the check or other record system, so that it can be used time after time.

The similarities between IBM's task orientation and these aspects of Taylor's scientific management are manifold:

- Both were a method of analysis as well as a way of communication,
- Both focused on tasks and used a means of measurement to determine how to optimize task performance,
- Both replaced humans—the IBM salesman of the 1940s and early 50s and the driving foreman of the 19th century—with documents,
- Both separated worker actions from the thinking and designing done by the writer of the task-oriented instructions or the Planning Department on instruction cards,

- Both used the rhetorical asymmetry of the imperative mood approach for the instructions. Writers became teachers, and readers supposedly became students. For example, one advocate of scientific management wrote that:[61]

> In performing their duties, these teachers of the men are guided by the written instructions of the Planning Department just as much as the workmen are. They, however, study all of the work in advance of the workmen, and in attending to their duties act not as driver and task master, but rather as guide, philosopher and friend. Their work is that of teaching and helping the men, watching over them, and finding their weak points, and then correcting them by example and by actually doing the work themselves while the workmen look on, as well as by talking to them.

- Both focused on audiences that were fairly inexperienced. For example, when Taylor wished to show how a worker might be instructed, he often used a example of the pig-iron handler named Schmidt:[62]

> "Well [says Taylor to Schmidt] if you are a high-priced man, you will do exactly as this man tells you to-marrow, from morning till night. When he tells you to pick up a pig and walk, you pick it up and you walk, and when he tells you to sit down and rest you sit down. You do that right straight through the day. And what's more, no back talk. Do you understand that? When this man tells you to walk, you walk; when he tells you to sit down, you sit down, and you don't talk back at him. Now you come on to work to-marrow morning and I'll know before night whether you are really a high-priced man or not."
>
> This seems to be rather rough talk. And indeed it would be if applied to an educated mechanic, or even an intelligent laborer. With a man of the mentally sluggish type of Schmidt it is appropriate and not unkind, since it is effective in fixing his attention on the high wages which he wants and away from what, if it were called to his attention, he probably would consider impossibly hard work.

- Both saw human performance in terms of a machine model: In the IBM articles, humans were called performers and their actions measured and modeled on those of electrical circuits with AND and NAND gates; in the Taylor system, stopwatches, slide rules, and stop-action photography measured and remeasured with the "precision" of science.[63]

Essentially, Taylor's scientific management task analysis and instruction cards reincorporated the asymmetric role of expert teacher and novice student that was initiated as far back as the first sewing machine instruction manuals. Furthermore, such task analysis and instruction cards were primarily methods of getting workers to accommodate to the work. Nearly a century after Taylor promulgated his ideas, his methods of analysis and of communication are open to inspection and evaluation. However, there was also an almost immediate reaction and evaluation of his methods. Moreover, neither the immediate reaction or evaluation was positive.

DESCRIBE THE EVALUATION OF THE DECISION MAKER'S SOLUTION IN THE PAST—SCIENTIFIC MANAGEMENT WAS "AUTOCRATIC IN TENDENCY"

Scientific management resulted not only in a management realignment but in worker realignment as well. For example, instruction cards were so integral to Taylor's system that even when workers were told at the Norfolk Navy shipyard in Virginia that a new British efficiency plan was coming into the yard, and when only the instruction forms were shown to them, the workers went on strike immediately, recognizing the plan as Taylor's and not any so-called British system.[64] Furthermore, the Norfolk strike was just one of many strikes that occurred throughout the Army and Navy arsenals and shipyards as Taylor's time-and-motion studies and instruction forms were brought onto the factory floor seeking to transfer the knowledge and skills of the workforce into the hands of management. This transfer of knowledge occasioned an emotional loss for workers:[65]

> Those rule-of-thumb ways of doing things which were anathema to Taylor had at least given a man on the job the feeling that he was doing what he should. To abolish the rule of thumb in factory work would excise a part of every worker's emotional investment and personal satisfaction. Could a worker now fail to feel that he was doing somebody else's job, or a job dictated by the machine? Rule of thumb was personal rule. Scientific management, which made the worker into a labor unit and judged his effectiveness by his ability to keep the technology flowing, had made the worker himself into an interchangeable part.

In the early 1910s Taylor and scientific management were actually brought before a Congressional subcommittee and accused of being anti-labor. *The Final Report and Testimony* of the U.S. Commission on Industrial Relations in 1912 rendered this evaluation of scientific management:[66]

> To sum up, scientific management in practice tends to weaken the competitive power of the individual worker and thwarts the formation of shop groups and weakens group solidarity. . . . In practice scientific management must, therefore, be declared autocratic in tendency...

Perhaps because the mechanician-entrepreneurs had hoped to make their workers and their activities into a smooth running machine factory with interchangeable human parts, the accusations were not too far fetched. Nobel pointed out in his 1984 critique of automation:[67]

> Taylor and his disciples tried to change the production process itself, in an effort to transfer skills from the hands of the machinist to the handbooks of management. Once this was done, they hoped, management would be in a position to prescribe the details of production tasks, through planning sheets, and instruction cards.

Later in the same book, Nobel pointed out that:[68]

> Taylor preached that productivity required that "doing" be divorced from "planning". . . . Once operations have been analyzed as if they were machine operations and organized as such...they should be capable of being performed by machines rather than by hand.

Many contemporary programmers and analysts have similarly objected to a task-oriented approach in their computer documentation because the style does not give them enough discretion to make independent decisions. Furthermore, the farmers in their mower-reaper manuals also objected for the same reasons to a similar approach in "instruction" books of the 1890s—they even turned their backs on the "experts." Thus, was the replacement of an 1880s driving foreman much different from the driving imperative mood of a Douglas Aircraft manager whose readers still flinched when they heard footsteps in a computer room?[69]

Perhaps it was only a very small step from the positive side of NCR's directive that "the mechanical side of the register might remain a profound mystery to the salesmen,"[70] or Bell's suggestion to Chapline of UNIVAC in 1949 that "with the relatively simple maintenance instructions which are contemplated by Northrop, it may be possible to materially simplify the discussion of the theory behind the maintenance instructions"[71] to the contemporary deskilling criticized by many in a world communicated to users in the asymmetric, imperative mood of task orientation.[72]

Earlier it was noted that some of the IBM commentators in their task-oriented style guide books saw a connection between their design principles of the 1980s and earlier military guide books. So, too, there was a connection between Taylor's scientific management and the military drill and the standard operating procedures manuals of his days. The connection was "soldiering" that Taylor saw as one of his single biggest foes in the workplace:[73]

> Workers paced themselves for many reasons: to keep some time for themselves, to avoid exhaustion, to exercise authority over their work, to avoid killing so-called gravy piece-rate jobs by overproducing and risking a rate cut, to stretch out available work for fear of layoffs, to exercise their creativity, and last, but not least, to express their solidarity and their hostility to management.

Moreover, Taylor went beyond such a passing allusion to explicitly describe the militaristic element of scientific management in a Harvard lecture in 1909:[74]

> As you know, one of the cardinal principles of the military type of management is that every man in the organization shall receive his order directly through the one superior officer who is over him. The general superintendent of the works transmits his order on tickets or written cardboards [instruction cards] through the various officers to the workmen in the same way that orders through a general in command of a division are transmitted.

Giedion (1969) specifically commented on the militaristic aspects of this speech by Taylor:[75]

> The hierarchy from general superintendent to worker, the soldierly discipline for efficiency's sake, doubtless offers industrial parallels to military life. But let there be no mistake: Taylorism and military activity are essentially unlike. The soldier indeed has to obey. But when under greatest stress, he faces tasks which demand personal initiative, his mechanical weapon becomes useless as soon as there is no moral impulse behind it. In the present situation where the machine is not far enough developed to perform certain operations, Taylorism demands of the mass of workers not initiative but automization. Human movements become levers in the machine.

WAS THERE, IS THERE, AN ALTERNATIVE?

Taylor's scientific management approach—like the task-oriented design principles—quickly passed into the mainstream of American thought. Few would use either in their entirety. Many times they were amended and adapted, and sometimes the specific words of the instruction card simply became embodied in the physical workstations of an assembly line. Nevertheless, there was an alternative approach to this adaptation of the worker to the work and that was to adapt the work to the worker.[76] For example, John Patterson at NCR in the mid-1890s created not only the *Silver Dollar Demonstration* (see Chapter 3) but also one of the first, if not the first, Personnel Department. Patterson's Personnel Department[77]

> [h]andled grievances and discharges, promoted sanitation and shop safety, kept management informed of legislation and court decisions pertained to labor matters, and conducted foremen's meetings to bring the supervisors more into sympathy with its aims and purposes.

This alternative to Taylor and his view of the factory and workers as a mechanism focused on the idea of[78]

> [i]mproving the external conditions of daily life, on the assumption that better living and working conditions would render him more cooperative, loyal, and content, and, thus, more efficient and "level-headed."

This alternative essentially sought to adjust the industrial reality to the worker through the new tools of psychology and sociology rather than vice versa.[79] This alternative was adapted later by Henry Ford in the initial creation of his Sociology Department in 1914:[80]

> It was doubtless to avoid the unctuously condescending term "welfare" that John R. Lee named his new agency the "Sociological Department." Under great pressure of time he worked out a practical plan by which the company might establish friendly and mutually respectful relations with its employees . . .

Just as American management was offered an alternative to scientific management, so, too, IBM and technical communication in general were also offered an alternative in the 1980s to task orientation. This alternative was called *minimalism*. Like NCR's Personnel Department and Ford's initial design for his Sociology Department, minimalism sought to adjust the instruction to the worker rather than demand the worker adjust to the work.

Minimalism and Task orientation

In the 1980s, research and development work on minimalism at the T. J. Watson Research Center of IBM culminated in a book written by John Carroll,

(1990).[81] Carroll began his book by pointing out that "adults resist explicitly addressing themselves to new learning."[82] Moreover, he explained that resistance to learning did not mean that adults burned books or cursed manuals as soon as they received them but that they performed a number of actions that could, according to the minimalist research, only be explained by such resistance. For example, how else could one explain that when given well-written and well-designed documentation, adult readers constantly skipped ahead and began using the system without reading the whole manual? How else could one explain the fact that adults were constantly guessing about what should and should not happen with a new system as soon as they began learning? Finally, how could one explain that "learners at every level of experience try to avoid reading?"[83]

One traditional explanation of these behaviors was that the readers were lazy or insubordinate. However, Carroll offered an alternative explanation that consisted of two paradoxes; the paradox of motivation and the paradox of assimilation:[84]

> A motivational paradox arises in the "production bias" people bring to the task of learning and using computer equipment. Their paramount goal is throughput. This is a desirable state of affairs in that it gives users a focus for their activity with a system, and it increases their likelihood of receiving concrete reinforcement from their work. But, on the other hand, it reduces their motivation to spend any time just learning about the system, so that when situations appear that could be more effectively handled by new procedures, they are likely to stick with the procedures they already know regardless of their efficacy.

He explained the paradox of assimilation as follows:[85]

> A second cognitive paradox devolves from the "assimilation bias": People apply what they already know to interpret new situations. This bias can be helpful, when there are useful similarities between the new and the old information. . . . But irrelevant and misleading similarities between the new and old information can also blind learners to what they are actually seeing and doing, leading them to draw erroneous comparisons and conclusions, or preventing them from recognizing possibilities for new functions.

These two paradoxes became problematic in the design of documentation that took a "systems" approach such as the task-oriented design principles. Such a design focused on step-by-step procedures in which the reader was expected simply to be passive and follow along. The problem with this "systems" design philosophy, however, was summed up by Carroll:[86]

> It is surprising how poorly the elegant scheme of systems-style instructional design actually works. . . . .Everything is laid out for the learner. All that needs to be done is to follow the steps, one, two, three. But, as it turns out, this is both too much and too little to ask of people. The problem is not that people cannot follow simple steps; it is that they do not. People are thrown into action; they can only understand through the effectiveness of their actions in the world. People are situated in a world more real to them than a series of steps, a world that provides rich context and convention for everything they do. People are always already trying things out, thinking things through, trying to relate what they already know to what is going on, recovering from errors. In a word, they are already too busy learning to make much use of the instruction.

What is Minimalism?

Carroll has never offered a set of specific minimalist guidelines. In fact, Carroll concluded his book by rejecting the notion that there even were a set of minimalist design techniques. Rather, Carroll embraced the notion of eclecticism and, in the development process, emphasized continuously examining how adults learn in particular situations. However, what seems consistent in the various examples in the book are the following alternative design principles:

- Cut secondary features of manuals and online documents—overviews, introductions, summaries, and so on.
- Focus on what readers need to know in order to immediately apply it to productive work.
- Test repeatedly during design; testing replaces any hard and fast rules and guidelines in the minimalist design philosophy.
- Make it easy for the reader of a page to coordinate the documentation with the screen information via pictures of screens or other graphics.
- Support error recognition and recovery because much important learning goes on during the correction of mistakes.
- Use what the readers already know by continuously linking new information to it (e.g., "given-new" metaphor approach).
- Encourage active exploration of a system via intentionally incomplete information.

To see how this design philosophy was put into action, consider the following case and the accompanying theoretical underpinnings.

An Example: IBM Word Processor Tutorial—The Minimal Manual

In the first application of this design philosophy, Carroll and his colleagues redesigned an IBM word processor's tutorial document. Specifically, they used the eight design elements and rationales (see Table 9.2). Experimental results from Carroll (1990) comparing the IBM word processor minimal manual design to a standard commercial self-instruction (task-oriented) version of the manual found that by using the minimal manual:

- there was 40% less learning time;
- 58% more tasks were completed;
- 93% more tasks were completed per unit of time;
- there was 20% fewer errors and 10% less time in recovering from errors; and
- there was 80% more exploratory episodes.

In short, this case showed minimalism to be a resounding success even when contrasted with a well-written, task-oriented manual.

Nevertheless, numbers and a sound theoretical grounding are insufficient for choosing between task orientation and minimalism as two paradigms for technical communication. After all, there were a number of published shortcomings of minimalism:

- Learning by self-discovery was less predictable and gaps or lack of depth in learning could appear.[95]
- This design assumed a motivated audience. However, all the other document designs such as playscript and structured writing assumed just the opposite. Which was a more accurate assessment of readers?
- In allowing the free choice of goals, some users picked unattainable or ineffective goals.
- How did one know when concision became cryptic? Would minimalist designers be more open to liability challenges than systematic, comprehensive, task-oriented documenters?
- Would designers miss the all-important testing element—the hard work—and just employ the easiest step, that is, cut out words.
- Finally—and the most difficult to reckon with—minimalism was quite different from the ways documenters had traditionally been encouraged to write. From sewing machines to automobiles to the IBM style guide, there appeared to be a nice consistency of approach. Why change instruction manual paradigms after 200 years?

Table 9.2. Design Elements and Rationales.

Minimalist Design Elements	**Theoretical Rationales**
Slash 75% of the pages by cutting previews, reviews, index, practice exercises, troubleshooting, "welcome to this application" introductions, and most pictures of screens.	*"The rule of thumb in minimalist design is to try first to cut and condense text and other passive components but the goal of this is to enrich the training experience. The trick is to give the learner more to think about, but less to overcome."*[87]
Make iterative testing an integral part of the document design, not just design verification.	*"The process of tuning a design by iterative user testing has been found extremely effective, for example, in the development of the Apple Lisa and Macintosh systems, which profited from very frequent user tests conducted by the applications software manager himself."*[88]
Leave procedural details deliberately incomplete to encourage learner exploration but give "enabling hints."	*[R]eaders who learned procedures by working exercises "that forced them to independently apply the information in the manual performed significantly better."*[89]
Continually try to exploit readers' prior knowledge of the task implemented by the computer or analogous noncomputerized activities.	*"Many benefits come from encouraging users to utilize their existing knowledge through metaphors. For example, users will tend to know what is happening when the system gives them a certain type of feedback; they will tend to know how to respond to that feedback; they will tend to know how to plan their tasks better; and they will tend to be able to generalize from known, already encountered situations, to new and unknown ones."*[90]
Use open-ended exercises called "On Your Own."	*"Procedural knowledge is difficult or impossible to write down and difficult to teach. It is best taught by demonstration and best learned through practice."*[91]

Table 9.2. Design Elements and Rationales (con't).

Minimalist Design Elements	Theoretical Rationales
Include manual and system coordination information in such ways as "Can you find this prompt on the display: Enter name of document?" or show a figure representing what the display should look like if all is well	*"What is required is the incorporation of richer implicit and explicit linkages into the training materials."*[92]
Inventory principal user errors and then use them in developing specific error recognition and recovery sections.	*"A model for giving instructions is the directions you might give someone on how to find a restaurant in the country. . . . All directions should have in them the indications that you have gone too far."*[93]
Keep chapters brief (averaging less than three pages).	*"The units must be very streamlined so that learners are not likely to skip around within them; the organization of the units must be very simple so that learners can more successfully skip over units or skip among units."* ... *"We hoped that this would make it more tractable for learners to turn pages, to skip from topic to topic as they deemed necessary, but also be able to skip back when they wanted."*[94]

The Difference Between Task Orientation and Minimalism

Task-oriented design principles and minimalist design principles were often confused. For example, the author of an article on task orientation at a 1988 technical communication conference stated that he was writing on that topic, yet he wound up quoting liberally from two articles by Carroll.[96] Furthermore, in a 1984 paper on task orientation, the author described how she herself got lost as she wrote her task-oriented procedures:[97]

> As I wrote procedures, I found the trace information very difficult to use. You had to leave the procedure and to use information simultaneously in three or four other sections of the book. The reader shouldn't have that difficulty.

Thus, in this article purportedly advocating task-oriented design principles, the author instead described how she developed a kind of minimal manual, a "template that can be laid on the records that contains summary information to explain the hexadecimal codes in the records."[98] This material with its summary information was kept separate from the rest of the manual, and the first customers to use the author's task-oriented manual "singled out the template as an excellent idea; they wished more IBM products had templates."[99]

Moreover, these two sets of design principles may have been confused because both focused on the goals of the reader, the user, or the audience rather than on the parts of the product. Also, both sets of design principles at IBM continued to enact the directives passed down through the decades from Tom Watson, Sr.'s mentor at his first company, John Patterson of NCR, "Don't talk machines; don't talk cash registers; talk the prospect's business."[100]

Additionally, both sets of design principles had an integral element of iterative testing as part of their development process. About this testing, Carroll wrote:[101]

> Design is a highly creative process. The approach taken here tried to respect more seriously the fact that the designer is inventing something and cannot be both effective and mechanical. There are three things that are useful to inventors: having a goal, having an inventory of good examples, and having empirical techniques for assessing process.

Finally, both sets of design principles examined how actions were initially performed and then reexamined the efficacy of instructions on user performance.

Nevertheless, minimalism with its inherent incompleteness, dependence on reader-independent exploration, careful use of prior reader knowledge, use of user errors, and its focus on getting readers immediately productive, recognized that there was a vast context outside the instructions on a page or a screen. Minimalism recognized that it was the real-world context that motivated readers to read, allowed readers to co-create meaning based on their own memories and experience, checked their reading with the performance of the directions, and gave an explicit intrinsic payoff—they got their work done.

Minimalism also re-incorporated the role of ancillary illustrations first used in early preparadigmatic mower-reaper manuals. For example, minimalism design techniques suggested that instruction manual authors offer readers pictures of the screens and allow them to draw their own links between the text, the pictures, and their tasks. On the other hand, in the task-oriented style guide, there were ample examples of applying the design concepts to words, but nowhere any mention of illustrations, graphics, or pictures.[102] This was especially noticeable in one instance in which a task-oriented design advocate had earlier presented a paper promoting the use of graphics in instruction manuals. Yet, two years later when the author's article on task-oriented design was published, all mention of illustrations and pictures disappeared.[103]

By cutting away the bulk and very presence of instructions, minimalism embodied a tradition that had been present throughout the history of American technical communication. This tradition occurred in the mower-reaper manuals which were a fraction of the size of the sewing machine manuals and which refused to take on the moniker of "instructions" or "directions" for nearly the whole of the 19th century. This tradition led to the apologetic attitude of the automobile manuals which replaced the finger-pointing words of the *Directions for Using the New Under-feed Sewing Machine* (1870):

☞ You Cannot use the Machine until you thoroughly understand the Directions.

with the following in the *Instructions for the Operation and Care of Chevrolet Cars (1919—FB Model)*:[104]

> But unless the operator of a Chevrolet knows how to properly handle and care for his or her car, the maximum of satisfaction and enjoyment, which our painstaking efforts have provided, will not be realized. It is to give this desirable, if not entirely necessary, knowledge, that this Instruction Book has been prepared.

Moreover, the tradition of symmetric instruction that desired to "get out of the way of the learner learning his or her tasks" was IBM's own previous internal design tradition embodied in the Principles of Operation manuals—a tradition that evidently had been lost within the company by the time of the advent of the PC. Lida in 1953 included a "Programmer's Card" as perhaps one of the first published examples of a minimal manual (see Figure 8.13). Moreover, rather than imperative mood directions, Lida and the other Principles of Operation authors consistently offered examples to explore and study:[105]

> This section gives several examples of programming that progressively use more and different operations and so become lengthier and more complicated. These examples can be studied in sequence to understand how different operations work together to create a meaningful program.

A minimalist design philosophy was an alternative to the deskilling possibilities inherent in task-oriented design principles. For example, by letting readers use what they already knew (given-new principle), minimalism avoided addressing them as beginners or as "students," and rapidly empowered them as "colleagues." Additionally, minimalism gave back to the readers a span of discretion by remaining intentionally incomplete and encouraging active exploration as well as trial-and-error learning. Minimalism also tried to provide intrinsic motivation to read the material by linking all documentation to work readers actually needed to do. In short, minimalism attempted to correct the autocratic tendencies of task-oriented design principles by diminishing the very presence of the instruction manual.

A DEBATE NEVER HELD

No record of a debate within IBM concerning these competing design methodologies has ever come to light. Indeed, no formal debate was held in the professional journals or at the professional conventions until a decade later. Only in the *Journal of Computer Documentation* (1991) were the design principles of task orientation critically examined and discussed.[106] Indeed, only in the *Journal of Computer Documentation* was there much substantive discussion of Carroll's book.[107] Only as the avalanche of articles on task orientation diminished did 12 articles on minimalism begin appearing in the professional journals and conference proceedings between 1988 and 1994.[108] Probably precedence played a large role in how one rather than the other design methodology has affected the profession of technical communication. After all, the work on minimalism did take place at IBM after the publication of the *Publications Guidelines* and after the guidelines were made an official IBM standard with all the attendant prestige and leadership function that such a standard has had in the profession of technical communication in the United States.

In 1994, John Carroll, the director of research in and the primary proponent of minimalism, left IBM's T. J. Watson Research Center to join the faculty of the Department of Computer Science at Virginia Polytechnic Institute. A question-and-answer session was held on the occasion of Carroll's reception of the SIGDOC RIGO Award in 1994 for significant lifetime achievement[109] (the same award given four years earlier to Joseph Chapline for his work, see Theme 3 Introduction). During this Q and A, Carroll mentioned that many of the research projects in minimalism were unfinished at the time of his departure and that the best new work on minimalism was no longer being done in the United States, but at the University of Twente in the Netherlands.[110]

Without an historical perspective, any comparison between task orientation and minimalism might well rest on dollars and cents. For example, Odescalchi at IBM did indeed publish research in 1986 demonstrating a 41% overall increase in the productivity of those readers who used a task-oriented style of documentation.[111] However, in 1903, Frederick Taylor also wrote of an 82% increase in productivity with the application of his task instruction cards:[112]

> The instruction card for [overhauling and cleaning the boilers] filled several typewritten pages, and described in detail the order in which the operations should be done and the exact details of each man's work, with the number of each tool required, piece work prices, etc.
>
> The whole scheme was much laughed at when it first went into use, but the trouble taken was fully justified, for the work was better done than ever before, and it cost only eleven dollars to completely overhaul 300 H. P. boilers by this method, while the average cost of doing the same work on day work without an instruction card was sixty-two dollars.

However, was the true cost to readers measured in either case? Moreover, even if the true cost was measurable, is it only objective research that maintains the sway of task orientation in the profession of technical communication? To change from one set of design principles such as task orientation which has now passed unquestioned into the fabric of technical communication, to a minimalist approach would take some "letting go" and overcoming of fears:[113]

> Fears of "letting go" of standard procedures techniques when developing GE+P [minimal manual] prototypes was our toughest problem to overcome. As technical writers, we were trained to write specific, clear and complete procedures for end-users. The GE [minimal manual] approach ran contrary to our training in that it required us to write truncated "hints" which often left information purposefully incomplete.

It was, it is very hard to "let go." Perhaps technical communicators are all too often caught in the same psychological vise that held Joseph Chapline of UNIVAC in 1950 from wholeheartedly affirming a new documentation style as well as a new method of documentation management. Norman called that psychological vise "cognitive hysteresis:"[114]

> [Cognitive hysteresis] was derived from the phenomenon in magnetism called hysteresis, which refers to the property that once a material has been magnetized in one direction, it is difficult to make it change to being magnetized in the opposite direction. Simple magnetic fields that earlier would have caused the unmagnetized substance to be magnetized in either direction no longer have any effect. Once the magnetic state has been set, it is stable and resists further change. . . . There are other names for this phenomenon in the psychological research community: functional fixedness, cognitive narrowing, tunnel vision. Whatever the name, they all describe pretty much the same phenomenon: People tend to focus upon the active hypothesis and, once focused, find it very difficult to change, even in the face of contradictory information.

ENDNOTES

1. Portions of this chapter appeared earlier as "Protecting Technical Communication Innovations?," a RIGO Award address given at the *SIG-DOC '86 Conference*, Ann Arbor, Michigan, November 1–2, and as "The Why, Where and How of Minimalism." in *Proceedings of SIGDOC '90 Conference, Little Rock, Arkansas November 1–2 (SIGDOC** 14 (4)): 111–9.
2. Matsushita, K.: "The Secret is Shared." *Manufacturing Engineering*. 100 (2) February 1988.
3. Carroll, Paul: *Big Blues: The Unmaking of IBM*. New York: Crown Publishers, Inc., 1993: 218.
4. *Calculating Machine Operator: Practice Tests for Civil Service Examinations*. Milwaukee, WI: Pergande Publishing Company, 1942: 4.
5. *IBM Operator's Guide*. New York: IBM, 1951, 1955: "Preface."
6. Carroll, Paul, op. cit., 1993: 52.
7. *IBM Operator's Guide*. New York: IBM, 1951, 1955: "Preface."
8. IBM Corporation. *IBM System/360 Principles of Operation*. Form A22-6821-3. Poughkeepsie, NY: IBM Corporation, Customer Manuals Dept., May 1966: 1.
9. *IBM Corporation. IBM System/370 Principles of Operation*. Publication Number GA22-7000-10. Poughkeepsie, NY: IBM Corporation, Central Systems Architecture Department, September 1987 (11th edition): iii.

10. Carroll, Paul, op. cit., 1993: 63–4.
11. Carroll, Paul, op. cit., 1993: 41.
12. qtd. in Chposky, James, and Lensis, Ted: *Blue Magic: The People, Power and Politics Behind the IBM Personal Computer*. New York: Facts on File, 1988: 134–5.
13. Carroll, John: *The Nurnberg Funnel: Designing Minimalist Instruction for Practical Computer Skill*. Cambridge, MA: MIT Press, 1990: 17.
14. Chposky and Lensis, op. cit., 1988: 124–5.
15. I S and T G Customer and Service Information, Document # ZC28-2525-2. Poughkeepsie, NY, 1980, Second Edition, 1982.
16. See Horn, Robert E. and Kelley, John: July 1981 *SIGDOC* * newsletter, 7 (4): 4–25; Horn, Robert E. and Kelley, John: "Structured Writing and Text Design." In (Jonassen, David, ed.). *The Technology of Text*. Englewood Cliffs, NJ: Educational Press, 1982: 341–367; Horn, Robert E. and Kelley, John: "Results with Structured Writing Using the Information Mapping Writing Service Standards." In (Duffy Thomas M. and Waller, Robert, eds.) *Designing Usable Texts*. Orlando, FL: Academic Press, 1985: 179–212.
17. Mathies, Leslie H.: *The New Playscript Procedure, Management Tool for Action*. Stamford, CT: Office Publications Inc., 1977.
18. Hughes Aircraft. *STOP: How to Achieve Coherence in Proposals and Reports*. Fullerton, CA: 1964.
19. In *Proceedings of the 29th International Technical Communication Conference, May 1982*. Washington, DC: Society for Technical Communication, 1982: Hadley, B. H.: "Designing Task-Oriented Documentation Using A Universal Architecture," C-38–C-40; and Mitchell, Georgina E.: "Solving Problems of Information Gathering with Task-Oriented Library Design," C-76–C-78; In *The Proceedings of the 30th International Technical Communication Conference, May 1983*. Washington, DC: Society for Technical Communication, 1983: Mitchell, Georgina E.: "Need to Know: Bringing Task Analysis to the People Who Need It—Writers!" W&E-102–W&E-103; Henderson, Allan: "Need to Know: Defining A Computer System's User Interface via Scenarios—A Methodological Cookbook, W&E-104–W&E-105; and Cortwright, Jarey L.: "Workshop: Need to Know: The Way to Go, Experiences in Task Analysis," W&E-159; In *The Proceedings of the 31st International Technical Communication Conference, May 1984*. Washington, DC: Society for Technical Communication, 1984: Mitchell, Georgina E.: "Task Analysis (Need to Know: The Way to Go)," WE-84–WE-85; Sederston, Candace: "Task Analysis: Applying Composition Theory in an Industrial Forum," WE-89–WE-92; Waite, Robert G.: "Organizing Computer Manuals on the Basis of User Tasks," WE-38–WE-40; Ward, Bob: "A Task Analysis Primer for Technical Communicators," WE-86–WE-88;

Wight, Eleanor G.: "Need to Know: The Way to Go—Case Study in Task Analysis," WE-93–WE-96; and Cortwright, Jarey Lee: "Need to Know: Way to Go—Two Forms of Task-Oriented Information," WE-100–WE-101, In *The Proceedings of the 33rd International Technical Communication Conference, May 1986*. Washington, DC: Society for Technical Communication, 1986; Bowman, George T.: "Applying Theory to Develop Task-Oriented Documents," 174–176; Odescalchi, Esther Kando: "Productivity Gain Attained by Task-Oriented Information," 359–362, and Simpson, Amy: "Task Oriented Writing: Using Action Sentences To Get The Job Done," 447–449; In *The Proceedings of the 35th International Technical Communication Conference, May 1988*. Washington, DC: Society for Technical Communication, 1988: Oppenheimer, Richard: "Introduction to the Task Analysis Panel," WE-78–WE-79; Bradford, Annette N.: "What is a Task Analysis Matrix," WE-80–WE-83. Girill, T. R.: "Ease of Use and the Richness of Documentation Adequacy." *The Journal of Computer Documentation* 15(2) September 1991: 22—"According to the Society for Technical Communication's comprehensive, annotated bibliography of technical writing papers published between 1966 and 1980, only four publications-related articles in that 15-year period had task analysis as their primary topic. Just six years later, the *Proceedings of the 34th International Technical Communication Conference* contained four papers on task analysis in its first 50 pages alone (Society for Technical Communication, 1987). It included dozens on that topic altogether."

20. The publication's ordering number was ZC28-2525-2, and the "Z" prefacing the number meant the document was for "IBM Internal Use Only."
21. Brockmann, R. John. *Writing Better Computer User Documentation, Version 2.0*. New York: John Wiley and Sons, 1986: 53–4.
22. Girill, T. R., Luk, Clement H., and Norton, Sally: "Reading Patterns in Online Documentation: How Transcript Analysis Reflects Text Design, Software Constraints, and User Preferences." In *The Proceedings of the 34th International Technical Communication Conference, May 1987*. Washington, DC: Society for Technical Communication, 1987: RET-114—"Reference retrieval of focused passages to answer particular questions clearly dominates the use of our on-line documentation. . . . Hierarchical, example-rich, *task-oriented text* markedly increases this trend toward reference reading online."
23. New York: John Wiley and Sons, 1994.
24. Thus, the final line of Bethke et al.'s paper introducing task-orientation to IBM in 1981 functions today quite ironically: "Finally, we expect task orientation concepts and their application to be widely accepted and grow far beyond their genesis in publications." Bethke, F. J., Dean, W. M., Kaiser, P. H., Ort, E., and Pessin, F. H.: "Improving the Usability of

Programming Publications." *IBM Systems Journal* 20(3) 1981 reprinted in *The Journal of Computer Documentation* 15(2) September 1991: 16.

25. Hadley, op. cit., 1982: C-38.
26. Hadley, ibid.: C-40.
27. Bradford, op. cit., 1988: WE-80–WE-83; see also Waite, op. cit., 1984: WE-88.
28. Bradford, op. cit., 1988: WE-83.
29. For further information see Jonassen, David H., Hannum, Wallace H., and Tessmer, Martin: *The Handbook of Task Analysis*. New York: Praeger, 1989.
30. Wight, op. cit., 1984: WE-93.
31. Mitchell, op. cit., 1982: C-78. Jonassen, et al., op. cit., 1989: 4—"Task analysis, as a process, seeks to reduce ambiguity in instruction by scientifically defining the parameters of any performance or learning situation."
32. Simpson, op. cit., 1986: 449. See also I S and T G, op. cit., 1982: 21–2.
33. Simpson, op. cit., 1986: 449.
34. Waite, op. cit., 1984: WE-39. See also I S and T G, op. cit., 1982: 23–105.—These guidelines offer many examples of applying task-orientation to a variety of "universal software tasks:" evaluation, planning, installation, resource definition, and so on.
35. Hadley, op. cit., 1982: C-39.
36. Odescalchi, op. cit., 1986: 359–362.
37. Bethke, et al., op. cit.: 1981. Reprinted in reprint and discussion series of *The Journal of Computer Documentation* 15(2) September 1991: 6.
38. Mitchell, op. cit., 1984: WE-84.
39. Simpson, op. cit., 1986: 449.
40. Hauke, R. N., Kern, R. P., Sticht, T. G., and Welty, D.: *Guidebook for the Development of Army Training Literature*. U.S. Army Research Institute for Behavior and Social Sciences, November 1977.
41. Of course, this style of writing described in the 1977 military guidelines drew on a long tradition of military drill and standard operating procedures manuals. For instance, one can find the following procedure in the War Department: *Rules for the Management and Cleaning of the Rifle Musket, Model 1863 for The Use of Soldiers*. Washington, DC: Government Printing Office, 1863: 19— "Rules for Dismounting The Rifle Musket (the numbers within the parentheses refer to pictures spread out over some 14 previous pages):

    1st. Unfix the bayonet (15.)
    2nd. Put the tompion (45) into the muzzle of the barrel.
    3rd. Draw the ramrod (5.)
    4th. Take out the tange-screw (3)

> 5th. Take off the lock (16;) to do this first put the hammer at half-cock, then unscrew partially the side-screws (19 a, b,) and, with a slight tap on the head of the screw with a wooden instrument, loosen the lock from its bed in the stock, then turn the side-screw and remove the lock with the left hand. . . .

From such Civil War drill manuals, one can follow the military "task-oriented" vision to R. B. Miller's work in the 1950s which systematized the basic procedures of task analysis and began to explicitly use such terms as "job," "function," and "task." His work helped shape what was known later as "human factors engineering." (Bailey, Richard W.: *Human Performance Engineering: A Guide for System Designers*. Englewood Cliffs, NJ: Prentice-Hall, Inc., 1982: 39.) Also in the 1950s, task orientation was widely used by the U.S. Air Force for training in weapons systems.

42. See also the *Proceedings of the Conference on Uses of Task Analysis in the Bell System*. Conducted at the Western Electric Corporate Education Center, Hopewell, NJ, October 18–20, 1972.
43. *IBM Operator's Guide*. New York: IBM, 1951, 1955: "Preface." Part of the Herkimer County Community College Library bound collection of early "IBM Machine Manuals." Herkimer, NY: Library of Congress No. QA 76.I25.
44. Killingsworth and Gilbertson also explored some interesting links between "scientific management" and technical communication. See *Signs, Genres, and Communities in Technical Communication*. Amityville, NY: Baywood Publishing Company, Inc., 1992: 102: Chapter 9 ("Management and the Writing Process").
45. qtd. in Norman, Donald A.: *Things That Make Us Smart: Defending Human Attributes in the Age of the Machine*. Reading, MA: Addison-Wesley Publishing Co., 1993: 155.
46. For more information on this early stage of American management, see Chandler, Alfred D.: *The Visible Hand: The Managerial Revolution in American Business*. Cambridge, MA: Harvard University Press, Belknap Press, 1977: 50–67, and Nelson, Daniel.: *Managers and Workers: Origins of the New Factory System in the United States, 1880–1920*. Madison, WI: University of Wisconsin Press, 1975: 3–4.
47. For more information on the role of foreman, see Chandler, ibid., 1977: 67–72; Nelson, ibid., 1975: 3–4; and Tucker, Barbara M.: "The Merchant, the Manufacturer, and the Factory Manager: The Case of Samuel Slater." *Business History Review* 55 (Autumn 1981): 297–313; and Beniger, James R.: *The Control Revolution: Technological and Economic Origins of the Information Society*. Cambridge, MA: Harvard University Press, 1986: 242.
48. qtd. in Nelson, op. cit., 1980:8.

49. Nelson, op. cit., 1980: 7.
50. JoAnne Yates and Wanda J. Orlinkowski: "Genres of Organizational Communication: A Structurational Approach to Studying Communication and Media." *Academy of Management Review* 17(2) (1992): 313—"To improve efficiency and regain control from the workers and foremen, managers developed new approaches and techniques that coalesced into a new managerial philosophy, later labeled systematic management."
51. Taylor, Frederick Winslow.: *Shop Management* (New York: American Society of Mechanical Engineers, 1903) New York: Harper and Row, 1947: 63–4.
52. Taylor, Frederick Winslow.: *Principles of Scientific Management.* (New York: Harper Brothers, 1911) Harper and Row, 1947: 59, 120.
53. Parkhurst, Frederick A.: *Applied Methods of Scientific Management.* New York: John Wiley and Sons, Inc., 1917.
54. Parkhurst, op. cit., 1917: 158.
55. Taylor, op. cit., 1947: 49.
56. Gillespie, James. and Wolle, Harley C.: "Report on Establishment of Piece Work in Connection with Loading of Pig Iron at the Works of the Bethlehem Iron Company," June 17, 1899. qtd. in Nelson, op. cit., 1980: 92. See also Taylor's famous story of analyzing the handling of pig-iron and his training of Schmidt, Copley, op. cit., 1923, 42–6.
57. Gilbreth, Frank B.: Motion Study: *A Method for Increasing the Efficiency of the Workman.* New York: D. Van Nostrand Company, 1911: 3—"Our duty is to study the motions and to reduce them as rapidly as possible to standard sets of least in number, least in fatigue, yet most effective motions. This has not been done perfectly as yet for any branch of the industries. In fact, so far as we know, it has not, before this time, been scientifically attempted. It is this work, and the method of attack for undertaking it, which it is the aim of this book to explain."
58. Noble, David F.: *America By Design: Science, Technology, and the Rise of Corporate Capitalism.* New York: Alfred A. Knopf, 1977: 269–70.
59. Gilbreth, Frank B.: *Primer of Scientific Management.* New York: D. Van Nostrand Company, 1921. The table of contents for Chapter II, "Laws, or Principles, of Scientific Management," lists time study, standards, instruction cards, functional foreman, rate of compensation, and prevention of soldiering as the six major principles.
60. Taylor qtd. in Gilbreth, ibid., 1921: 16.
61. Copley, op. cit., 1923, Vol. 1: 323.
62. Copley, op. cit., 1923, Vol. 2: 42–6.
63. Beniger, op. cit., 1986: 294—"Taylor's carefully refined methods for subordinating the human processor to the industrial process can be found in his six-step prescription for a proper time study: (1)'Find, say 10 to

15 different men . . . especially skillful in doing the particular work'; (2) 'study the exact series of elementary operations or motions which each of these men uses'; (3) 'study with a stop watch the time required to make each of these elementary movements'; (4) 'eliminate all false movements, slow movements, and useless movements'; and (6) substitute 'this new method (for the) inferior series which were formerly in use.'"

64. Nelson, op. cit., 1980: 160.
65. Boorstin, Daniel J.: *The Americans: The Democratic Experience*. New York: Random House, Vintage, 1973: 369.
66. Qtd. in Chandler Alfred, and Tedlow, Richard S.: *The Coming of Managerial Capitalism*. Homewood, IL: Irwin, 1985: 136, 138.
67. *Forces of Production: A Social History of Industrial Automation*. New York: Alfred A. Knopf, 1984: 33. See also Noble, 1977.
68. Nobel, ibid., 1984: 231.
69. Ibid.: 191—"Those who have worked for John Lowe will quickly recognize the tone that is set here. His three rubrics—neatness, timeliness, accuracy—are engraved indelibly in all our minds, and, lest we forget, are hammered home on the very next page of the manual."
70. Johnson and Lynch, ibid., 1932: 197–8.
71. Eltgroth to Chapline, 6/21/49, Hagley Museum and Archives, Wilmington, DE: Accession 1825, Box 2.
72. van Oss, Joseph E.: "Documenting Systems: Changing Products, Tools, and Theory." In *Proceedings of the 31st International Technical Communications Conference, 1984*. Washington, DC: Society for Technical Communication: WE-150–3. See also, Shelton, J. A.: "The Individual 'Work-to-Rules': Reducing Determinism in Taylor-made Expert Systems." *Interacting with Computers: The Interdisciplinary Journal of Human-Computer Interaction* 1(3) December 1989: 338–42.
73. Nobel, op. cit., 1984: 33.
74. Copley, op. cit., 1923: 282.
75. Giedion, Seigfried: *Mechanization Takes Command*. New York: W. W. Norton and Company, Inc., 1969: 99.
76. Killingsworth and Gilbertson (ibid., 1992) also explored the human relations alternative to "scientific management" as it applied to technical communication.
77. Nelson, op. cit., 1980: 18.
78. Noble, op. cit., 1977: 290.
79. See also Yates, JoAnne: *Control through Communication: The Rise of System in American Management*. Baltimore, MD: Johns Hopkins University Press, 1989: 16.
80. Nevins, Allan: *Ford: The Times, the Man, the Company*. New York: Charles Scribner's Sons, 1954: 554.
81. Carroll, John, op. cit., 1990: 17.

82. Carroll, John and Rosson, M B.: "The Paradox of the Active User." In *Interfacing Thought: Cognitive Aspects of Human-Computer Interaction*. Cambridge, MA: MIT Press/Bradford Books, 1987: 101.
83. Carroll and Rosson, ibid., 1987: 83.
84. Carroll and Rosson, ibid., 1987: 80–1.
85. Carroll and Rosson, ibid., 1987: 81.
86. Carroll John, op. cit., 1990: 74.
87. Carroll, John, 1990, op. cit., Chap. 4. See also Charney, Davida H., Reder, Lynne M., and Wells, Gail W.: "Studies in Elaboration in Instructional Texts." In (Doheny-Farina, Stephen, ed.). *Effective Documentation: What We Have Learned from Research*. Cambridge, MA: MIT Press, 1988: 55.
88. Laudauer, T. K.: "Psychology as a Mother of Invention." In *Proceedings of CHI + GI'87: Human Factors in Computing Systems and Graphics Interface (Toronto, April 5–9)*. New York: Association for Computing Machinery, 1987: 333–5.
89. Charney, et al., 1988, op. cit.: 63.
90. Hampton, James A.: "Principles from the Psychology of Language." In (Gardiner, Margaret M. and Christie, Bruce, eds.). *Applying Cognitive Psychology to User-Interface Design*. Chichester, England: John Wiley and Sons, 1987: 229.
91. Norman, Donald.: *The Psychology of Everyday Things*. New York: Basic Books, 1988: 58–9.
92. Carroll, John, 1990, op. cit.: 86.
93. Wurman, Richard Saul.: *Information Anxiety*. New York: Doubleday, 1989: 327; see also Ramey, Judith: "How People *Use* Computer Documentation: Implications for Book Design." In (Doheny-Farina, Stephen, ed.). *Effective Documentation: What We Have Learned from Research*. Cambridge, MA: MIT Press, 1988: 151.
94. Carroll, John, 1990, ibid.: 85, 149.
95. Carroll and Rosson, 1987, op. cit.: 92.
96. Lillies, Phillip: "Some Guidelines for Making a Computer Manual More Task-Oriented." In The *Proceedings of the 35th International Technical Communication Conference, May 1988*. Washington, DC: Society for Technical Communication, 1988: WE-46–WE-48.
97. Wight, op. cit., 1984: WE-96.
98. Wight, op. cit., 1984: WE-96. See also Cortwright, op. cit., 1984: W-102–3. In this article Cortwright also offers a poster and quick reference card both of which could be considered minimalist documentation within her task-oriented article, .
99. Wight, op. cit., 1984: WE-96. It might be interesting to note that Sidney Lida working within the Principles of Operations style of IBM manuals did include a Programmer's Quick Reference Card (see Figure 8.13).

100. Ibid., 1932: 199.
101. Carroll, John, op. cit., 1990: 304.
102. I S and T G, op. cit., 1982: 21–2.
103. Only Mitchell in 1983 ("Need to Know: Bringing Task Analysis to the People Who Need It—Writers!" W&E-102–W&E-103) has any mention of illustrations in the 16 task-orientation presentations by IBM, and then only suggests that an author should work with a graphic artist to provide effective illustrations.
104. *Instructions for the Operation and Care of Chevrolet Cars (1919—FB Model)*: 5.
105. *Principles of Operation: Type 701 and Associated Equipment, 1953*: 86.
106. *The Journal of Computer Documentation* 15(2) September 1991. In this first of the reprint and discussion series in JCD, the seminal task-orientation article by Bethke, Dean, Kaiser, Ort, and Pessin, op. cit., 1981 was reprinted (3–17). Then five articles discussed and analyzed its design principles: Girill, T. R.: "Ease of Use and the Richness of Documentation Adequacy" (18–22); Mirel, Barbara: "Toward a More Precise Definition of Task-Sufficient Information: Reconsidering Bethke et al. on Task Orientation and Usability" (22–33); Prigge, David: "A Response to Bethke et al." (33-37); Ramey, Judith: "Studying Document Usability: Grasping the Nettles" (37-40); and Bethke, Fredrick J: "Usability Revisited" (41–43).
107. *Journal of Computer Documentation* (1992) 16(1) January 1992: Horn, Robert E.: "Commentary on *The Nurnberg Funnel*" (3–10); Hallgren, Chris: "*The Nurnberg Funnel*: A Minimal Collection," (11–16); Carlson, Patricia A.: "From Document to Knowledge Base: Intelligent Hypertext as Minimalist Instruction," (17–31).
108. The 12 articles in the publications of the Society for Technical Communication include: in The *Proceedings of the 34th International Technical Communication Conference, May 10–13 1988*. Washington, DC: Society for Technical Communication, 1988: Vanderlinden, Gay, Cocklin, Thomas G., McKita, Martha: "Testing and Developing Minimalist Tutorials," RET-196–RET-199; in *The Proceedings of the 37th International Technical Communication Conference, May 20–23, 1990, Santa Clara, California*. Washington, DC: Society for Technical Communication, 1990: Duin, Ann Hill: "Minimal Manuals vs. Elaboration in Documentation: Centering on the Learner," RT-77–RT-80; in *The 1992 Proceedings of the 39th Annual Conference, May 10–13, 1992*. Arlington, VA: Society for Technical Communication, 1992: Chiang, Alice and McBride, Kevin: "TIRS Tutorial: New Techniques for Guided Exploration", 327–330; Chiang, Alice, McBride, Kevin and Payne, Edmund: "TIRS Tutorial: New Techniques for Guided Exploration," 331–334; in *The 1993 Proceedings of the 40th Annual Conference, June 6–9,*

1993. Arlington, VA: Society for Technical Communication, 1993: Elser, Arthur G.: "Minimalist Strategies for Improving User Documentation," 11–14; Smart, Karl: "Accommodating Active Learners in Software Documentation Decisions," 15–17; Bringhurst, Robert: "Concrete Methods that Promote Active Learning in Software Manuals," 18–20; and Mirel, Barbara: "A Study of Instructions for Information Systems: Variations on a Minimalist Theme," 363–366; in *The 1994 Proceedings of the 40th Annual Conference, May 15–18, 1943*. Arlington, VA: Society for Technical Communication, 1994: Miles, Terri J.: "Writing to Reduce Information," 29–31; and Elser, Arthur G.: "Designing Minimalist Principles into User Interfaces," 95. See also, Mirel, Barbara, Feinberg, Susan, and Allmendinger, Leif: "Designing Manuals for Active Learning Styles" *Technical Communication* 38(1) (First Quarter 1991): 75–87.

109. SIGDOC '94, ACM 12th Annual International Conference on Systems Documentation. October 2–5, Banff, Alberta, Canada.
110. See the work of van der Meji, H.: "Catching the User in the Act." In (Steehoudern, M., Jansen, C., van der Poort, P., and Verheijen, R., eds.) *Quality of Technical Documentation*. Amsterdam, Netherlands, 1994: 201–10, and "Principles and Heuristics for Designing Minimalist Instruction." *Technical Communication* 42(2) (Second Quarter, May 1995): 243–61. See also, Lazonder, A. W.: *Minimalist Computer Documentation: A Study on Constructive and Corrective Skills Development*. (dissertation, Enschede, The Netherlands) 1994.
111. Odescalchi, 1986, op. cit.: 359–62.
112. Taylor, 1903–1947, op. cit.: 187.
113. Vanderlinden, Cocklin, and McKita, ibid., 1988: RET-198.
114. Norman, op. cit., 1993: 133–4. See also Brockmann, R. John: "Minimalism-A Case of Information Transfer in Technical Communication." In (Carroll, J., ed.) *Reconsidering Minimalism*. Boston, MA: MIT Press, forthcoming.

Name/Subject Index

B

F

G

M

Y

Z